Block Copolymer Nanocomposites

Block Copolymer Nanocomposites

Editor

Emeka Arendse

scitus
academics

Block Copolymer Nanocomposites
Edited by **Emeka Arendse**

ISBN: 978-1-68117-714-4
Library of Congress Control Number: 2016958815

SCITUS Academics LLC,
616, Corporate Way, Suite 2, 4766,
Valley Cottage, NY 10989

www.scitusacademics.com

Notice

Printed in United States of America on Acid Free Paper

Preface

Polymers are much more than ever under pressure to be affordable and offer customized property profiles in this era of portability and rapid technological advances. The key is often the design of alloys and mixtures carefully structured from current polymers at the apposite scale (preferably under a micrometer). Block copolymers are thought to offer a solution for a long time. Local segregation of the different polymer blocks yields nanometer-sized molecular aggregates. The latest advances in synthetic chemistry have revealed extraordinary opportunities to prepare tailored block copolymers at reasonable cost. Over 20 years of intensive academic research and the development of powerful statistical theories and computer techniques should help to predict the balance and non-balance behavior of copolymers and their mixtures with other polymers. Polymer nanocomposites extend beyond the initial scope of polymer nanocrystals for clay-filled homo-polymers or refractive index tuning, mainly for mechanical reinforcement, to a wide range of applications.

Highlighting recent research and trends in the field of structural formation in nanocomposite materials based on block copolymers, this book shows opportunities for new materials based on the inclusion of different types of nanoparticles. The majority of block copolymer applications derive from their power to form microdomains in solution and in large quantities. Current applications of block copolymers include thermoplastic elastomers and compatibility of polymer blends. There are enormous potential uses of block copolymers in immersive technologies such as nanotechnology, nanolithography, photonics and controlled drug delivery. The interactions between the chemical structure and physical properties of block copolymers and their current and future applications are emphasized. With this in mind, the book highlights recent advances in the synthesis and characterization of block copolymers, discusses a number of key material opportunities for block copolymers and highlights several of the challenges that currently limit the further realization of block copolymers in promising nanoscale applications. It provides a comprehensive overview of the synthesis and the characterization of nanocomposites based on block copolymers. It is ideal for students, practitioners and researchers, with extensive research leading to nanotechnology and nanoscience.

TABLE OF CONTENTS

1

CHAPTER

Routes to Nanoparticle-Polymer Superlattices

Sara Mehdizadeh Taheri, Steffen Fischer and Stephan Förster *

Physikalische Chemie I, Universität Bayreuth, D-95447 Bayreuth, Germany

ABSTRACT

Nanoparticles can self-assemble into highly ordered two- and three-dimensional superlattices. For many practical applications these assemblies need to be integrated into polymeric matrices to provide stability and function. By appropriate co-assembly of nanoparticles and polymers it has become possible to tailor the nanoparticle superlattice structure via the length and stiffness of the polymer chains. The present article outlines and discusses established routes to nanoparticle-polymer superlattices. Recent progress has been remarkable so that the integration into functional devices has become the next challenge.

Keywords: nanoparticles; block copolymers; self-assembly

1. INTRODUCTION

Nanoparticles can self-assemble into ordered superlattices leading to the formation of two-dimensional highly ordered monolayers [1] or three-dimensional microcrystals (Figure 1) [2,3]. For many applications the use of bare nanoparticle assemblies is, however, unsuitable. Although free-standing nanoparticle monolayers over dimensions of micrometers have been shown to be stable [4], the stability generally needs to be considerably improved. To provide stability and function, it would be highly desirable to integrate nanoparticle superlattices into polymeric matrices [5]. By appropriate assembly of nanoparticles and polymers, it should additionally be possible to tailor the superlattice structure by the length and stiffness of the polymer chains. Although distance control for nanoparticles with attached low molecular weight ligands is to some degree possible [6], polymers would make a far broader range of distances accessible.

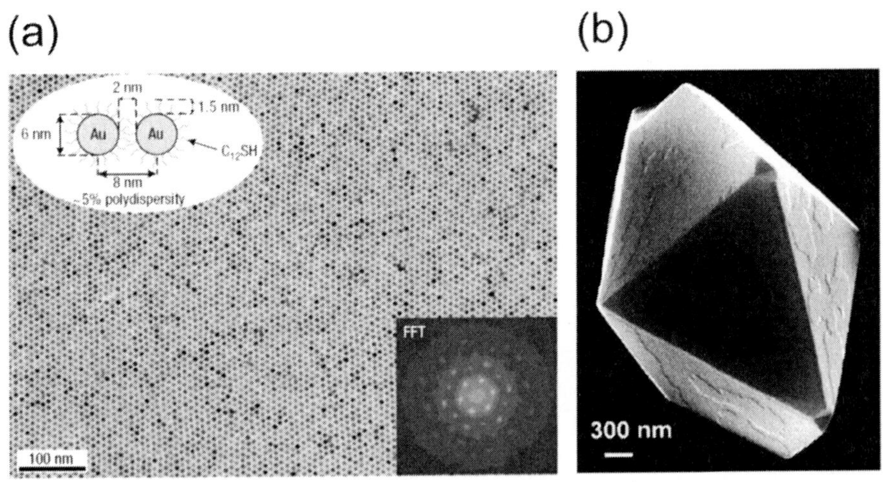

Figure 1. (**a**) TEM image of a two-dimensional array of gold nanoparticles [1] and (**b**) SEM-image of a three-dimensional, highly ordered FePt nanoparticle microcrystal [3]. From Ref. [1] and [3].

The preparation of such ordered nanoparticle-polymer assemblies, however, represents a considerable challenge. The three most important issues are

a. to incorporate nanoparticles of sufficient quality and regularity,

b. to provide compatibility with the polymer matrix,

c. and to establish a superlattice of high q uality.

2. ROUTES TO POLYMER-NANOPARTICLE SUPERLATTICES

To solve, overcome or circumvent these issues, different routes to polymer nanocomposites have been devised, which are summarized in Figure 2.

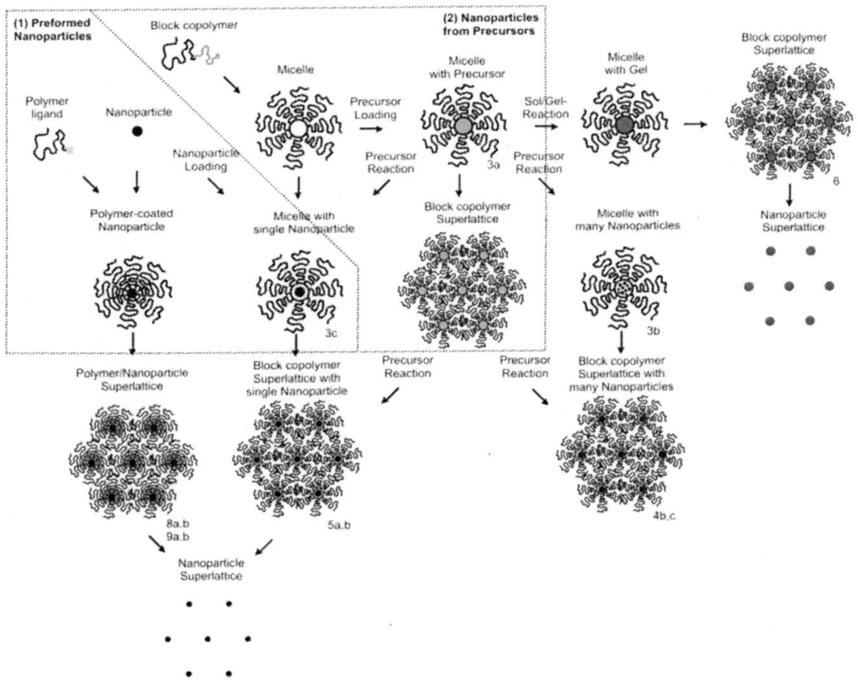

Figure 2. Various routes to nanoparticle superlattices. The two major routes involve the use of preformed nanoparticles that are subsequently integrated into a polymer matrix (1) or the use of polymer microdomains to synthesize nanoparticles from precursors within the polymer matrix (2). Both routes are designed to finally lead to the controlled assembly of nanoparticle-polymer superlattices. The numbers indicate the corresponding figures in the manuscript.

Two major routes may be distinguished:

(1) The incorporation of preformed nanoparticles into polymeric matrices.

This route has the advantage that the best synthesis routes to high quality nanoparticles can be employed. Since methods such as hot-injection techniques have become available, the preparation of nanoparticles with narrow size distribution and high crystallinity has become straightforward, even on larger scales. These synthetic procedures allow one to prepare nanoparticles with different chemical compositions (metals, chalcogenides) and properties (catalytic, conducting, semi-conducting, fluorescent, magnetic). A major difficulty when incorporating these nanoparticles into homopolymer and block copolymer matrices is the incompatibility of the nanoparticles with the polymer matrix. In mixtures with polymers, the nanoparticles will generally aggregate into large clusters, which often deteriorate the properties of the nanoparticles and the nanocomposite. They prohibit the formation of ordered nanoparticle lattices, even if the size distribution of the primary nanoparticles or the block copolymer microdomains is sufficiently narrow.

(2) The use of precursors to synthesize nanoparticles within polymer microdomains

The advantage of this method is that low molecular weight precursors can readily be solubilized into polymer microdomains, which can self-assemble into ordered lattices. The structure of these superlattices can be well tailored via the lengths of the polymer blocks. A major issue is how to control the size of the nanoparticles during growth in the polymer matrix. Also for this route, there is the issue of the incompatibility of synthesized nanoparticles with the polymer matrix [7,8].

For both approaches (1) and (2), the problem of incompatibility with the polymer matrix needs to be overcome. This is a major challenge that currently limits the success of nanocomposites in general. For the preformed nanoparticles (route 1) compatibility can be achieved *thermodynamically* by proper compatibilization of the nanoparticle surface which is possible by coating with appropriate compatibilizers (stearic acid, copolymers, *etc.*) before mixing with the polymer matrix. When nanoparticles are synthesized from polymer-miscible precursors inside the polymer matrix (route 2), they are usually *kinetically* entrapped. For any kinetic entrapment without thermodynamic compatibilization, long term stability is an issue. Eventually, the nanoparticles will phase separate, which leads to agglomeration in homopolymer matrices or relocalization and aggregation of the nanoparticles to the interface of the block copolymer microdomains, where their surface energy is minimized.

3. NANOPARTICLE ARRAYS FROM PRECURSORS

The kinetic entrapment of nanoparticles by reaction of polymer-compatible precursors inside block copolymer microdomains is a strategy that simultaneously takes advantage of the capability of block copolymers to assemble into well-ordered superlattices while providing compatibility during, and possibly after, the entrapment of the nanoparticles. The synthesis of nanoparticles within polymer microdomains has been termed the "microreactor" concept, because the chemical reaction from the precursor to the nanoparticle takes place within the microdomains, which themselves can also confine the size of the formed nanoparticles. For this, the microdomains are loaded with a suitable precursor, which can be chemically transformed into the desired nanoparticles either by addition of further reagents or simply by heating. Miscibility or compatibility of precursor and polymer can be provided by the use of secondary interactions (H-bond, coordinative, acid-base) which can be mediated by functional groups in the matrix polymer. Acid-base interactions with precursors such as gold acid ($HAuCl_4$) with basic groups such as polyvinylpyridines, or coordinative interactions between transition metal complexes with amine-groups in polymers, or just the miscibility of hydrophobic precursors in hydrophobic polymers enable to load sufficient precursor material into the polymeric matrix. The precursors are then converted into the nanoparticles by redox-reactions or acid-base reactions [9-11]. Since the addition of reagents into polymeric matrices involves slow diffusion processes, often gaseous reagents or simply decomposition by heating are used. The microreactor domains can either be block copolymer micelles that are dispersed in solvents, which can later be assembled into superlattices, or they can be microdomains within ordered block copolymer phases (see Figure 2).

The direct integration into preformed block copolymer lattices has been first explored by the group of Cohen, where precursors have been loaded into spherical, cylindrical or lamellar microdomains of block copolymers [12,13]. After reaction into the desired nanoparticles, the nanoparticles were shown to be selectively located in the original microdomains, but not in the matrix. However, no nanoparticle superlattices could be established because the nanoparticle growth is difficult to control leading to a large number of small, polydisperse nanoparticles. This is similarly the case for the synthesis of nanoparticles inside block copolymer micelles.

The number and size of the nanoparticles depends on the relative rates of nucleation and growth. The time for the formation of a critical nucleus depends on the precursor concentration and the interfacial energy of the nucleus in the polymer matrix. The growth of the critical nucleus in polymeric matrices is diffusion limited by the transport of precursor molecules through the polymer matrix. If the critical nucleation time, *i.e.*, the time for the formation for the first critical nucleus, is much smaller than the diffusion time to transverse the microdomain, then throughout the microdomain nuclei are formed, resulting in a large number of small nanoparticles per microdomain (see Figure 3).

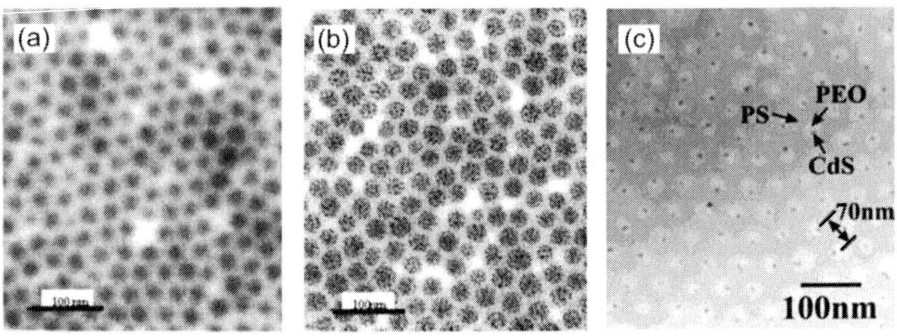

Figure 3. Block copolymer micelles filled with precursor (**a**), micelles with many gold nanoparticles after fast reduction (**b**), and preformed CdS nanoparticles solubilized into block copolymer microdomains (**c**). From Refs. [14] and [19].

Although by variation of reaction conditions the mean diameter of the nanoparticles within the polymer microdomain can be varied within a certain range by using different amounts of precursor, the polydispersity is generally large. While this approach is not useful for the preparation of nanoparticle superlattices, it is, however, suitable for the preparation of catalytically active nanoparticles, where it is the high specific surface area, the small size of the nanoparticles together with the stabilization in the polymer microdomain that provides high efficiency and the possibility for catalyst recycling in catalytic reactions [15].

The volume fraction of nanoparticles that can be generated within a microdomain is usually limited by the solubility of precursors. If the amount of nanoparticles in the microdomains is sufficiently large, nanoparticles can be made to fuse inside the polymer microdomains.

This can be achieved by chemical vapor deposition of metals onto microphase separated block copolymer films [16]. Since growth is limited by the microdomain size, this can lead to the formation of wires as shown in Figure 4.

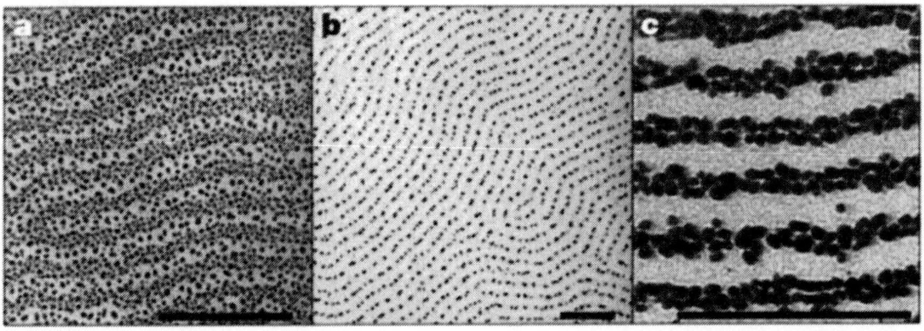

Figure 4. Nanoparticles (**a**) and after phase separation (**b**), and after growth and fusion (**c**) in block copolymer superlattices. The large volume fraction of nanoparticles allows the fusion into nanowires. From Ref. [16].

In the best case that the critical nucleation time is much larger than the diffusion time, only one nucleus is formed, which then grows until all precursor material within the micellar core has been taken up. If during this time no further nucleus is formed, only one nanoparticle per microdomain is generated. To achieve this goal, *i.e.*, to grow nanoparticle single crystals within a microdomain, requires optimization of precursor concentration, precursor mobility (by addition of solvents or variation of temperature) and interfacial energy (by the presence of stabilizing ligands). Although difficult, this has been achieved for gold nanoparticles in polyvinylpyridine-microdomains of polystyrene-polyvinylpyridine block copolymers (see Figure 5) on surfaces and in bulk. It was shown that a control of the nanoparticle size and distance was independently possible via the amount of precursor in the micellar core and the block length of the block copolymers forming the micelles.

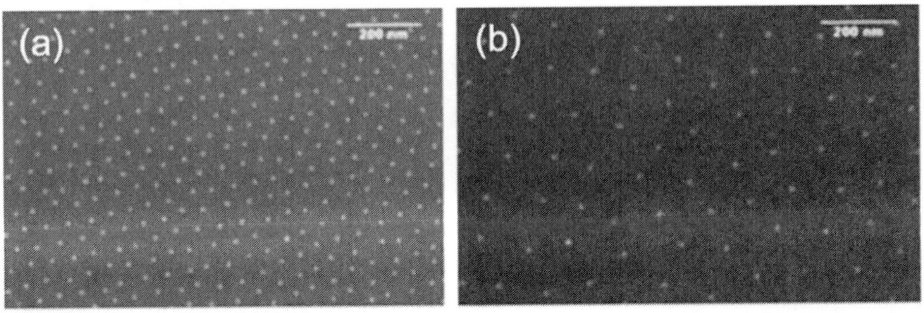

Figure 5. SEM-images (**a, b**) of a monolayer of gold-nanoparticles obtained by precursor reduction in block copolymer micelles. The distance between the nanoparticles can be controlled via the polymer block length. From Ref. [11].

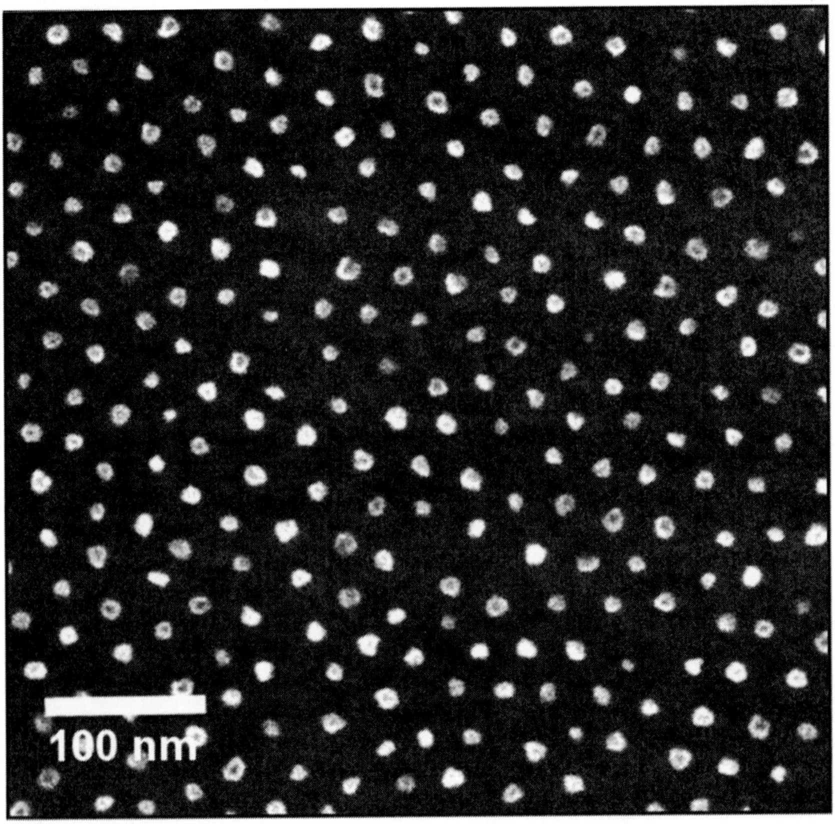

Figure 6. SEM-image of silica nanoparticles obtained from a sol/gel synthesis in micellar cores. From Ref. [17].

For sol/gel-reactions where the gel is homogeneously formed within the microdomain, proper conditions for the formation of one gel particle per microdomain are easier to find. This extends the use of this concept to prepare nanoparticles of a large class of inorganic oxides such as silica, titania and vanadia via sol/gel-routes (see Figure 6).

4. NANOPARTICLE ARRAYS FROM PREFORMED NANOPARTICLES

The direct integration of preformed nanoparticles into polymeric microdomains offers the advantage of using high quality nanoparticles. These nanoparticles can be selectively loaded into microdomains of ordered block copolymers [18,19]. For the formation of regular superlattices the challenge is to load exactly one nanoparticle per microdomain, which is only possible by optimizing loading conditions (seeFigure 3(c)). To avoid this problem, preformed nanoparticles can be directly coated with polymer chains, which make them structurally similar to block copolymer micelles, where the nanoparticle represents the micellar core. These could then be similarly assembled into superlattices. In thin films, preformed nanoparticles and block copolymers can exhibit interesting co-assembly behavior, where domain size and orientation can be tailored [20].

For preformed nanoparticles to be miscible with the polymer matrix, the nanoparticle surface has to be compatibilized. Compatibility of nanoparticles with polymers is often achieved by coating the nanoparticles with low molecular weight hydrophobic molecules that have end groups that bind to the nanoparticle surface. Frequent examples are stearic acid that is used for oxidic (basic) nanoparticle surfaces such as ZnO or $CaCO_3$. Also reactions of oxidic nanoparticles with silylation reagents have been used, e.g., for the compatibilization of silica nanoparticles in polymer membranes [21]. If possible, the nanoparticles are covered with a layer of a matrix-miscible polymer, ideally with the matrix polymer itself.

Generally, simple coating of the nanoparticle surface with a compatibilizing layer is still insufficient to provide complete miscibility, because of the loss of conformational entropy of the matrix polymers close to the nanoparticle surface [22]. Due to the large specific surface area of nanoparticles, this purely entropic effect alone leads to immiscibility and aggregation, a phenomenon that in colloidal solutions is known as depletion flocculation. In order to provide sufficient conformational freedom for the matrix polymers close to the nanoparticles, the polymer layer covering the nanoparticles should have a low segment density at its periphery to provide sufficient conformational freedom for the matrix polymers. This type of layer structure is ideally realized in spherical polymer brushes [23].

This requires the polymers to be bound to the surface of the nanoparticle with only one chain-end. This is chemically possible by grafting-to or grafting from techniques, where the grafting-from technique offers the advantage of high surface coverage at the cost of more synthetic effort. This has been realized for silica-polystyrene nanoparticles [24]. A more versatile method uses polymer chains containing a ligand end-group that binds to the nanoparticle surface. The attachment of brush-like polymer layers onto nanoparticles has recently been demonstrated for polyethylene glycol and polystyrene by using a ligand exchange procedure [25]. The attachment of polymer chains leads to an increased separation of nanoparticles as shown in Figure 7.

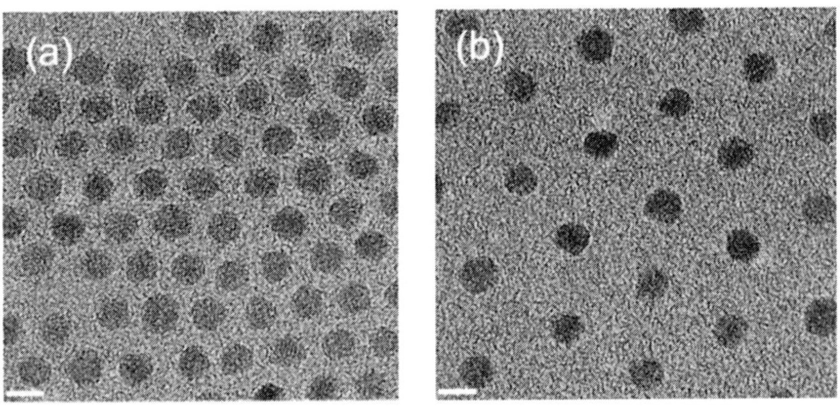

Figure 7. TEM-images of bare nanoparticles covered with low molecular weight ligand (oleic acid) (**a**) and polystyrene coated nanoparticles (**b**). The distance between the nanoparticles can be controlled via the polymer chain length. From Ref. [25].

The attachment of a brush-like polymer layer provides complete miscibility of nanoparticles and polymer matrix over the entire range of nanoparticle volume fractions without aggregation [26]. The route is successful for many chemically different nanoparticles such as metals, oxides or selenides, with a variety of different nanoparticle properties (magnetic, catalytic, semiconducting). It so far provides a very versatile way to two- and three-dimensional nanoparticle superlattices in polymeric matrices.Figure 8 shows this for the two-dimensional case of a magnetite nanoparticle monolayer in a polystyrene matrix, and for the three-dimensional case of a CdSe-superlattice in polystyrene. In both cases, the distance between the nanoparticles can be controlled via the length of the attached polymer chains.

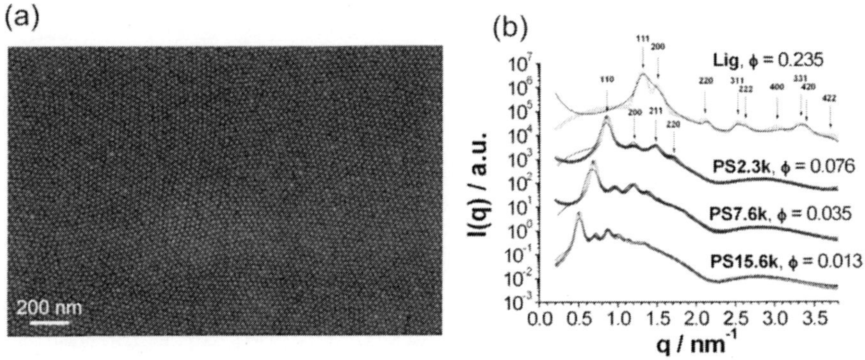

Figure 8. SEM-image of a highly ordered two-dimensional magnetite/polystyrene monolayer (**a**) and SAXS-curves of ordered three-dimensional CdSe/polystyrene nanocomposites (**b**); where the distance between the nanoparticles was controlled via the chain length of the polystyrene chains. From Ref. [26].

5. DNA-NANOPARTICLE ARRAYS

A sophisticated method to tailor nanoparticle self-assembly into superlattices is based on the work of Mirkin [27] and Alivisatos [28] who recognized the potential to control nanoparticle assembly via the attachment of DNA-strands. While this method is mainly limited to gold nanoparticles, where a stable attachment to the DNA-strands is achieved via a DNA-SH-Au link, it illustrates the potential and the quality of nanoparticle superlattices that can be assembled with polymers. Figure 9 shows examples for the two-dimensional case of a gold nanoparticle/DNA monolayer and the three-dimensional assembly of gold nanoparticle/DNA, where the distance between the nanoparticles is controlled via the number of DNA base pairs. By selective assembly of gold nanoparticles on specific sites on the DNA-strands, more complex "nanogrids" can be assembled [29,30].

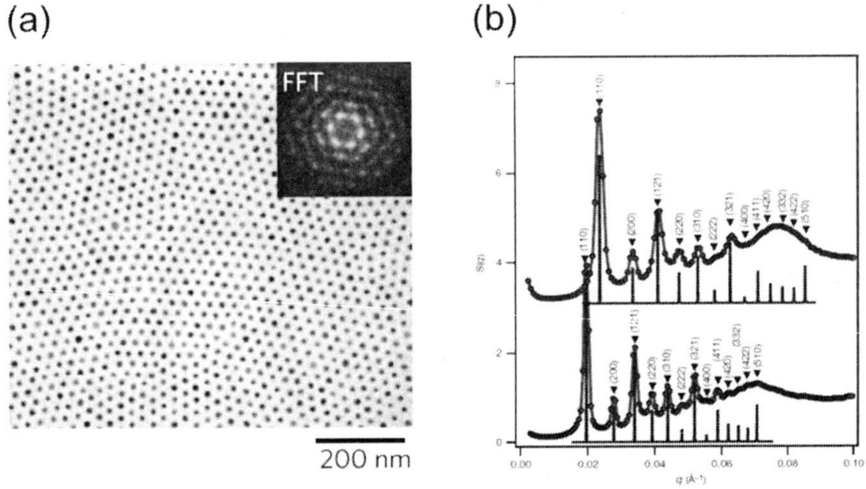

Figure 9. TEM-image of highly ordered two-dimensional gold/DNA monolayers [31] (**a**) and SAXS-curves of ordered three-dimensional gold/DNA nanocomposites (**b**); where the distance between the nanoparticles was controlled via the number of base pairs of the DNA-chain [32]. From Refs. [31] and [32].

6. CONCLUSION AND OUTLOOK

Extensive research on the assembly of nanoparticle/polymer composites has shown that the assembly of nanoparticles into superlattices in polymeric matrices is a particularly demanding challenge. It was found that the most critical issue was to prepare and conjugate nanoparticles of sufficient quality with a polymer layer that provided good miscibility with the polymer matrix. Various routes to solve this issue have been developed, of which at least two, the micellar precursor route, and the attached polymer brush route, have successfully led to the formation of nanoparticle superlattices in polymer matrices. These routes enable control of the interparticle distance via the length of the attached polymer chain on the 2–50 nm length scale. This is a feature that allows one to address critical design concepts in magnetic storage and hybrid solar cell applications, where the interparticle distance has to be carefully adapted to the length scale of magnetic dipolar interactions or the excition diffusion length. The possibility to translate structure into function is thus a currently exciting research area where progress will have profound influences for the use of polymer nanocomposites in functional devices.

Current challenges include assembling nanoparticles in a regular fashion at particular positions in two-phase nanostructures. This could be within the interface to provide electronic excitation and charge separation for the use in hybrid solar cells (Figure 10(a)), or the inclusion of two or more chemically different nanoparticles for directed energy transfer in light harvesting assemblies, the emission of white light in organic light emitting diodes, or the development of multiple contrast agents for medical diagnostics (Figure 10(b)).

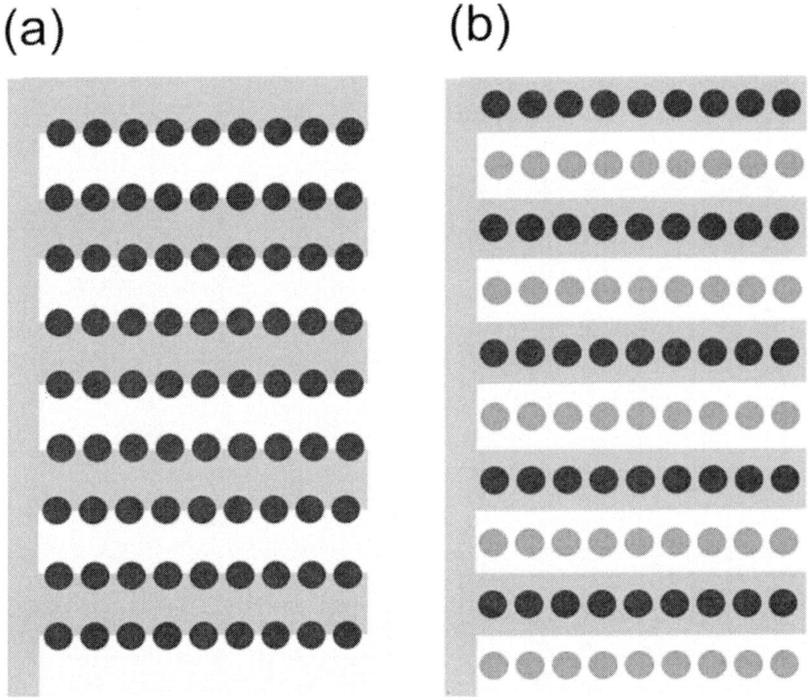

Figure 10. Regular assembly of nanoparticle in the interface of a two-phase device (**a**) or assembly of two different nanoparticles in two-phase systems for dual or coupled functions (**b**).

ACKNOWLEDGMENTS

The author would like to thank Frank Caruso for his hospitality during the sabbatical at the University of Melbourne, where the article has been written.

REFERENCES

1. Bigioni, T.P.; Lin, X.-M.; Nguyen, T.T.; Corwin, E.I.; Witten, T.A.; Jaeger, H.M. Kinetically driven self-assembly of highly ordered nanoparticles monolayers. *Nat. Mater.* **2006**, *5*, 265–270.
2. Murray, C.B.; Kagan, C.R.; Bawendi, M.G. Self-organization of CdSe nanocrystallines into 3-dimensional quantum-dot superlattices. *Science* **2008**, *270*, 1335–1338.
3. Shevchenko, E.V.; Talapin, D.; Kornowski, A.; Wiekhorst, F.; Kötzler, J.; Haase, M.; Rogach, A.L.; Weller, H. Colloidal crystals of monodisperse FePt nanoparticles grown by a three-layer technique of controlled oversaturation. *Adv. Mater.* **2002**, *14*, 287–290.
4. Mueggenburg, K.E.; Lin, X.-M.; Goldsmith, R.H.; Jaeger, H.M. Elastic membranes of close-packed nanoparticle arrays. *Nat. Mater.* **2007**, *6*, 656–660.
5. Haryono, A.; Binder, W. Controlled arrangement of nanoparticle arrays in block copolymer domains. *Small* **2006**, *2*, 600–611.
6. Martin, J.E.; Wilcoxon, J.P.; Odinek, J.; Provencio, P. Control of the interparticle spacing in gold nanoparticle superlattices. *J. Phys. Chem. B* **2000**, *104*, 9475–9486.
7. Förster, S.; Antonietti, M. Amphiphilic block copolymers in structure-controlled nanomaterials hybrids. *Adv. Mater.* **1998**, *10*, 195–217.
8. Förster, S.; Plantenberg, T. From self-organizing polymers to nanohybrid and biomateirals. *Angew. Chem. Int. Ed.* **2002**, *41*, 688–714.
9. Antonietti, M.; Förster, S.; Hartmann, J.; Oestreich, S. Novel amphiphilic block copolymers by polymer analogous reactions and their use for solubilization of metal salts and metal colloids.*Macromolecules* **1996**, *29*, 3800–3806.

10. Spatz, J.P.; Herzog, T.; Mössmer, S.; Ziemann, P.; Möller, M. Inorganic-polymer micellar hybrid systems—A tool for nanolithography. *Adv. Mater.* **1999**, *11*, 149–153.

11. Spatz, J.P. Nano- and micropatterning by organic-inorganic templating of hierarchical self-assembled structures. *Angew. Chem. Int. Ed.* **2002**, *114*, 3359–3362.

12. Clay, R.T.; Cohen, R.E. Synthesis of metal nanoclustes within microphase-separated diblock copolymers: Sodium carboxylate *vs.* carboxylic acid functionlization. *Supramol. Sci.* **1998**, *5*, 41–48.

13. Abes, J.I.; Cohen, R.E.; Ross, C.A. Selective growth of cobalt nanoclusters in domains of block copolymer films. *Chem. Mater.* **2003**, *12*, 1125–1131.

14. Förster, S.; Konrad, M. From self-organizing polymers to nano- and biomaterials. *J. Mater. Chem.***2003**, *13*, 2671.

15. Klingelhöfer, S.; Heitz, W.; Greiner, A.; Oestreich, S.; Förster, S.; Antonietti, M. Preparation of palladium colloids in block copolymer micelles and their use for the catalysis of the Heck reaction. *J. Am. Chem. Soc.* **1997**, *119*, 10116–10120.

16. Lopez, W.A.; Jaeger, H.M. Hierarchical self-assembly of metal nanostructures on diblock copolymer scaffolds. *Nature* **2001**, *414*, 735–738.

17. Frömsdorf, A.; Kornowski, A.; Pütter, S.; Lee, L.T. Highly ordered nanostructured surfaces obtained by silica filled diblock copolymer micelles as templates. *Small* **2007**, *5*, 880–889.

18. Fogg, D.E.; Radzilowski, L.H.; Blanski, R.; Schrock, R.R.; Thomas, E.L.; Bawendi, M.G. Fabrication of quantum dot-polymer composites: Semiconductor nanoclusters in dual-function polymer matrices with electron-transporting and cluster-passivating properties. *Macromolecules* **1997**, *30*, 8433–8439.

19. Yeh, S.-W.; Wu, T.-L.; Wei, K.-H. Spatial position control of pre-synthesized CdS-nanoclusters using a self-assembled diblock copolymer template. *Nanotechnology* **2005**, *16*, 683–687.

20. Lin, Y.; Böker, A.; He, J.; Sill, K.; Xiang, H.; Abetz, C.; Li, X.; Wank, J.; Emrick, T.; Long, S.; Wang, O.; Balasz, A.; Russell, T.P. Self-directed self-assembly of nanoparticle/copolymer mixtures. *Nature***2004**, *434*, 55–59.

21. Merkel, T.C.; Freeman, B.D.; Spontak, R.J.; He, Z.; Pinnau, I.; Meakin, P.; Hill, A.J. Ultrapermeable, reverse-selective nanocomposite membranes. *Science* **2002**, *296*, 519–522.

22. De Gennes, P.G. *Scaling Concepts in Polymer Physics*; Cornell University Press: Ithaca, NY, USA, 1979.

23. Milner, S.T. Polymer brushes. *Science* **1991**, *251*, 905–914.

24. Pyun, J.; Matyjaszewski, C. Synthesis of nanocomposites organic/inorganic hybrid materials using controlled/ "living"polymerization. *Chem. Mater.* **2001**, *13*, 3436–3448.

25. Nikolic, M.A.; Alexandrovic, V.; Krack, M.; Kornowski, A.; Förster, S.; Weller, H. Taylor-made ligands for biocompatible nanoparticles. *Angew. Chem. Int. Ed.* **2006**, *45*, 6577–6580.

26. Fischer, S.; Salcher, A.; Kornowski, A.; Weller, H.; Förster, S. Completely miscible nanocomposites.*Angew. Chem. Int. Ed.* **2011**. in press.

27. Mirkin, C.A.; Letsinger, R.L.; Mucic, R.C.; Storhoff, J.J. A DNA-based method for rationally assembling nanoparticles into macroscopic materials. *Nature* **1996**, *382*, 609–611.

28. Alivisatos, A.P.; Johnsson, K.P.; Peng, X.; Wilson, T.E.; Loweth, C.J.; Bruchez, M.P., Jr.; Schultz, P.G. Organization of 'nanocrystal molecules' using DNA. *Nature* **1996**, *382*, 609–611.

29. Zhang, J.P.; Liu, Y.; Ke, Y.G.; Yan, H. Periodic square-like gold nanoparticle arrays templated by self-assembled 2D DNA nanogrids on a surface. *Nano Lett.* **2006**, *6*, 248–251.

30. Zheng, J.W.; Constantinou, P.E.; Micheel, C.; Alivisatos, A.P.; Kiehl, R.A.; Seeman, N.C. Two-dimensional nanoparticle arrays show the organizational power of robust DNA motifs. *Nano Lett.***2006**, *6*, 1502–1504.

31. Cheng, W.; Campolongo, M.J.; Cha, J.J.; Tan, S.J.; Umbach, C.C.; Muller, D.A.; Luo, D. Free-standing nanoparticle superlattice sheets controlled by DNA. *Nat. Mater.* **2009**, *8*, 519.

32. Nykypanchuk, D.; Maye, M.M.; van der Lelie, D.; Gang, O. DNA-guided crystallization of colloidal nanoparticles. *Nature* **2008**, *451*, 549–552.

2

CHAPTER

Block Copolymer Nanostructures for Technology

Yu-Chih Tseng * and Seth B. Darling

Center for Nanoscale Materials, Argonne National Laboratory, 9700 South Cass Avenue, Argonne, IL 60439, USA

ABSTRACT

Nanostructures generated from block copolymer self-assembly enable a variety of potential technological applications. In this article we review recent work and the current status of two major emerging applications of block copolymer (BCP) nanostructures: lithography for microelectronics and photovoltaics. We review the progress in BCP lithography in relation to the requirements of the semiconductor technology roadmap. For photovoltaic applications, we review the current status of the quest to generate ideal nanostructures using BCPs and directions for future research.

Keywords: block copolymer; lithography; photovoltaics

1. INTRODUCTION: APPLICATIONS USING BLOCK COPOLYMER NANOSTRUCTURES

Block copolymers (BCPs), where dissimilar polymer blocks are joined by covalent bonds, draw tremendous interest for their potential applications. This is especially true for thin films of BCPs, which can be deposited inexpensively on a wide variety of substrates using solution processing techniques. The dissimilar blocks can separate into distinct domains with controllable dimensions and functionalities. The degree of long-range order and orientation of these domains can be controlled by a variety of factors, such as the interaction of the BCP molecules with the substrate, the film thickness, and the post-deposition annealing procedures [1,2]. Most studies on BCPs have focused on coil-coil diblock copolymers, where each block is a fully flexible polymer chain. Structures generated by these BCPs are well understood. For virtually all coil-coil BCPs, the various equilibrium morphologies can be completely described by a phase diagram such as that depicted in Figure 1a [3]. The self-assembled domain shapes can be tuned by adjusting the relative volume fraction of each block (f), Flory-Huggins interaction parameter χ, and the degree of polymerization N. In a typical BCP, the dimension of the domains range from 10 nm to 100 nm and can be controlled by changing the overall molecular weight of the macromolecule. BCP structures described in the phase diagram are schematically shown in Figure 1b [1]. This set of properties enables BCPs to be used as a general route for patterning a variety of materials into periodic structures. Transferring the self-assembled BCP pattern into the supporting substrate constitutes the most common example [4,5], while templating of magnetic materials [6], nanoparticles [7,8,9,10,11], and continuous thin metal films [12] have all been reported.

Among the plethora of potential applications for BCPs, two topical technologies will be addressed here: microelectronic manufacturing and photovoltaic cells. The former is already approaching commercialization, whereas the latter is just beginning to show its promise. Potential for these specific applications derives primarily from the fact that BCPs are amenable to solution-based processing methods and the nanostructures they generate can be applied inexpensively and quickly over a large area. This property leads to high throughput and is fundamentally important for both cases. At present, photolithography presents a bottleneck for throughput in the microelectronics manufacturing process, and the success of any competing patterning technology would need to have a similar throughput; not all lithography techniques are suitable [13]. In this respect, BCP-based lithography holds an advantage compared to other high-resolution patterning technologies such as electron beam lithography. For wide-scale global photovoltaic applications, the cost of generating electricity from any solar cell technology needs to be equal to or less than the cost of using conventional electricity sources such as fossil fuels, and current solar panels based on silicon are too expensive and too slow to produce. Polymer-based photovoltaic cells present an inexpensive and more scalable alternative, presuming efficiency and stability requirements can be met [14].

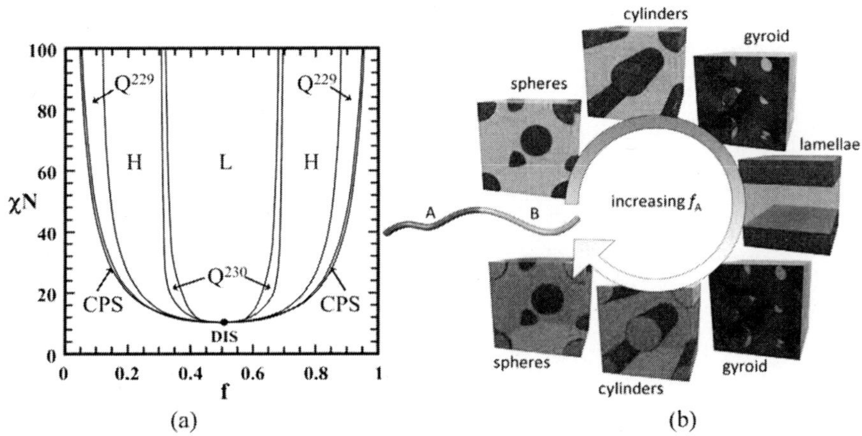

Figure 1. (a) Typical phase diagram of a coil-coil diblock copolymer.*f*: Volume fraction of one block. χ: Flory-Huggins interaction parameter. *N*: degree of polymerization. L: lamellae, H: hexagonally packed cylinders, Q^{230}: double-gyroid phase, Q^{229}: body-centered spheres, CPS: closed-packed spheres, DIS: disordered. Reprinted with permission from Reference [3]. Copyright 2006 American Chemical Society. **(b)** Structures of the different phases described in (a). f_A is the volume fraction of block A. Reprinted from Materials Today, Vol. 13, I. Botiz and S.B. Darling, Optoelectronics using block copolymers [25]. Copyright 2010 with permission from Elsevier.

The most important near-term application of BCPs is their use in the microelectronics industry. An impressive array of research has shown that BCPs can enhance the current capabilities of optical lithography. The dominant technology for producing fine patterns, optical lithography is fundamentally constrained by the wavelength of the light source. The microelectronics industry, the main driver of making patterns with dimensions far below the optical diffraction limit, has not reduced the wavelength of the illumination source of lithography tools to less than 193 nm since 2001, due to the difficulty of finding a suitable material for the optical elements in the lithography tools below this wavelength. In this context, templates based on block copolymers, which are high controllable, uniform and inexpensive to generate, present an attractive option to extend the semiconductor technology roadmap [15]. The magnetic recording industry has a similar impetus to make small, dense and regular patterns in magnetic materials to increase information storage density [16,17,18]. Pioneering work by Nealey *et al.* [19,20] and Ross *et al.* [21,22,23] shows the potential to generate such structures to enhance the capabilities of production-level lithography methods, by means of density multiplication.

Beyond generating fine patterns, BCPs may also be useful for photovoltaic applications. The solution-processible property of polymers is exemplified by large-area organic light-emitting diodes (OLEDs). The same property should enable large-area organic photovoltaic (OPV) cells to be produced inexpensively, although OPVs still suffer from considerable drawbacks, chiefly their low energy conversion efficiency (currently 8.13% at best [24]). In a typical OPV cell, two organic materials are present in the active layer (Figure 2), an electron donor (D) and an electron acceptor (A). The energy levels between the donor and acceptor material are offset as depicted in Figure 3 [25]. When light is absorbed in the active material, it generates bound electron-hole pairs (excitons), which are then separated at the donor-acceptor interface, leading to electric current. However, the photo-generated excitons do not diffuse further than about 10 nm before recombining. Moreover, charge carrier mobility is generally low in polymers, only on the order of 0.1 $cm^2/V{\cdot}s$ at best [26,27], and many separated charge carriers generated by solar illumination recombine before diffusing to the electrodes. Consequently, to convert light efficiently into mobile carriers, the D-A interface needs to be close to the location where the excitons are generated. At the same time, there should be a direct pathway for the separated carriers to reach the electrodes. Together, these requirements lead to the idealized design depicted in Figure 4b. The most efficient OPV cells demonstrated to date do not employ the structure shown in Figure 4b. Rather, they contain a disordered blend of the donor and acceptor materials that are phase-separated on a broadly distributed length scale. Not all photo-generated excitons are effectively separated, and the pathway to transport the separated carriers to the electrodes is convoluted and not continuous (Figure 4a). It has been proposed that employing the ideal structure in Figure 4b should improve the conversion efficiency.

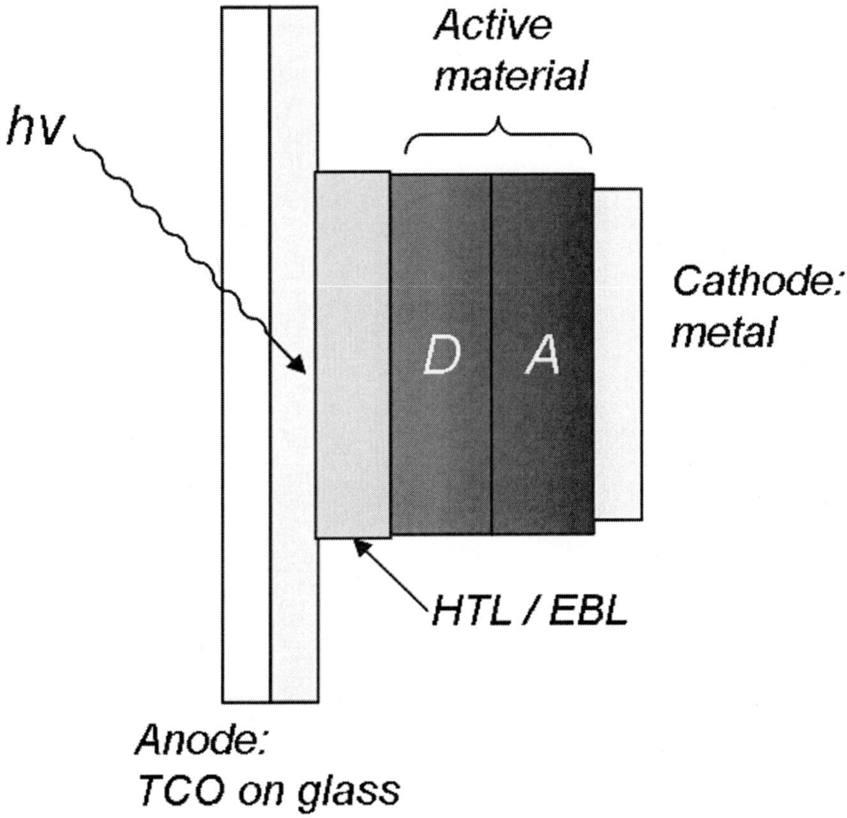

Figure 2. Structure of a bilayer organic photovoltaic cell. The material used for the anode is frequently a transparent conducting oxide (TCO) on glass. Exciton separation is encouraged by using a hole-transporting layer (HTL), which serves also as an electron-blocking layer (EBL). This layer allows holes to flow to the anode but not the electrons.

BCPs provide a promising way to generate a structure similar to that in Figure 4b, while allowing solution-processibility at the same time. Most proposals call for using a diblock copolymer with a donor block and an acceptor block. When processed in films, the donor and the acceptor blocks form separate domains, leading to structures approaching the idealized vision. Structure-based enhancement of OPV properties has been recently demonstrated [28], although the structure was formed using nanoimprint lithography, in which case the area of the PV cell is limited by the size of the stamp and achieving a characteristic length scale of the domains comparable to the exciton diffusion distance remains a challenge. Self-assembled structures using BCPs may be more easily fabricated, thereby enabling unprecedented studies of relationships between nanostructure and photophysical properties while also being potentially amenable to roll-to-roll, truly large-scale processing.

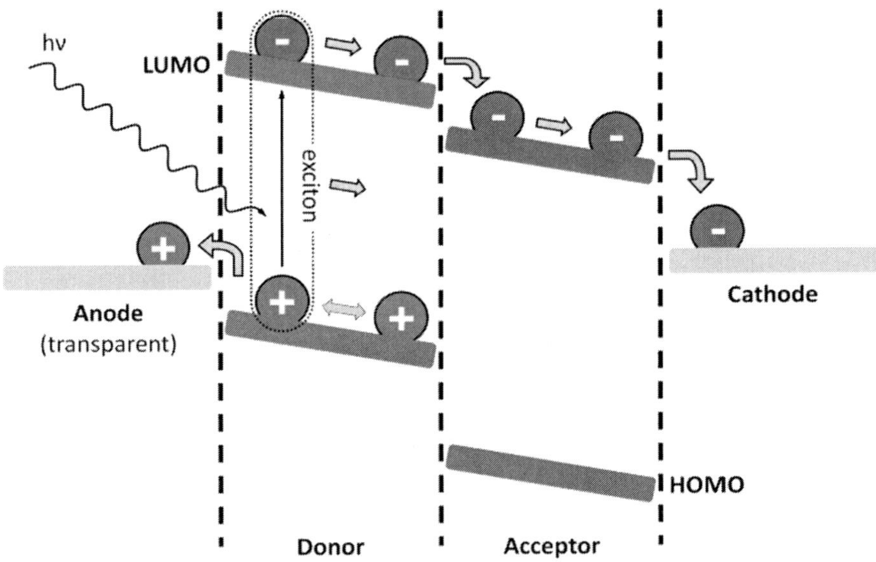

Figure 3. Schematic depiction of energy conversion in an organic photovoltaic cell. Absorption of light leads to the generation of excitons. The offset between the donor and acceptor energy levels (LUMO and HOMO) causes the excitons to separate. Reprinted from Materials Today, Vol. 13, I. Botiz and S.B. Darling, *Optoelectronics using block copolymers* [25]. Copyright 2010, with permission from Elsevier.

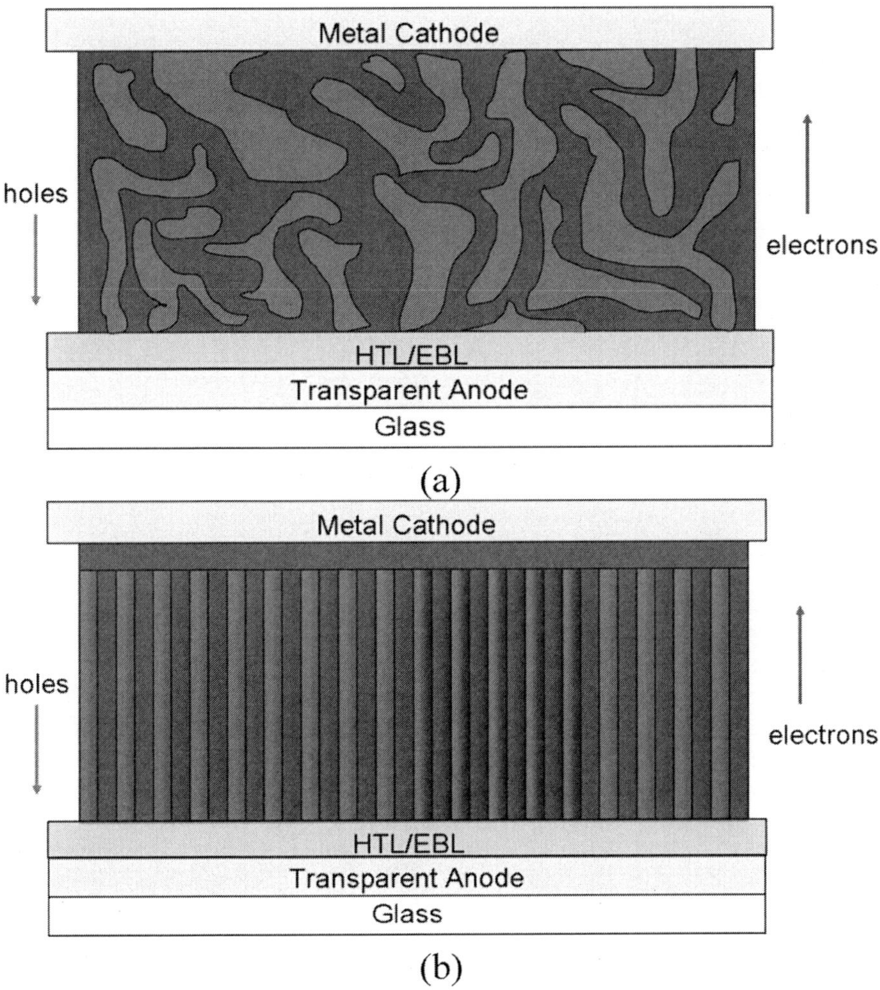

Figure 4. (a) Commonly depicted morphology of a bulk heterojunction OPV cell with disordered and discontinuous charge transport networks. Blue: acceptor domains. Red: donor domains. **(b)** Commonly proposed ideal morphology for an OPV cell. Blue: acceptor domains. Red: donor domains. The width of each domain is comparable to the exciton diffusion length and the active layer thickness is sufficient to absorb the incident sunlight.

In this review, we examine BCP-based nanostructures in the context of these two important applications: lithography for the microelectronics industry and solar energy conversion. Specifically, we examine the achievements and remaining challenges associated with BCPs in making high resolution patterns for the microelectronic industry. We review as well the applications of BCPs in photovoltaics, the current status of polymer assembly, the achieved energy conversion efficiency, and future directions of research. As related reviews exist in each of these areas, here we focus only on some key recent developments.

2. CURRENT STATUS OF DENSITY MULTIPLICATION USING GUIDED ASSEMBLY OF BLOCK COPOLYMERS

Optical lithography has been enormously successful in making fine patterns. Over a span of four decades, this technology has met the microelectronic industry's demand for resolution, uniformity, throughput, multi-layer registry and cost. The reduction of the illumination source's wavelength has largely enforced Moore's law for over 40 years. However, there are fundamental difficulties with further reducing the wavelength of the illumination source. Instead, working with an illumination wavelength of 193 nm, lithography engineers in the semiconductor industry employ sophisticated techniques, such as phase-shifting masks [29], immersion lithography [30,31], and increasingly, double-exposure methods [32] to meet the resolution requirements of the semiconductor technology roadmap.

BCPs can further advance optical lithography by ways of density multiplication. This is achieved by making BCP self-assemble on a pre-patterned substrate, as first demonstrated by Rockford *et al.* [33]. The patterns provide either topographical or chemical constraints that guide the assembly of the BCP domains and can be of much larger dimensions than the domains of the BCP. In general, they need to be an integer multiple of the period of the BCP structure, or commensurate, to generate defect-free patterns.

Nealey's group has done extensive work in assembling BCPs on surfaces with chemically modified patterns, showing the possibility of replicating faithfully the underlying patterns [34]. Using chemical contrast between SiO_2 and hydroxyl-terminated polystyrene, ordered poly(styrene-*block*-methyl methacrylate) (PS-*b*-PMMA) domains can be formed, with long-range ordering and a four-fold density multiplication [9]. Work from Cheng *et al.* [35] introduced the concept of inducing density multiplication by using a neutral surface to both BCP blocks, with constraints provided by chemical modification (Figure 5) [35]. Guided assembly of BCPs is sufficiently flexible to generate fairly complicated patterns that occur frequently in integrated circuits (Figure 6) [20,36], and beyond the periodic BCP patterns predicted by the phase diagram in Figure 1b. A substantial body of work has shown the possibility to constrain and order BCPs, especially PS-*b*-PMMA, using topographical features. However, to generate a truly periodic pattern over a large area using density multiplication, the guiding pattern needs to have the same orientation as the BCP domains (*i.e.*, posts to guide out-of-plane cylinders, trenches to guide lamellae or in-plane cylinders). This requires the thickness of the BCP layer to be appropriate for the desired patterns, and precise thickness control is critical [37]. Nevertheless, high density multiplying factor can be achieved, as exemplified by Tada *et al.*, where the density can be increased up to nine times [16] using pre-patterned and etched surfaces. By exploiting the effect of commensurability, recent work by Yang *et al.* [38] demonstrates the possibility of generating complex patterns in poly(styrene-*block*-dimethylsiloxane) (PS-*b*-PDMS) using only a sparse guiding pattern.

It has been consistently shown that guided self-assembly of BCPs is robust with respect to the imperfections in the underlying pattern. As an example, work by Tada *et al.* [17] clearly shows that order in the BCP film can be immune to defects in substrate topography, and that the commensurability of the guiding pattern has the strongest effect on the final BCP lateral order. BCP domains are also immune to non-idealities in the lithographic process, such as line-edge roughness (LER) in the topography [39] and even broken lines [35].

Even without using a guiding substrate, BCP domains can be ordered over a fairly large area [40,41]. Most experimental characterization of BCP structures relies on AFM imaging, a technique suitable for only relatively small area. Methods more applicable to large area examination such as Moiré patterns, GISAXS measurements, and SEM image-based orientation correlation have been demonstrated by Park*et al.* [42], and show that lateral order over 10 μm is possible.

The range of ordering can be greatly enhanced by the pattern of the underlying substrate [43,44,45] without interrupting the polymer lattice. Park *et al.* used a textured sapphire surface in combination with solvent annealing to generate perfect ordering of PS-*b*-PEO over an area of 4 cm^2 [46]. Recent work by Jung *et al.* [47] also shows that, in a lithography-guided assembly, the uniformity in the lateral ordering can be extended over an area of several cm^2.

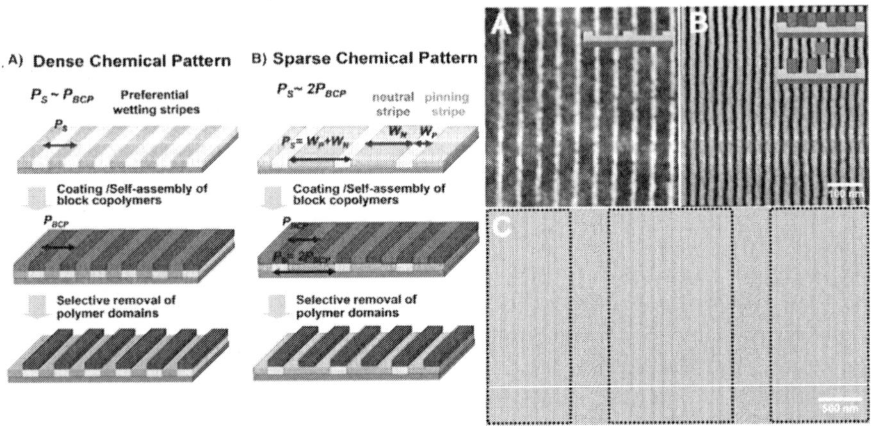

Figure 5. Density multiplication and directed assembly using chemical patterns. E-beam lithography defines sparse pinning stripes. (Left) Comparison of schemes for directing BCP self-assembly. Scheme B) was used to induce density multiplication. (Right) A) shows the sparse e-beam resist lines defining the pinning stripes. B: resulting PS-*b*-PMMA pattern showing density multiplication. Reprinted with permission from Reference [35]. Copyright 2008, Wiley-VCH.

3. SOME CHALLENGES AND POTENTIAL LIMITATIONS OF BCP LITHOGRAPHY

3.1. Pattern Transfer

Patterns generated using any lithography technique are usually transferred to the underlying materials using etching processes. For very small features, dry etching processes such as reactive ion etching (RIE) or ion milling are the only methods capable of transferring them faithfully. Since the polymer blocks in most BCPs are based on carbon, the etch selectivity between the blocks is generally poor in technologically relevant RIE gases. In the most commonly studied BCP

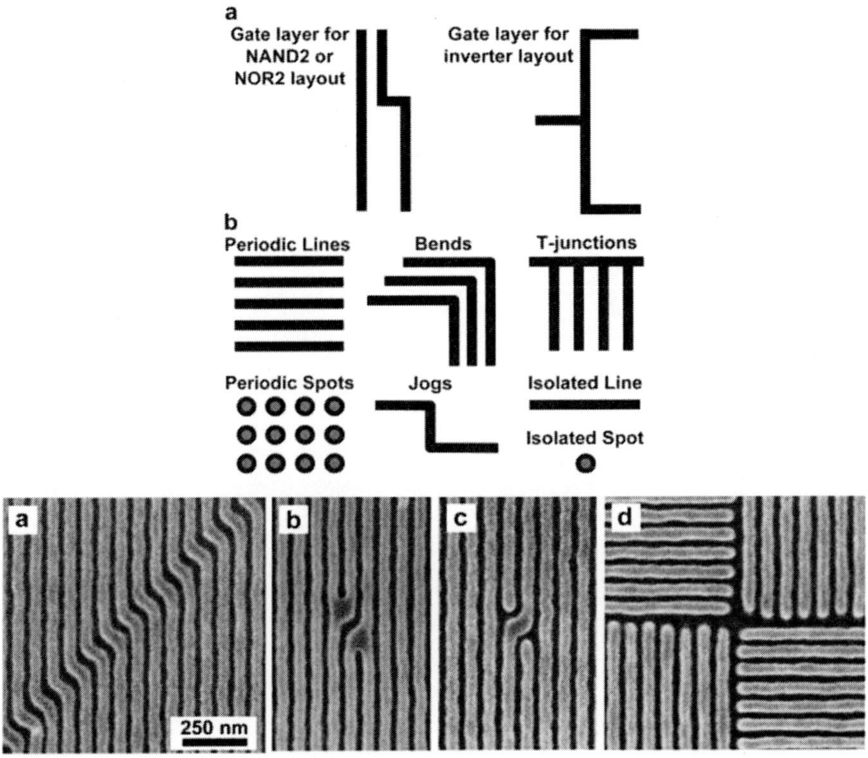

Figure 6. Common lithographic patterns for microelectronics and realization of some of the geometries using block copolymer self-assembly. Reprinted with permission from Reference [36]. Copyright 2007 American Chemical Society.

for lithography, PS-*b*-PMMA, the etch rate of PMMA blocks is only about twice that of PS for most RIE gases [48]. A BCP film used for lithography is typically thinner than 50 nm because of the critical dependence of the film morphology on the thickness, and is therefore too thin to withstand the etching process. Only a few polymer blocks have shown promising etch resistance, and most pattern transfer processes following BCP lithography necessitate a hard mask. Polymer blocks containing metals such as polyferrocenylsilanes (PFS) [49,50,51,52] or silicon such as PDMS [53,54,55] have enhanced resistance to plasma etching, and therefore a large etch contrast relative to organic blocks. However, since iron degrades transistor performance, its presence in semiconductor processing is generally undesirable, especially for front-end processes where the high-resolution capability of BCP lithography is most relevant.

3.2. Line-Edge Roughness

As mentioned in the Introduction, optical lithography has been successfully applied for microelectronics manufacturing because it satisfies many stringent requirements of the industry. Achieving high resolution is only one of the many requirements. For future technology nodes, the industry also needs better *quality* in the lithographic pattern, quantified namely by the line-edge roughness (LER) of a printed line in the photoresist (Figure 7) [56]. This roughness, if transferred into the smallest printed feature, which corresponds usually to the gate length of a transistor, can change the performance of the device significantly and in a complex manner. For an integrated circuit containing more than one billion such transistors, the results are mostly negative. Some consequences are: increased transistor leakage [57] (large stand-by power consumption), spread in threshold voltage [58] (decreased distinction between the digital 1s and 0s), and reduced read/write margin in a memory cell [59]. As a result, the requirements on LER are stringent, being $3\sigma < 1.4$ nm for technology nodes more advanced than 22 nm [13].

Guided BCP assembly can reduce LER of the guiding pattern [23], although the studies performed to date began with exaggerated roughness, and it is not clear whether BCPs can improve on the currently achievable LER using optical lithography. In fact, the BCP lithography community has yet to demonstrate a solution that meets the industry's requirement on LER. Intrinsic roughness of each BCP domain is the limiting factor, which arises from a non-vanishing interface width between two dissimilar domains (Figure 8). Simulation studies [60] suggest that increasing the Flory-Huggins parameter χ is the most direct way to reduce LER, since the interface width between the blocks is proportional to $\chi^{-1/2}$[61]. Highly dissimilar blocks such as PS and PDMS ($\chi = 0.26$) should improve considerably the edge sharpness of the domains [62], although there is a dearth of confirming reports. The LER also depends on the degree of polymerization N of each block, with larger N leading to smoother lines [63]. However, this is at odds with achieving high spatial resolution using small domain sizes, which would require low-molecular weight blocks. The optimization of LER is only beginning, and experimental results on quantification and reduction of the LER in BCP patterns are unfortunately still lacking. Recent measurement using soft x-ray diffraction [64] seems to provide a general pathway in characterizing this important parameter, although the reported quantity, $3\sigma \sim 4.9$ nm, is large for industrial lithographic purposes.

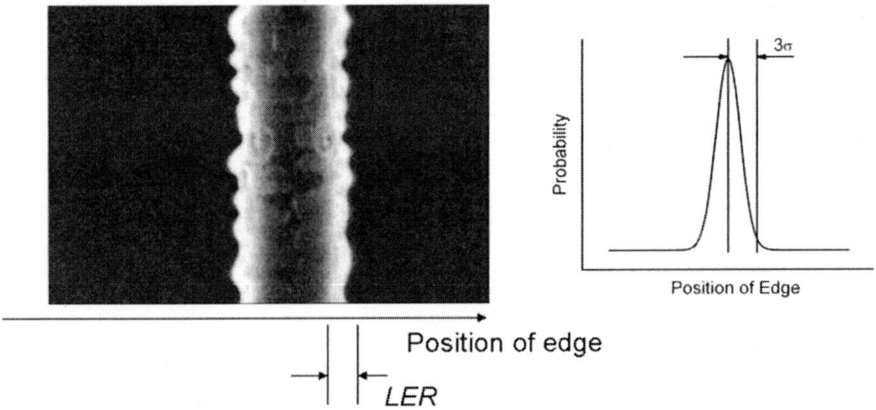

Figure 7. Definition of line edge roughness (LER). A normal distribution (right) of the line edge position is compiled from an SEM image (left), and LER is defined as 3 × the standard deviation σ. SEM image was reprinted with permission from Reference [56]. Copyright 1999, American Vacuum Society.

4. REQUIREMENTS ON THE NANOSTRUCTURE OF CONJUGATED BCPS FOR PHOTOVOLTAIC PURPOSES

Another emerging technological application of BCPs is their use in photovoltaic cells. The justification for applying BCPs to photovoltaics was explained in the Introduction, and much of the work was reviewed recently [25,65,66]. The polymers used for photovoltaic applications usually have a conjugated backbone on which the molecular orbitals are delocalized. This leads to an energy gap that matches parts of the solar spectrum, generally towards the bluer wavelengths for commonly studied materials. The rigid, conjugated backbone leads to different phase-separation behavior from the more frequently studied coil-coil BCPs. Conjugated blocks are often described as "rods" since they are rigid compared to more flexible coil blocks. Rod-coil BCPs form various kinds of lamellar phases [67] depending on the relative fraction of the rod. These phases are better described using the theory for liquid crystals, especially in the limit where the coil block constitutes only a small volume fraction. Olsen *et al.* [68,69] used a variety of techniques, including x-ray diffraction and optical transmission, to map out the phase diagram of a poly(phenylenevinylene-*block*-isoprene) PPV-*b*-PI rod-coil polymer.

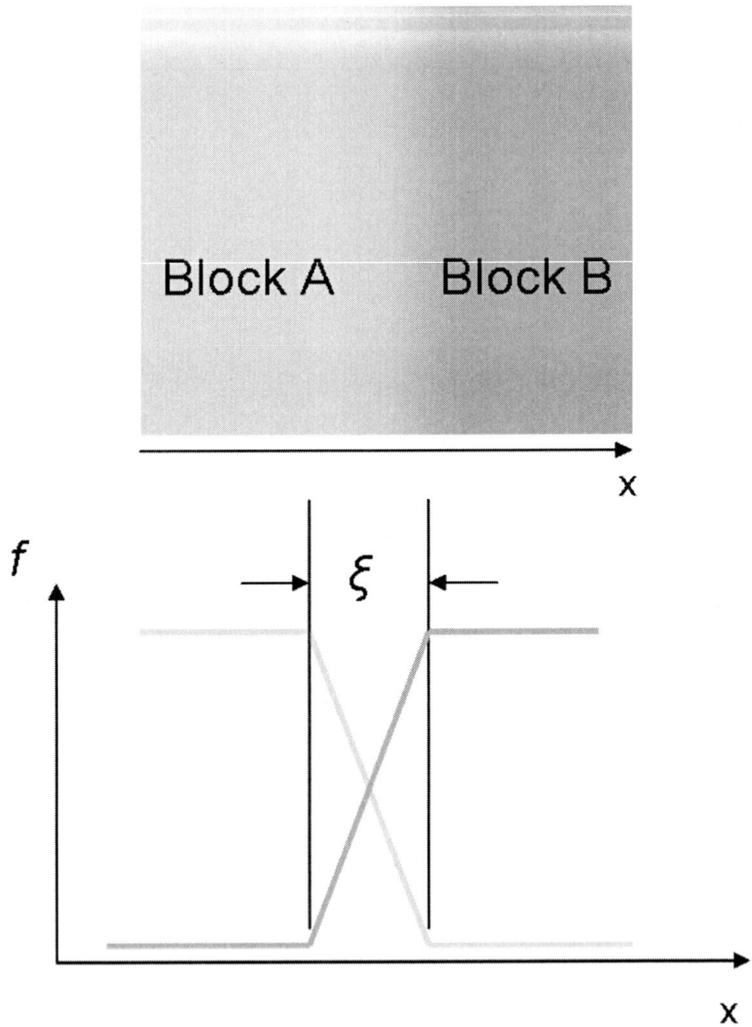

Figure 8. Illustration of ξ, the interface width between two polymer domains. f: volume fraction of each polymer block. Block A and B are randomly mixed inside the interface region.

For photovoltaic applications, the domain sizes need to match approximately the exciton diffusion length (~10 nm). Since the rod blocks are substantially dissimilar from the coil blocks, due to the stiffness of the rods, conjugated BCPs can easily separate into small (<10 nm) domains [67]. Because the conjugated blocks are rigid, it is not necessary to have a large block to form lamellar domains with width smaller than 10 nm. This is in contrast to coil-coil BCPs where a larger minimum degree of polymerization N is required to produce phase separation. Thus, for making the idealized OPV structure illustrated in Figure 4b, rod-coil BCPs can satisfy the requirements in dimension and structure. The spacing between the conjugated blocks, defined by the coil block, is also important, since this determines the dimensions of the other phase. This parameter can also be controlled, and it was shown by Tao *et al.* [70] that adding rod homopolymer to the BCP can change the spacing between the lamellar domains.

Detailed requirements on the conjugated BCP structure for PV purposes are, however, not limited to the size and orientation of the domains. Numerical simulation by Shah [71] shows that the anisotropic mobility in the polymer and a non-vanishing interface width between the donor/acceptor domains can both strongly influence the conversion efficiency (Figure 9) [71]. The interface between the rod phase and the coil phase can be sharp, but only at high rod fractions [72], and there is a paucity of measurements on the interface quality. Additionally, the ideal OPV structure depicted in Figure 4b can be realized only if the lamellae of the rod blocks align normally to the substrate. For a film that is thick enough to absorb a significant amount of light, this means that the lamellae of length comparable to the film thickness need to align normally to the substrate. In thick films of rod-coil BCPs, this is not always the case. Studies by Olsen *et al.* [73,74] on PPV-*b*-PI show that the orientation of the lamellae depends strongly on the affinity of the blocks to the substrate surface, and that the mutual alignment of the "grains" of lamellae becomes poor at large film thicknesses (Figure 10) [74]. This is analogous to coil-coil BCPs, where the orientation of (for example) cylindrical domains is influenced strongly by the substrate surface and the film thickness. In the context of making efficient OPV cells, the orientation of the lamellae is critical. Using a brush layer that presents a neutral surface to both blocks may help, although it would have to be a good conductor at the same time. Other ordering techniques, such as using magnetic fields [75] may also be adequate.

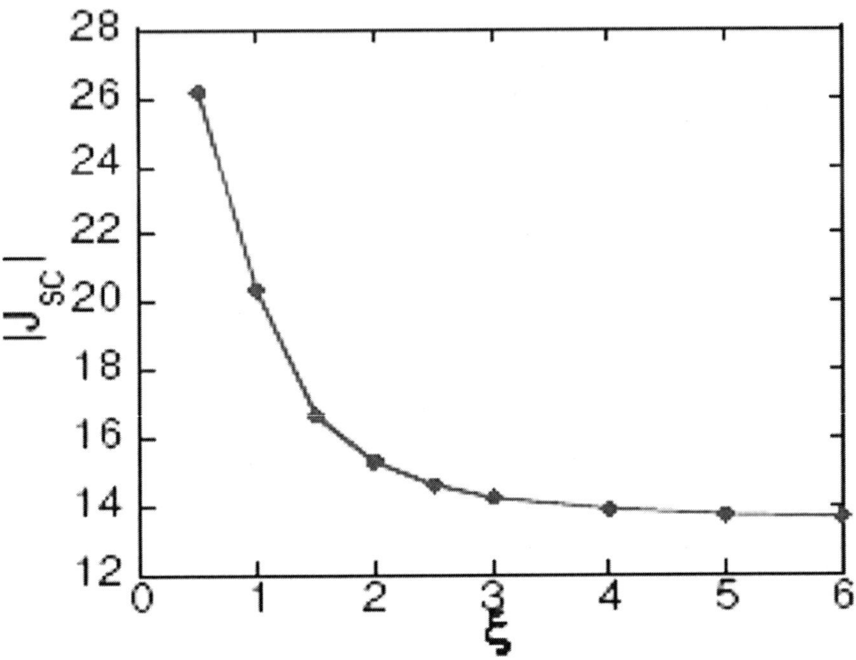

Figure 9. Calculated short-circuit current density J_{sc} (mA/cm^2) in an OPV cell *vs.* the interface width ξ (nm) between the donor and acceptor domain. The assumed structure contains donor and acceptor domains *perpendicular* to the plane of electrode. Reprinted with permission from Reference [71]. Copyright 2010 American Chemical Society.

5. CURRENT PROGRESS USING DONOR-ACCEPTOR BCPS

The highest-efficiency OPV cell reported to date is a homogeneous blend of electron-donating and -accepting materials without an ordered, continuous pathway for charge extraction. It does not guarantee uniform D/A interfaces for exciton separation throughout the active layer volume either. Phase separation of a copolymer consisting of a donor block and an acceptor block may be a straightforward way to generate an ideal OPV

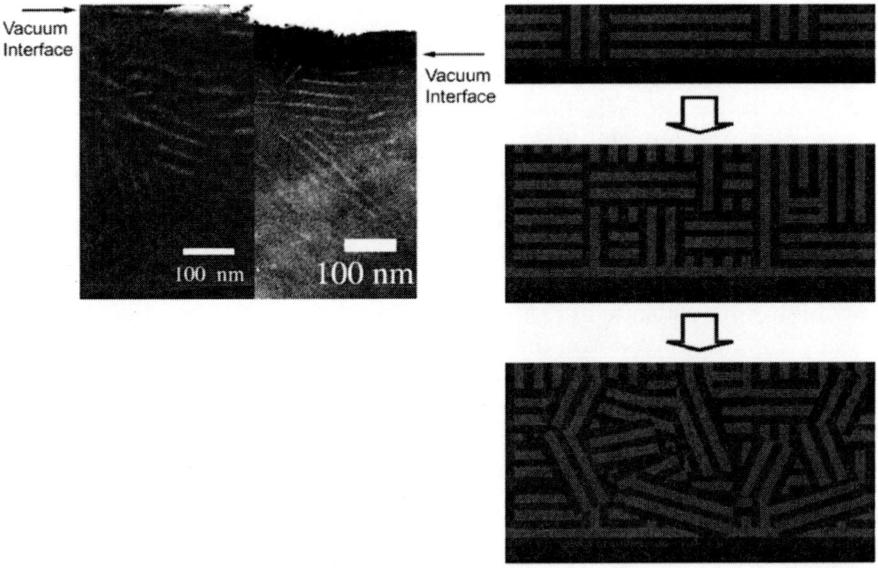

Figure 10. (Left) Cross-sectional TEM image of a rod-coil BCP film of PPV-*b*-PI. (Right) Schematic of morphology of a rod-coil BCP film as a function of film thickness. Reproduced with permission from Reference [74]. The Royal Society of Chemistry. Hyperlink:http://dx.doi.org/10.1039/B809092K.

morphology. Reported energy conversion efficiencies using this approach have not been high: Hazdiioannou's group has several reports on synthesizing BCPs containing PPV and C_{60}, with the latter grafted onto polystyrene [76]. A small photovoltaic effect under monochromatic illumination was demonstrated [77]. Similar work using a poly(3-hexylthiophene) (P3HT) donor block and C_{60}-grafted PMMA [78] shows improved photoluminescence quenching compared to a simple blend of P3HT and phenyl-C_{61}-butyric acid methyl ester (PCBM), indicating better exciton dissociation, although photovoltaic cells made from such material demonstrated only a modest photovoltaic effect. Recent work by Tao *et al.* [79] using P3HT as the donor and perylene as the acceptor, and by King *et al.* [80] using PvTPA-perylene, also demonstrate phase separation of the blocks, although the PV effect is decidedly modest in both cases. Zhang *et al.* [72] synthesized a P3HT-*b*-perylene diimide acrylate (P3HT-*b*-PPDA) block copolymer, taking advantage of the high carrier mobility in both phases. The films show complete PL quenching and the PV device shows a respectable conversion of efficiency of 0.49%, although no clear relationship was established between the structure of the film and the efficiency, since the device was fabricated on an as-spun thin film that did not show the same structure as a post-annealed film. A recent report validates this approach, however, by showing a significant increase in efficiency in an inverted device structure, using poly(3-hexylthiophene-*block*-4-vinyl pyridine) P3HT-*b*-P4VP copolymer, with PCBM acceptor grafted to the P4VP blocks [81].

While the donor-acceptor diblock copolymer methodology can enable fabrication of the ideal OPV structure, the covalent link between these two blocks may also lead to rapid charge carrier recombination. Several recent reports have examined the use of a spacer or "bridge" block between the donor and the acceptor to improve PV device performance. The demonstration by Sun [70] using PPV-based blocks shows that this approach can significantly improve the open-circuit voltage, although the photocurrent is low. Nevertheless, photoluminescence quenching is observed [82]. A similar attempt was made using polycarbazole (PCz) donors and perylenetetracarboxydiimide (PDI) acceptor with a saturated spacer [83], but a conversion efficiency of only 0.004% was demonstrated. Phase separation in a P3HT-*b*-PS-*b*-(PS/C_{60}) triblock copolymer, with ungrafted PS chains as spacer, has also been reported [84], although no device performance was presented.

Another approach to fabricate the structure in Figure 4b would be a multi-step process that employs a rod-coil BCP containing only the donor (rod) phase, and a coil block that can be removed after phase separation. The coil block can be etched and the space refilled with the acceptor material. Botiz *et al.*[85] and Boudouris *et al.*[86] both demonstrated the possibility of generating structures in poly(3-hexylthiophene-*block*-lactide) (P3HT-*b*-PLA) by the removal of the PLA domains, with the former work also showing subsequent re-filling using fullerene acceptors and concomitant PL quenching. The wet procedures used to etch the PLA blocks introduce the problem of pattern collapse [87], which commonly occurs in the fabrication of high-aspect ratio nanostructures. A three-dimensional structure, such as that generated using the gyroid phase [88] as in Figure 1b, may provide the structural rigidity that allows the pattern to withstand the capillary forces encountered during the fabrication procedures, although the pathway for charge carrier transport is less direct.

So far, donor-acceptor BCPs appear to have the most positive impact on OPV devices when they are used as a "compatibilizer", or surfactant to facilitate the blending of the donor and acceptor materials. The addition of BCP to PV devices containing P3HT and fullerenes improves the initial energy conversion efficiency of bulk heterojunction devices [89], and also significantly enhances their thermal stability [90,91,92]. Under prolonged thermal annealing, the addition of BCP prevents the macroscopic aggregation of P3HT and fullerenes [91,92]. This preserves the large exciton dissociation interface in a bulk heterojunction structure that is responsible for the high energy conversion efficiency. In a blend of polythiophene and polyquinoline, previously shown to have significant PL quenching [93], the improvement on morphology is also demonstrated [94], where the addition of the copolymer strongly reduces macroscopic phase separation.

6. SUMMARY AND FUTURE DIRECTIONS

There is enough knowledge on block copolymer synthesis and assembly to warrant serious consideration for important technological applications. For lithography, BCPs can provide uniform and small feature size, long-range order, robust etch resistance, and can generate diverse patterns. For photovoltaic applications, BCP self-assembly is a promising way to create, over a large area, nanoscopically ordered structure of functional materials with dimensions suitable for efficient extraction of photo-generated carriers.

Overcoming the remaining challenges in BCP lithography for industrial purposes requires engineering on the molecular level. Very dissimilar blocks that have large Flory-Huggins parameter would facilitate phase segregation for small molecules (enhancing resolution) and improve on the LER. By incorporating inorganic components, these blocks would also conceivably have larger etch contrast. Indeed, recent research on BCP lithography using polystyrene-*block*-polyethyleneoxide (PS-*b*-PEO) [95,96,97] shows the possibility of making dense sub-10 nm features, dimensions that are difficult to produce even for electron-beam lithography. Methods such as multi-layer BCPs [47], and inorganic polymer blocks containing polyhedral oligomeric silsesquioxane (POSS) [98] have also clearly demonstrated sub-10 nm feature sizes.

As to improving pattern transfer, polymer blocks containing inorganic blocks, such as PDMS, or the recently-reported POSS, can significantly enhance the etch contrast. Other issues concerning pattern transfer include the detailed three-dimensional geometry of the pattern. In photolithographically generated photoresist features, the side-wall profile strongly influences the final dimensions and geometry of the transferred pattern, and it is generally desired to have straight side walls that are perpendicular to the substrate plane. In diblock copolymer lithography, the side-wall angle is not independently controllable from the shape of the domains. In-plane cylinders have curved walls and would not do well for transferring the patterns vertically. Although out-of-plane cylinders and lamellar domains have vertical side walls, using these phases exclusively would place additional restriction the design of the polymer.

Beyond the challenging synthetic chemistry needed to make BCPs with functional blocks suitable for PV applications, many issues need to be resolved before BCPs can be a viable route for producing useful PV cells. Controlling the film morphology, such that the polymer domains orient perpendicularly to the substrate plane, stands out as a basic problem to be solved. This is analogous to the case of coil-coil BCPs used in lithography, where the domain orientation can be controlled by manipulating the boundary conditions of the BCP film. The interface sharpness between the blocks is also a fundamental obstacle, since it may limit the exciton separation efficiency. This is also a fundamental problem for BCP lithography.

Many of the challenges in improving the optoelectronic properties of BCPs are the same as those in organic optoelectronic devices in general. Organic PV materials usually absorb higher energy photons because of a large HOMO-LUMO gap. Since the optimal band-gap energy for all PV materials is about 1.4 eV [99], lowering the HOMO-LUMO gap is generally desired. Reducing the optical gap would also imply stronger optical absorption from a thinner film, and controlling the morphology of a thinner film using boundary conditions may be easier. Moreover, carrier mobility is generally low in organic semiconductors, leading to small exciton diffusion length. Improving carrier mobility would relax many requirements on the BCP geometry, especially the size of the domains. Nevertheless, despite these challenges, the fundamental advantage of BCPs—the possibility for inexpensive large-scale deployment remains a valid and compelling motivation for further research.

ACKNOWLEDGEMENTS

Use of the Center for Nanoscale Materials was supported by the U.S. Department of Energy, Office of Science, Office of Basic Energy Sciences, under Contract No. DE-AC02-06CH11357.

REFERENCES

1. Darling, S.B. Directing the self-assembly of block copolymers. *Progr. Polym. Sci.* **2007**, *32*, 1152–1204.

2. Kim, H.C.; Park, S.M.; Hinsberg, W.D. Block copolymer based nanostructures: Materials, processes, and applications to electronics. *Chem. Rev.* **2010**, *110*, 146–177.

3. Cochran, E.W.; Garcia-Cervera, C.J.; Fredrickson, G.H. Stability of the gyroid phase in diblock copolymers at strong segregation. *Macromolecules* **2006**, *39*, 2449–2451.

4. Guarini, K.W.; Black, C.T.; Zhang, Y.; Kim, H.; Sikorski, E.M.; Babich, I.V. Process integration of self-assembled polymer templates into silicon nanofabrication. *J. Vac. Sci. Technol. B* **2002**, *20*, 2788–2792.

5. Black, C.T.; Ruiz, R.; Breyta, G.; Cheng, J.Y.; Colburn, M.E.; Guarini, K.W.; Kim, H.C.; Zhang, Y. Polymer self assembly in semiconductor microelectronics. *IBM J. Res. Dev.* **2007**, *51*, 605–633.

6. Hamley, I.W. Nanostructure fabrication using block copolymers. *Nano Technol.* **2003**, *14*, R39–R54.

7. Aizawa, M.; Buriak, J.M. Block copolymer templated chemistry for the formation of metallic nanoparticle arrays on semiconductor surfaces. *Chem. Mater.* **2007**, *19*, 5090–5101.

8. Shenhar, R.; Norsten, T.B.; Rotello, V.M. Polymer-mediated nanoparticle assembly: Structural control and applications. *Adv. Mater.* **2005**, *17*, 657–669.

9. Darling, S.B. Mechanism for hierarchical self-assembly of nanoparticles on scaffolds derived from block copolymers. *Surf. Sci.* **2007**, *601*, 2555–2561.

10. Darling, S.B.; Hoffmann, A. Tuning metal surface diffusion on diblock copolymer films. *J. Vac. Sci. Technol. A* **2007**, *25*, 1048–1051.

11. Lopes, W.A.; Jaeger, H.M. Hierarchical self-assembly of metal nanostructures on diblock copolymer scaffolds. *Nature* **2001**, *414*, 735–738.

12. Park, S.; Kim, B.; Cirpan, A.; Russell, T.P. Preparation of metallic line patterns from functional block copolymers. *Small* **2009**, *5*, 1343–1348.

13. The International Technology Roadmap for Semiconductors. Available online: http://www.itrs.net(accessed on 18 October 2010).

14. Krebs, F. *Polymeric Solar Cells: Materials, Design, Manufacture*, 1st ed.; DEStech Publications, Inc.: Lancaster, PA, USA, 2010; pp. 119–169.

15. Craig, G.S.W.; Nealey, P.F. Directed assembly of block copolymers on lithographically defined surfaces. *Proc. SPIE* **2010**, *7637*, 76370L.

16. Tada, Y.; Akasaka, S.; Takenaka, M.; Yoshida, H.; Ruiz, R.; Dobisz, E.; Hasegawa, H. Nine-fold density multiplication of hcp lattice pattern by directed self-assembly of block copolymer. *Polymer* **2009**, *50*, 4250–4256.

17. Tada, Y.; Akasaka, S.; Yoshida, H.; Hasegawa, H.; Dobisz, E.; Kercher, D.; Takenaka, M. Directed self-assembly of diblock copolymer thin films on chemically-patterned substrates for defect-free nano-patterning. *Macromolecules* **2008**, *41*, 9267–9276.

18. Tada, Y.; Akasaka, S.; Chen, F.; Yoshida, H.; Takenaka, M.; Hasegawa, H. Density multiplication by directed self-assembly of block copolymer binary blends. *J. Photopolym. Sci. Technol.* **2009**, *22*, 229–233.

19. Ruiz, R.; Kang, H.M.; Detcheverry, F.A.; Dobisz, E.; Kercher, D.S.; Albrecht, T.R.; de Pablo, J.J.; Nealey, P.F. Density multiplication and improved lithography by directed block copolymer assembly. *Science* **2008**, *321*, 936–939.

20. Liu, G.L.; Thomas, C.S.; Craig, G.S.W.; Nealey, P.F. Integration of density multiplication in the formation of device-oriented structures by directed assembly of block copolymer-homopolymer blends. *Adv. Funct. Mater.* **2010**, *20*, 1251–1257.

21. Cheng, J.Y.; Ross, C.A.; Chan, V.Z.H.; Thomas, E.L.; Lammertink, R.G.H.; Vancso, G.J. Formation of a cobalt magnetic dot array via block copolymer lithography. *Adv. Mater.* **2001**, *13*, 1174–1178.

22. Ross, C.A.; Smith, H.I.; Savas, T.; Schattenburg, M.; Farhoud, M.; Hwang, M.; Walsh, M.; Abraham, M.C.; Ram, R.J. Fabrication of patterned media for high density magnetic storage. *J. Vac. Sci. Technol. B* **1999**, *17*, 3168–3176.

23. Bita, I.; Yang, J.K.W.; Jung, Y.S.; Ross, C.A.; Thomas, E.L.; Berggren, K.K. Graphoepitaxy of self-assembled block copolymers on two-dimensional periodic patterned templates. *Science* **2008**, *321*, 939–943.

24. Solarmer Energy, Inc. Available online: http://www.solarmer.com (accessed on 18 October 2010).

25. Botiz, I.; Darling, S.B. Optoelectronics using block copolymers. *Mater. Today* **2010**, *13*, 42–51.

26. Sirringhaus, H.; Brown, P.J.; Friend, R.H.; Nielsen, M.M.; Bechgaard, K.; Langeveld-Voss, B.M.W.; Spiering, A.J.H.; Janssen, R.A.J.; Meijer, E.W.; Herwig, P.; de Leeuw, D.M. Two-dimensional charge transport in self-organized, high-mobility conjugated polymers. *Nature* **1999**, *401*, 685–688.

27. Horowitz, G. Organic field-effect transistors. *Adv. Mater.* **1998**, *10*, 365–377.

28. He, X.M.; Gao, F.; Tu, G.L.; Hasko, D.; Huttner, S.; Steiner, U.; Greenham, N.C.; Friend, R.H.; Huck, W.T.S. Formation of nanopatterned polymer blends in photovoltaic devices. *Nano Lett.***2010**, *10*, 1302–1307.

29. Levenson, M.D.; Viswanathan, N.S.; Simpson, R.A. Improving resolution in photolithography with a phase-shifting mask. *IEEE Trans. Electron. Devices* **1982**, *29*, 1828–1836.

30. Hinsberg, W.; Wallraff, G.; Larson, C.; Davis, B.; Deline, V.; Raoux, S.; Miller, D.; Houle, F.; Hoffnagle, J.; Sanchez, M.; Rettner, C.; Sundberg, L.; Medeiros, D.; Dammel, R.; Conley, W. Liquid immersion lithography--Evaluation of resist issues. In *Liquid immersion lithography--Evaluation of Resist Issues*; Sturtevant, J.L., Ed.; SPIE-International Society Optical Engineering: Bellingham, WA, USA, 2004; pp. 21–33.

31. Lin, B.J. Immersion lithography and its impact on semiconductor manufacturing. In *Immersion Lithography and its Impact on Semiconductor Manufacturing*, 1st ed.; Smith, B.W., Ed.; SPIE-International Society Optical Engineering: Bellingham, WA, USA, 2004; pp. 46–67.

32. Byers, J.; Lee, S.; Jeri, K.; Zimmerman, P.; Turr, N.J.; Willson, C.G. Double exposure materials: Simulation study of feasibility. *J. Photopolym. Sci. Technol.* **2007**, *20*, 707–717.

33. Rockford, L.; Liu, Y.; Mansky, P.; Russell, T.P.; Yoon, M.; Mochrie, S.G.J. Polymers on nanoperiodic, heterogeneous surfaces. *Phys. Rev. Lett.* **1999**, *82*, 2602–2605.

34. Kim, S.O.; Solak, H.H.; Stoykovich, M.P.; Ferrier, N.J.; de Pablo, J.J.; Nealey, P.F. Epitaxial self-assembly of block copolymers on

lithographically defined nanopatterned substrates. *Nature* **2003**,*424*, 411–414.

35. Cheng, J.Y.; Rettner, C.T.; Sanders, D.P.; Kim, H.C.; Hinsberg, W.D. Dense self-assembly on sparse chemical patterns: Rectifying and multiplying lithographic patterns using block copolymers. *Adv. Mater.* **2008**, *20*, 3155–3158.

36. Stoykovich, M.P.; Kang, H.; Daoulas, K.C.; Liu, G.; Liu, C.C.; de Pablo, J.J.; Mueller, M.; Nealey, P.F. Directed self-assembly of block copolymers for nanolithography: Fabrication of isolated features and essential integrated circuit geometries. *ACS Nano* **2007**, *1*, 168–175.

37. Guarini, K.W.; Black, C.T.; Yeuing, S.H.I. Optimization of diblock copolymer thin film self assembly. *Adv. Mater.* **2002**, *14*, 1290–1294.

38. Yang, J.K.W.; Jung, Y.S.; Chang, J.B.; Mickiewicz, R.A.; Alexander-Katz, A.; Ross, C.A.; Berggren, K.K. Complex self-assembled patterns using sparse commensurate templates with locally varying motifs. *Nat. Nano Technol.* **2010**, *5*, 256–260.

39. Welander, A.M.; Nealey, P.F.; Cao, H.; Bristol, R. Impact of trench width roughness on the graphoepitaxial assembly of block copolymers. *J. Vac. Sci. Technol. B* **2008**, *26*, 2484–2488.

40. Kim, S.H.; Misner, M.J.; Russell, T.P. Solvent-induced ordering in thin film diblock copolymer/homopolymer mixtures. *Adv. Mater.* **2004**, *16*, 2119–2123.

41. Kim, S.H.; Misner, M.J.; Xu, T.; Kimura, M.; Russell, T.P. Highly oriented and ordered arrays from block copolymers via solvent evaporation. *Adv. Mater.* **2004**, *16*, 226–231.

42. Park, S.; Kim, B.; Xu, J.; Hofmann, T.; Ocko, B.M.; Russell, T.P. Lateral Ordering of cylindrical microdomains under solvent vapor. *Macromolecules* **2009**, *42*, 1278–1284.

43. Stein, G.E.; Kramer, E.J.; Li, X.; Wang, J. Single-crystal diffraction from two-dimensional block copolymer arrays. *Phys. Rev. Lett.* **2007**, *98*.

44. Segalman, R.A.; Yokoyama, H.; Kramer, E.J. Graphoepitaxy of spherical domain block copolymer films. *Adv. Mater.* **2001**, *13*, 1152–1155.

45. Fontana, S.M.; Dadmun, M.D.; Lowndes, D.H. Long-range order of cylinders in diblock copolymer thin films using graphoepitaxy. *Thin Solid Films* **2010**, *518*, 2783–2792.

46. Park, S.; Lee, D.H.; Xu, J.; Kim, B.; Hong, S.W.; Jeong, U.; Xu, T.; Russell, T.P. Macroscopic 10-Terabit-per-Square- Inch arrays from block copolymers with lateral order. *Science* **2009**, *323*, 1030–1033.

47. Jung, Y.S.; Chang, J.B.; Verploegen, E.; Berggren, K.K.; Ross, C.A. A path to ultranarrow patterns using self-assembled lithography. *Nano Lett.* **2010**, *10*, 1000–1005.

48. Ting, Y.H.; Park, S.M.; Liu, C.C.; Liu, X.S.; Himpsel, F.J.; Nealey, P.F.; Wendt, A.E. Plasma etch removal of poly(methyl methacrylate) in block copolymer lithography. *J. Vac. Sci. Technol. B* **2008**,*26*, 1684–1689.

49. Hempenius, M.A.; Lammertink, R.G.H.; Peter, M.; Vancso, G.J. Poly(ferrocenylsilanes) as etch barriers in nano and microlithographic applications. *Macromol. Sym.* **2003**, *196*, 45–56.

50. Chuang, V.P.; Gwyther, J.; Mickiewicz, R.A.; Manners, I.; Ross, C.A. Templated self-assembly of square symmetry arrays from an ABC triblock terpolymer. *Nano Lett.* **2009**, *9*, 4364–4369.

51. Ramanathan, M.; Nettleton, E.; Darling, S.B. Simple orientational control over cylindrical organic-inorganic block copolymer domains for etch mask applications. *Thin Solid Films* **2009**, *517*, 4474–4478.

52. Lammertink, R.G.H.; Hempenius, M.A.; van den Enk, J.E.; Chan, V.Z.H.; Thomas, E.L.; Vancso, G.J. Nanostructured thin films of organic-organometallic block copolymers: One-step lithography with poly(ferrocenylsilanes) by reactive ion etching. *Adv. Mater.* **2000**, *12*, 98–103.

53. Jung, Y.S.; Jung, W.; Ross, C.A. Nanofabricated concentric ring structures by templated self-assembly of a diblock copolymer. *Nano Lett.* **2008**, *8*, 2975–2981.

54. Chao, C.C.; Wang, T.C.; Ho, R.M.; Georgopanos, P.; Avgeropoulos, A.; Thomas, E.L. Robust block copolymer mask for nanopatterning polymer films. *ACS Nano* **2010**, *4*, 2088–2094.

55. Jung, Y.S.; Ross, C.A. Orientation-controlled self-assembled nanolithography using a polystyrene-polydimethylsiloxane block copolymer. *Nano Lett.* **2007**, *7*, 2046–2050.

56. Nelson, C.; Palmateer, S.C.; Forte, A.R.; Lyszczarz, T.M. Comparison of metrology methods for quantifying the line edge

roughness of patterned features. *J. Vac. Sci. Technol. B* **1999**, *17*, 2488–2498.

57. Xiong, S.Y.; Bokor, J.; Xiang, Q.; Fisher, P.; Dudley, I.; Rao, P. Study of gate line edge roughness effects in 50 nm bulk MOSFET devices. In *Study of Gate Line Edge Roughness Effects in 50 nm Bulk MOSFET Devices*, 1st ed.; Herr, D.J.C., Ed.; SPIE--International Society Optical Engineering: Bellingham, WA, USA, 2002; pp. 733–741.

58. Diaz, C.H.; Tao, H.J.; Ku, Y.C.; Yen, A.; Young, K. An experimentally validated analytical model for gate line-edge roughness (LER) effects on technology scaling. *IEEE Electron. Device Lett.* **2001**,*22*, 287–289.

59. Yu, S.M.; Zhao, Y.N.; Du, G.; Kang, J.F.; Han, R.Q.; Liu, X.Y. The impact of line edge roughness on the stability of a FinFET SRAM. *Semiconduct. Sci. Technol.* **2009**, *24*, 9.

60. Srinivas, G.; Swope, W.C.; Pitera, J.W. Interfacial fluctuations of block copolymers: A coarse-grain molecular dynamics simulation study. *J. Phys. Chem. B* **2007**, *111*, 13734–13742.

61. Bates, F.S.; Fredrickson, G.H. Block copolymer thermodynamics--Theory and experiment. *Ann. Rev. Phys. Chem.* **1990**, *41*, 525–557.

62. Ross, C.A.; Jung, Y.S.; Chuang, V.P.; Ilievski, F.; Yang, J.K.W.; Bita, I.; Thomas, E.L.; Smith, H.I.; Berggren, K.K.; Vancso, G.J.; Cheng, J.Y. Si-containing block copolymers for self-assembled nanolithography. *J. Vac. Sci. Technol. B* **2008**, *26*, 2489–2494.

63. Bosse, A.W.; Lin, E.K.; Jones, R.L.; Karim, A. Interfacial fluctuations in an ideal block copolymer resist. *Soft Matter* **2009**, *5*, 4266–4271.

64. Stein, G.E.; Liddle, J.A.; Aquila, A.L.; Gullikson, E.M. Measuring the structure of epitaxially assembled block copolymer domains with soft X-ray diffraction. *Macromolecules* **2010**, *43*, 433–441.

65. Darling, S.B. Block copolymers for photovoltaics. *Energ. Environ. Sci.* **2009**, *2*, 1266–1273.

66. Segalman, R.A.; McCulloch, B.; Kirmayer, S.; Urban, J.J. Block copolymers for organic optoelectronics. *Macromolecules* **2009**, *42*, 9205–9216.

67. Lee, M.; Cho, B.K.; Zin, W.C. Supramolecular structures from rod-coil block copolymers. *Chem. Rev.* **2001**, *101*, 3869–3892.

68. Olsen, B.D.; Segalman, R.A. Structure and thermodynamics of weakly segregated rod-coil block copolymers. *Macromolecules* **2005**, *38*, 10127–10137.

69. Olsen, B.D.; Segalman, R.A. Nonlamellar phases in asymmetric rod-coil block copolymers at increased segregation strengths. *Macromolecules* **2007**, *40*, 6922–6929.

70. Sun, S.S.; Zhang, C.; Ledbetter, A.; Choi, S.; Seo, K.; Bonner, C.E.; Drees, M.; Sariciftci, N.S. Photovoltaic enhancement of organic solar cells by a bridged donor-acceptor block copolymer approach. *Appl. Phys. Lett.* **2007**, *90*, 3.

71. Shah, M.; Ganesan, V. Correlations between morphologies and photovoltaic properties of rod-coil block copolymers. *Macromolecules* **2010**, *43*, 543–552.

72. Zhang, Q.L.; Cirpan, A.; Russell, T.P.; Emrick, T. Donor-acceptor poly(thiophene-block-perylene diimide) copolymers: Synthesis and solar cell fabrication. *Macromolecules* **2009**, *42*, 1079–1082.

73. Olsen, B.D.; Li, X.F.; Wang, J.; Segalman, R.A. Thin film structure of symmetric rod-coil block copolymers. *Macromolecules* **2007**, *40*, 3287–3295.

74. Olsen, B.D.; Li, X.F.; Wang, J.; Segalman, R.A. Near-surface and internal lamellar structure and orientation in thin films of rod-coil block copolymers. *Soft Matter* **2009**, *5*, 182–192.

75. Tao, Y.F.; Zohar, H.; Olsen, B.D.; Segalman, R.A. Hierarchical nanostructure control in rod-coil block copolymers with magnetic fields. *Nano Lett.* **2007**, *7*, 2742–2746.

76. Barrau, S.; Heiser, T.; Richard, F.; Brochon, C.; Ngov, C.; van de Wetering, K.; Hadziioannou, G.; Anokhin, D.V.; Ivanov, D.A. Self-assembling of novel fullerene-grafted donor-acceptor rod-coil block copolymers. *Macromolecules* **2008**, *41*, 2701–2710.

77. de Boer, B.; Stalmach, U.; van Hutten, P.F.; Melzer, C.; Krasnikov, V.V.; Hadziioannou, G. Supramolecular self-assembly and opto-electronic properties of semiconducting block copolymers. *Polymer* **2001**, *42*, 9097–9109.

78. Lee, J.U.; Cirpan, A.; Emrick, T.; Russell, T.P.; Jo, W.H. Synthesis and photophysical property of well-defined donor-acceptor diblock copolymer based on regioregular poly(3-hexylthiophene) and fullerene. *J. Mater. Chem.* **2009**, *19*, 1483–1489.

79. Tao, Y.F.; McCulloch, B.; Kim, S.; Segalman, R.A. The relationship between morphology and performance of donor-acceptor rod-coil block copolymer solar cells. *Soft Matter* **2009**, *5*, 4219–4230.

80. King, S.; Sommer, M.; Huettner, S.; Thelakkat, M.; Haque, S.A. Charge separation and recombination in self-organizing nanostructured donor-acceptor block copolymer films. *J. Mater. Chem.* **2009**, *19*, 5436–5441.

81. Sary, N.; Richard, F.; Brochon, C.; Leclerc, N.; Leveque, P.; Audinot, J.N.; Berson, S.; Heiser, T.; Hadziioannou, G.; Mezzenga, R. A new supramolecular route for using rod-coil block copolymers in photovoltaic applications. *Adv. Mater.* **2010**, *22*, 763–768.

82. Zhang, C.; Choi, S.; Haliburton, J.; Cleveland, T.; Li, R.; Sun, S.S.; Ledbetter, A.; Bonner, C.E. Design, synthesis, and characterization of a -donor-bridge-acceptor-bridge-type block copolymer via alkoxy- and sulfone-derivatized poly(phenylenevinylenes). *Macromolecules* **2006**, *39*, 4317–4326.

83. Yang, C. Donor/spacer/acceptor block copolymer containing poly(2,7-carbazole) and perylenetetracarboxydiimide subunits. *Macromol. Chem. Phys.* **2010**, *211*, 1446–1451.

84. Dante, M.; Yang, C.; Walker, B.; Wudl, F.; Nguyen, T.Q. Self-assembly and charge-transport properties of a polythiophene-fullerene triblock copolymer. *Adv. Mater.* **2010**, *22*, 1835–1839.

85. Botiz, I.; Darling, S.B. Self-assembly of poly(3-hexylthiophene)-block-polylactide block copolymer and subsequent incorporation of electron acceptor material. *Macromolecules* **2009**, *42*, 8211–8217.

86. Boudouris, B.W.; Frisbie, C.D.; Hillmyer, M.A. Nanoporous poly(3-alkylthiophene) thin films generated from block copolymer templates. *Macromolecules* **2008**, *41*, 67–75.

87. Botiz, I.; Martinson, A.B.F.; Darling, S.B. Minimizing lateral domain collapse in etched poly(3-hexylthiophene)-block-polylactide thin films for improved optoelectronic performance. *Langmuir* **2010**, *26*, 8756–8761.

88. Crossland, E.J.W.; Kamperman, M.; Nedelcu, M.; Ducati, C.; Wiesner, U.; Smilgies, D.M.; Toombes, G.E.S.; Hillmyer, M.A.; Ludwigs, S.; Steiner, U.; Snaith, H.J. A bicontinuous double gyroid hybrid solar cell. *Nano Lett.* **2009**, *9*, 2807–2812.

89. Yang, C.; Lee, J.K.; Heeger, A.J.; Wudl, F. Well-defined donor-acceptor rod-coil diblock copolymers based on P3HT containing C-60: the morphology and role as a surfactant in bulk-heterojunction solar cells. *J. Mater. Chem.* **2009**, *19*, 5416–5423.

90. Lee, J.U.; Jung, J.W.; Emrick, T.; Russell, T.P.; Jo, W.H. Synthesis of C_{60}-end capped P3HT and its application for high performance of P3HT/PCBM bulk heterojunction solar cells. *J. Mater. Chem.* **2010**, *20*, 3287–3294.

91. Lee, J.U.; Jung, J.W.; Emrick, T.; Russell, T.P.; Jo, W.H. Morphology control of a polythiophene-fullerene bulk heterojunction for enhancement of the high-temperature stability of solar cell performance by a new donor-acceptor diblock copolymer. *Nano Technol.* **2010**, *21*, 9.

92. Sivula, K.; Ball, Z.T.; Watanabe, N.; Frechet, J.M.J. Amphiphilic diblock copolymer compatibilizers and their effect on the morphology and performance of polythiophene: Fullerene solar cells. *Adv. Mater.* **2006**, *18*, 206–210.

93. Alam, M.M.; Jenekhe, S.A. Nanolayered heterojunctions of donor and acceptor conjugated polymers of interest in light emitting and photovoltaic devices: Photoinduced electron transfer at polythiophene/polyquinoline interfaces. *J. Phys. Chem. B* **2001**, *105*, 2479–2482.

94. Economopoulos, S.P.; Chochos, C.L.; Gregoriou, V.G.; Kallitsis, J.K.; Barrau, S.; Hadziioannou, G. Novel brush-type copolymers bearing thiophene backbone and side chain quinoline blocks. Synthesis and their use as a compatibilizer in thiophene-quinoline polymer blends. *Macromolecules* **2007**, *40*, 921–927.

95. Park, S.M.; Park, O.H.; Cheng, J.Y.; Rettner, C.T.; Kim, H.C. Patterning sub-10 nm line patterns from a block copolymer hybrid. *Nano Technol.* **2008**, *19*, 6.

96. Freer, E.M.; Krupp, L.E.; Hinsberg, W.D.; Rice, P.M.; Hedrick, J.L.; Cha, J.N.; Miller, R.D.; Kim, H.C. Oriented mesoporous organosilicate thin films. *Nano Lett.* **2005**, *5*, 2014–2018.

97. Cheng, J.Y.; Pitera, J.; Park, O.H.; Flickner, M.; Ruiz, R.; Black, C.T.; Kim, H.C. Rapid directed self assembly of lamellar microdomains from a block copolymer containing hybrid. *Appl. Phys. Lett.* **2007**, *91*, 3

98. Hirai, T.; Leolukman, M.; Liu, C.C.; Han, E.; Kim, Y.J.; Ishida, Y.; Hayakawa, T.; Kakimoto, M.; Nealey, P.F.; Gopalan, P. One-step

direct-patterning template utilizing self-assembly of poss-containing block copolymers. *Adv. Mater.* **2009**, *21*, 4334–4338.

99. Green, M.A. *Solar Cells, Operating Principles, Technology and System Applications*, 1st ed.; University of New South Wales: Sydney, Australia, 1998; pp. 88–89.

3

CHAPTER

Functionalization of Block Copolymer Vesicle Surfaces

Stefan Egli [1], Helmut Schlaad [2], Nico Bruns [1,*] and Wolfgang Meier [1,*]

[1]Department of Chemistry, University of Basel, Klingelbergstr. 80, 4056 Basel, Switzerland

[2]Max Planck Institute of Colloids and Interfaces, Department of Colloid Chemistry, Research Campus Golm, 14424 Potsdam, Germany

ABSTRACT

In dilute aqueous solutions certain amphiphilic block copolymers self-assemble into vesicles that enclose a small pool of water with a membrane. Such polymersomes have promising applications ranging from targeted drug-delivery devices, to biosensors, and nanoreactors. Interactions between block copolymer membranes and their surroundings are important factors that determine their potential biomedical applications. Such interactions are influenced predominantly by the membrane surface. We review methods to functionalize block copolymer vesicle surfaces by chemical means with ligands such as antibodies, adhesion moieties, enzymes, carbohydrates and fluorophores. Furthermore, surface-functionalization can be achieved by self-assembly of polymers that carry ligands at their chain ends or in their hydrophilic blocks. While this review focuses on the strategies to functionalize vesicle surfaces, the applications realized by, and envisioned for, such functional polymersomes are also highlighted.

Keywords: amphiphilic block copolymer; vesicle; polymersome; biomimetic membrane; surface functionalization; targeting ligand; conjugation; immobilization; functional blocks

1. INTRODUCTION

Membranes play a crucial role in several biological and chemical processes. Since Singer and Nicolson introduced the fluid mosaic model of the structure of cell membranes [1], which are viewed as two-dimensional solutions of lipids and globular proteins, polymer chemists and nanoscientists have developed new, artificial membranes that are superior regarding stability and functional versatility, but in most cases are more complex in structure [2,3]. The building blocks of such artificial membranes are, for example, amphiphilic block copolymers composed of hydrophobic and hydrophilic polymer chain blocks. Certain block copolymers self-assemble in an aqueous solution into membranes that, in turn, fuse into their entropically most stable morphology as hollow spheres, called polymer vesicles or polymersomes [4]. It has been shown that biological membrane processes can be successfully mimicked by, for example, reconstituting integral membrane proteins [5], fusing vesicles [6], or entrapping DNA [7]. Furthermore, polymersomes can be designed to be biocompatible [8]. With regard to potential applications of artificial membranes in life science, for instance in drug and gene delivery, cell surface recognition and nanoreactors, the development of so-called smart polymersomes that respond to internal or external stimuli such as temperature, pH, oxidative stress, light, magnetic fields and ultrasound has been highly promoted during the last few years [9-13]. One example of such smart materials are membranes consisting of biohybrid block copolymers that are inherently functional and respond to external stimuli [14]. Especially in the field of drug delivery, such triggers are of interest in order to release a hydrophilic agent at certain disease sites in an organism by changing the permeability or even the morphology of the vesicle membrane [10]. For efficient therapeutic applications, another essential issue is the targeting of specific sites, such as diseased cells or tissue, by an active receptor-ligand mechanism [15]. This targeting can be realized by attaching or introducing functional, in the majority of cases biological, ligands, such as sugars, aptamers, peptides and proteins, vitamins and antibodies [16] to the membrane surface. The specific immobilization of polymersomes to solid substrates and surfaces is also of high importance in other fields of application—in sensors and for surface modifications [17-19].

Various ways exist to generate a wide range of functionalities on the surface of the hydrophilic leaflets of amphipilic block copolymer membranes (Figure 1). In this review, we highlight the state of the art on how to functionalize block copolymer vesicle surfaces and show their applications in biomedicine, surface- and nanoscience. For this purpose, we focus on four different concepts of membrane modification: First, we discuss the conjugation of ligands to preformed vesicles, where the conjugation chemistries in aqueous solution will be highlighted. In the second part, we examine the formation of membranes from end-functionalized block copolymers. Then, we outline the formation of membranes from biohybrid block copolymers consisting of synthetic polymers and biologically relevant sugars, peptides, proteins and oligonucleotides. Finally, we show that membrane proteins that were reconstituted in block copolymer membranes, can act as specific recognition and attachment sites.

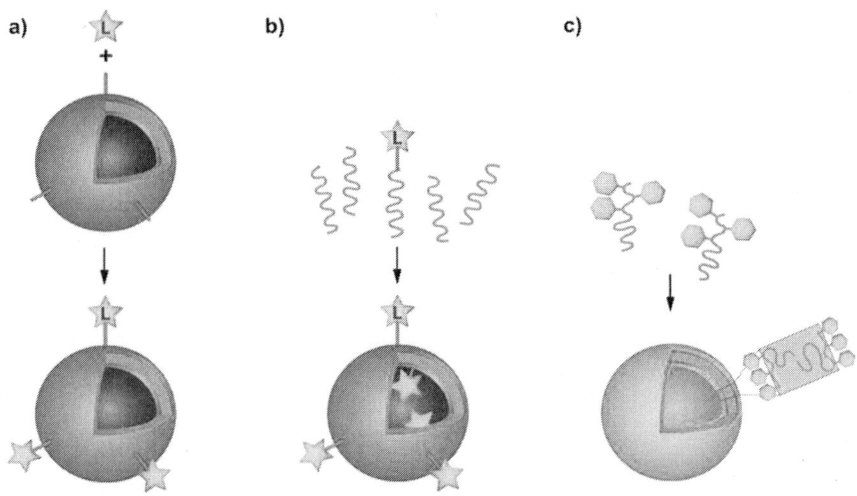

Figure 1. Routes to functional block copolymer vesicle surfaces. (**a**) Conjugation of a functional entity to preformed vesicles. (**b-c**) Self-assembly of functionalized and non-functionalized amphiphilic block copolymers into vesicles: (**b**) Endgroup functionalized block copolymers, and (**c**) block copolymers having a functional hydrophilic block. L = Functional entity or ligand: fluorophore, biomolecules, polymer tentacle, *etc.*

2. FUNCTIONALIZATION OF BLOCK COPOLYMER MEMBRANE SURFACES

2.1. Conjugation to Preformed Polymersomes

Polymersomes that have assembled from amphiphilic block copolymers have proven to be useful tools as drug delivery systems [8,20-22], nanoreactors and sensors [13,17,23-28]. The attachment of targeting ligands or enzymes to the polymersomes, as well as the immobilization of polymer vesicles on surfaces, is of crucial importance in most of the previously mentioned applications. In this chapter, we will review the chemical methods that have been used for conjugation of a variety of ligands to the membrane surfaces of preformed polymersomes (Figure 1(a)). The main requirements of such conjugation chemistries are their viability in an aqueous environment, the need to avoid cross-linking between polymersomes and between ligands, as well as stability and irreversibility of the resulting bond. Furthermore, it is desirable that the bonds be detectable and quantifiable in a simple, non-destructive manner. The state of the art in terms of common conjugation chemistry that fulfills all or some of these requirements is summarized in Table 1.

Table 1. Overview of methods used to conjugate ligands and other functional entities to preformed block copolymer vesicles.

Polymersome functional group	Ligand / substrate functional group	Resulting bond	Block copolymer	Ligand / substrate
Alkyne	Azide	1,4-disubstituted 1,2,3-triazole	PS_{40}-*b*-$PIAT_{50}$, or mixture of PS_{40}-*b*-$PIAT_{50}$ with PS_{40}-*b*-PEG_{68} [26,29]	Horseradish peroxidase [26] and an azidofunctionalized *Candida antarctica* Lipase B [29]
Azide	Alkyne	1,4-disubstituted 1,2,3-triazole	PBD-*b*-PEG [30,31], PS-*b*-PAA [32]	Mannose functional polyester dendron [30,31], dansyl probe [32], biotin [32], eGFP [32]
Biotin	Streptavidin	Biotin-Streptavidin interaction	PBD-*b*-PEG and PEG-*b*-PEE [33], PMOXA-*b*-PDMS-*b*-PMOXA [17,20,34,35]	Superavidin-coated microspheres [33], antibody [36], polyguanylic acid [20,34], streptavidin-modified glass surface [17], fluorescently labeled avidin [33,35]
Ni^{2+}-nitrilotriacetic acid (Ni^{2+}-NTA)	Oligohistidine sequence	Complexation	PDB-*b*-PEG [37,38]	Fluorescein labeled maltose binding protein [37], eGFP [37,38], RFP [38]
Cyclodextrin (CD)	Adamantane	Hydrophobic interactions	PS-CD [39], CD-PEI-CD [40]	HRP [39], PEG [40]
N-hydroxysuccinimidyl ester	Thiol	C-S bond	PCL-*b*-PEG [41]	Antibody OX26 [41]
Vinyl sulfone	Thiol	C-S bond	PEG-*b*-PMCL [42]	Cysteine-RGD peptide [42]
Amine	Succinimidyl ester	Amide	PDMS-*b*-PMOXA [43]	Alexa Fluor 633 succinimidyl ester [43]
Aldehyde	Amine	Imine	PCL-*b*-PEG, PLA-*b*-PEG and PI-*b*-PEG [19]	Aminated glass surface [19]
4-Formyl benzoic amide	6-Hydrazinonicotinic amide	Bis-arylhydrazone	PDMS-*b*-PMOXA [43]	eYFP, anti-biotin-IgG, trastuzumab [43]

Enhanced green fluorescent protein (eGFP), enhanced yellow fluorescent protein (eYFP), horseradish peroxidase (HRP), immunoglobulin G (IgG), poly(butadiene) (PBD), poly(caprolactone) (PCL), poly(dimethylsiloxane) (PDMS), poly(ethylethylene) (PEE), poly(ethylene glycol) (PEG), poly(ether imide) (PEI), poly(l-isocyanoalanine(2-thiophen-3-yl-ethyl)amide) (PIAT), poly(isoprene) (PI), poly(lactide) (PLA), poly(γ-methyl-ε-caprolactone) (PMCL), poly(2-methyl-2-oxazoline) (PMOXA), polystyrene (PS), red fluorescent protein (RFP).

The first method that was used to attach ligands to the surface of preformed polymersomes was the biotin-streptavidin conjugation approach [33], which allows an arbitrary variation of ligands bearing biotin groups [17,20,34,35]. Ligands are bound indirectly by biotin-streptavidin-biotin interactions to the polymersome surface. The modification of block copolymers with biotin has been performed either by*N*,*N'*-dicyclohexylcarbodiimide/4-(dimethylamino)pyridine-activated esterification [17,20,34,35] or by preactivation of the terminal hydroxyl functionality by tresyl chloride [44] or 4-fluoro-3-nitrobenzoic acid [36,45] and subsequent reaction with biocytin [33,36]. Both the esterification and the tresylation method resulted in a good yield of biotin modified polymer, which is essential for streptavidin and thus ligand binding after self-assembly of the polymersome.

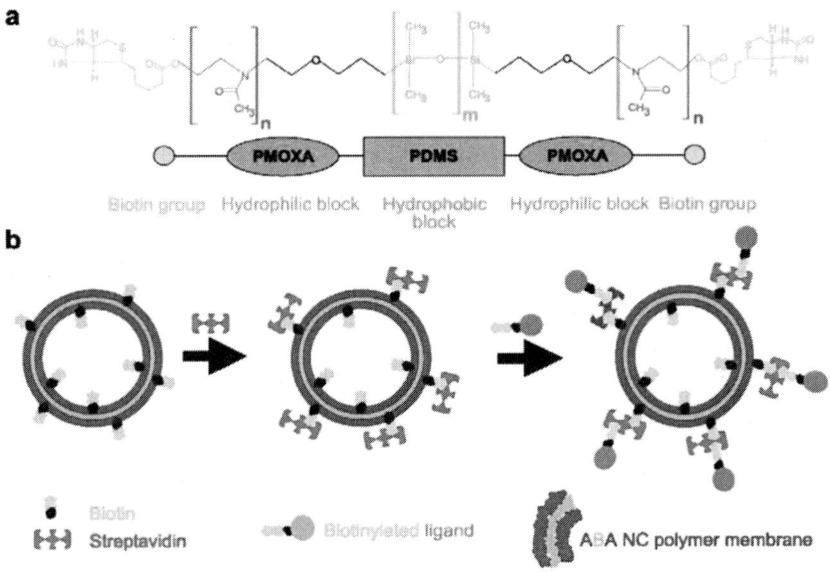

Figure 2. Surface-modification of polymersomes exploiting biotin-streptavidin binding. (**a**) Biotin- functionalized poly(2-methyl-2-oxazoline)-*block*-poly(dimethylsiloxane)-*block*-poly(2-methyl-2-oxazoline) (PMOXA-*b*-PDMS-*b*-PMOXA). This was synthesized from the unmodified block copolymer by an esterification. (**b**) Attachment of biotinylated ligands to the self-assembled polymersomes by a biotin-streptavidin binding approach. Reproduced by permission of Elsevier [20].

In order to obtain polymersomes that interact with biotin-modified ligands, biotin-modified block copolymers (Figure 2(a)) were hydrated in aqueous solution to form polymersomes that were then incubated in an excess of streptavidin (Figure 2(b)). The optimal concentrations of polymersomes and streptavidin need to be determined experimentally in order to avoid undesired cross-linking reactions. If the ratio of streptavidin to biotinylated polymersomes is too low, the probability of cross-linking between polymersomes increases. Once the surface of the biotin modified polymersome had been saturated with streptavidin, biotin modified ligands were then attached.

The modification strategy has some possible disadvantages. Although biotin and streptavidin form one of the strongest non-covalent interactions known (with dissociation constant around 10^{-15} M) [46], it is not a covalent bond. After binding, there is thus the possibility of ligand exchange with other biotinylated molecules in response to a change of ionic strength and/or temperature [47]. Furthermore, streptavidin is a macromolecule of considerable size (52.8 kDa). Therefore, when small ligands are intended for attachment to polymersomes via this strategy, the atom efficiency is not very favorable and steric problems might occur.

The binding of fluorescent dye-modified streptavidin to biotin modified polymersomes, with diameters between 140 and 172 nm, was demonstrated by fluorescence imaging [33], fluorescence correlation spectroscopy (FCS), as well as fluorescence cross-correlation spectroscopy (FCCS) [35]. A model system, investigated by FCS and FCCS and comprising rhodamine-green-biotin modified polymersomes and cyanin5-labeled streptavidin, was utilized in order to determine the number of binding sites and the dissociation constant, K_D, of the receptor interaction between biotin-modified polymersomes and streptavidin [35]. An average number of 1921 ± 357 streptavidin-Cy5 per polymersome and an intrinsic K_D of $(1.7 \pm 0.4) \times 10^{-8}$ M were determined [35]. The latter differs from the value of 10^{-15} M given in the literature for the biotin-streptavidin interaction, which was explained by entropic and steric effects.

The biotin-streptavidin binding approach was used to modify polymersomes with the oligonucleotide polyguanylic acid (polyG), which specifically targets the macrophage A1 scavenger receptor [20]. Moreover, the decorated polymersome has also proven to be suitable to target tumor cell lines [34]. The receptor-specific binding of these polymersomes in human and transgenic cell lines and in mixed cultures was followed by vesicular uptake. A major improvement in biologic efficacy was obtained for polymersomes (containing pravastatin) compared to free drug, whereas no increased cytotoxic effect was observed in muscle tissue [34].

The group of Hammer has conducted extensive work on the surface-modification of polybutadiene-*block*-poly(ethylene glycol) (PBD-*b*-PEG) polymersomes via biotin-avidin interactions with the aim of creating and investigating vesicles that stick to biological surfaces under flow conditions, thus mimicking the adhesive properties of leucocites (Figure 3). The first papers in this series investigated the adhesion of biotin-decorated vesicles to avidin-coated surfaces [33,48]. To this end, PBD-*b*-PEG conjugated to biotin-lysin (biocytin) was synthesized. The functionalized copolymer formed vesicles on its own, or when mixed with unmodified block copolymers. Two different unfunctionalized block copolmers were mixed with the biotinylated block copolymer. One was shorter, the other was the same length as the functionalized polymer. Micropipette aspiration allowed measurement of the critical tension required to peel the membrane away from a bead covered with avidin as a function of biotin surface density and of the length-ratio between unmodified and modified block copolymers. When biocytin was conjugated to a PEG block that was longer than the surrounding unmodified PEG blocks, the critical tension reached a maximum at a content of 55% functionalized blockcopolymer. However, when both copolymers were of the same length, the critical tension was maximal at 10%, after which it increased only marginally. Thus, the biotin end groups need to be accessible at the surface, while they tend to be buried in the hydrophilic polymer layer at higher concentrations.

Biotinylated vesicles were coated with neutravidin, by means of which various biotin-labeled ligands were able to be conjugated to the surface, these include Alexa Fluor 488-biocytin, biotinylated adhesion molecule-1 antibody (anti-ICAM-1) and biotin-sialyl Lewis^x (sLe^x; a selectin ligand) (Figure 3(a)) [36]. The surface density of the antibody was tuned by varying the ratio of anti-ICAM-1 to sLe^x during the conjugation. The adhesiveness was measured between the decorated vesicles and ICAM-1-coated beads. In contrast to previous studies, the adhesion strength did not depend on the amount of biotinylated block copolymer, but rather increased linearly with the surface-density of the binding ligand, because the flexible polymer chains are buried underneath the coat of neutravidin (Figure 3(b) and Figure 3(c)). Under physiological flow rates, such decorated vesicles do adhere to surfaces coated with inflammatory adhesion molecules P-selectin (to which sLe^x binds) and ICAM-1 [45,49], as well as to inflamed endothelium [49], indicating possible applications as a targeted drug-delivery system.

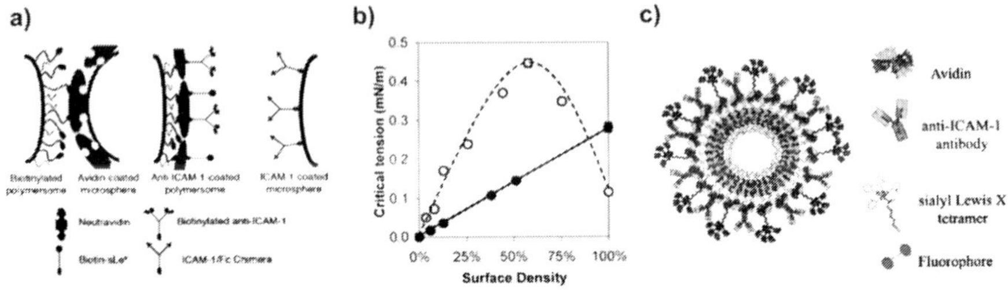

Figure 3. (**a**) Schematic illustration of biotin-coated polymersome and anti-ICAM-1-coated polymersome. This system was used to compare the critical tension behavior of biotinylated polymersomes to avidin-coated microspheres and anti-ICAM-1-coated polymersomes to ICAM-1-coated microspheres, respectively. (**b**) Comparison of the critical tension of those two different receptor-ligand pairs: biotin-avidin (open circle, dashed line) and anti-ICAM-1-ICAM-1 (closed circle, solid line). The divergence between the two critical tension curves is due to the variable manner by which the adhesion molecules were presented. (**c**) Schematic illustration of a leukocyte-mimicking polymersome. In order to mediate adhesion, biotinylated sialyl-Lewis X or biotinylated anti-ICAM-1, or both simultaneously, was (were) attached to the polymersome surface. Reproduced by permission of the American Chemical Society [36,49].

Although cell-specific targeting *in vitro* was shown with polymersomes that were surface-modified by a biotin-streptavidin approach [20,34], such polymersomes would not be of use *in vivo* studies because streptavidin is known to block essential immune reactions in the human body [50].

Another method of non-covalent but selective attachment of ligands to polymersomes and solid supported membranes was described by Nehring *et al.* [37,38]. Polymersomes were synthesized based on PBD-*b*-PEG diblock copolymer comprising carboxylic acid or lysine-nitilotriacetic acid (NTA) functional end groups. No disturbance of self-assembly behavior was observed upon mixing these two polymers. After vesicle formation, the peripheral NTA groups were able to form complexes with Ni^{2+} or Cu^{2+} ions. NTA-metal complexes are well known to selectively bind to oligohistidine sequences of proteins. Thus, maltose binding protein labeled with fluoresceine (His_{10}-MBP-FITC) [37], as well as His-tagged enhanced green fluorescent protein (His_6-eGFP) [37,38] and red fluorescent protein (His_6-RFP) [38], were able to be conjugated to the vesicle surface. After confirmation of the presence of the polymer-metal complex by UV-Vis spectroscopy and electron paramagnetic resonance (EPR), dissociation constants K_D of the different His-modified proteins to the Ni^{2+}-NTA functionalized polymersome were determined by FCS. A K_D of 7.0 ± 1.2 μM was measured for His10-MBP-FITC, 12.3 ± 1.2 μM for His_6-eGFP and 1.99 ± 0.42 μM for His_6-RFP [37]. These values are the same order of magnitude as the $K_D$values obtained with Ni^{2+}-NTA-modified liposomes (K_D = 4.3 μM) [51] and Cu^{2+}-NTA complexes [52].

Solid supported monolayers of the same block copolymers were prepared by Langmuir-Schaefer transfer and incubated with a His-tagged protein, with the aim of investigating protein binding to metal ion-complexing polymer surfaces. Atomic force microscopy (AFM) revealed structural details of the protein-decorated membranes, indicating that polymer monolayers may induce the formation of highly ordered protein arrays. Thus, the immobilization of densely packed proteins to such planar surfaces is a key step for 2D protein crystallization [53].

A further type of non-covalent interaction, the binding of adamantane in the cavity of cyclodextrins, can be exploited to functionalize polymersome surfaces. PS homopolymer terminated with a permethylated β-cyclodextrin (β-CD) group [39], as well as polyether imide (PEI) terminated at both chain ends with β-CDs [40] self-assembled into polymersomes, due to the hydrophilicity of the carbohydrate headgroups and the hydrophobicity of the polymer. The surface of these vesicles is a corona of β-CDs, which allowed the conjugation of adamantane-tagged horse radish peroxidase [39] and adamantane-terminated PEG [40]. Isothermal titration calorimetry (ITC) and static light scattering (SLS) revealed that both the inner and the outer polymersome surfaces were modified with the PEG. The enzyme-coated polymersomes were subjected to multiple washing steps. The amount of enzyme on the vesicles was found to decrease, indicating that the non-covalent interaction between β-CD and adamantane is too weak to withstand the washing procedure [39].

A much more efficient and—in contrast to the biotin-streptavidin, Ni-NTA-His-tag, and β-CD-adamantane interactions—covalent conjugation method is the Huisgen 1,3-dipolar cycloaddition of azides and alkynes, promoted by Sharpless and co-workers as a reaction that fulfills the criteria of click chemistry [54]. This reaction is usually carried out in a variety of organic solvents such as DMF, DMSO, THF, halogenated solvents [55], or in water-alcohol mixtures or pure water [56]. It is driven by heat or a catalytic amount of Cu(I). Since most polymersomes are destabilized in the presence of organic solvents, it is a prerequisite to perform this click chemistry conjugation reaction in aqueous media. Click chemistry with polymersomes in aqueous solution was introduced by van Hest and co-workers using poly(styrene)-*block*-poly(acrylic acid) (PS-*b*-PAA) diblock copolymers with terminal azide groups (Figure 4)

[32]. Different ligands, such as an alkyne bearing fluorescent dansyl probe, biotin and eGFP, have been conjugated to preformed polymersomes.

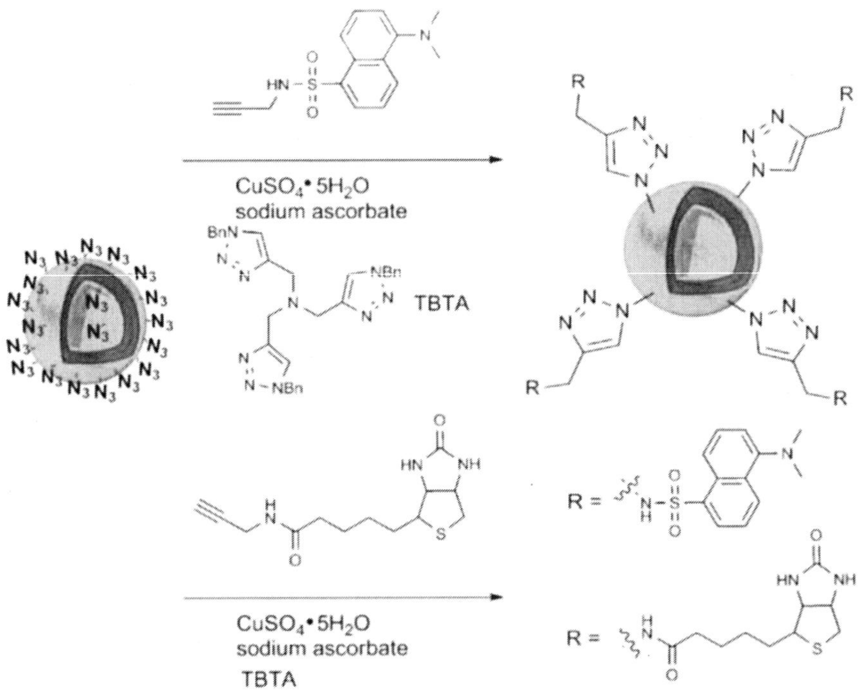

Figure 4. Modification of the outer polymersome surface by click chemistry. Alkyne bearing fluorescent dansyl probe and biotin were attached to azide functionalized polymersomes. Reproduced by permission of the Royal Society of Chemistry [32].

The Cu(I) species that is used to catalyze this click reaction is usually formed *in situ* by mixing sodium ascorbate with a copper sulfate solution. In order to reach high conversion, the reduced Cu(I) species should be kept active for about 24 hours. For this purpose, the Cu(I) was stabilized by tris-(benzyltriazolylmethyl)amine (TBTA) or sodium bathophenantroline disulfonate (BPhT). The degree of functionalization was investigated by measuring the fluorescence intensity of the alkyne bearing fluorescent dansyl probe "clicked" to azide functionalized polymersomes. In order to find optimal reaction conditions and to improve the degree of functionalization of the polymersome surface, different parameters, such as the Cu(I) stabilizing ligands TBTA and

BPhT, the copper concentration, the reaction temperature, and the pH of the reaction buffer, were varied. Degrees of functionalization between 24% and 26% were found under all conditions tested, which did not indicate a significant influence by the changed parameters. However, the relatively low degree of functionalization might be explained by functional group availability on the polymersome surface, with the majority of functional groups being buried in the hydrophilic layer of the block copolymer membrane. In addition to azide-functional polymersomes, van Hest and co-workers also introduced polymersomes based on alkyne-functionalized PS-*b*-PEG mixed with a high ratio of poly(styrene)-*block*-poly(l-isocyanoalanine(2-thiophen-3-yl-ethyl)amide) (PS-*b*-PIAT) [29].

An approach used to increase the number of surface-accessible functional groups on polymersomes was shown by Gillies and co-workers. By click chemistry, they attached dendritic and nondendritic displays of mannose, bearing an alkyne functionality, to azide functional polymersomes and compared their binding to the mannose binding protein, concanavalin A, by using a hemagglutination assay [31]. As shown in Figure 5, membranes functionalized with dendritic displays are expected to show a higher availability of the mannose ligands than the nondendritic ones. In fact, this hypothesis was verified by experiments that showed that the binding of polymersomes functionalized with dendritic mannose displays was increased by 1–2 orders of magnitude relative to the nondendritic polymersomes.

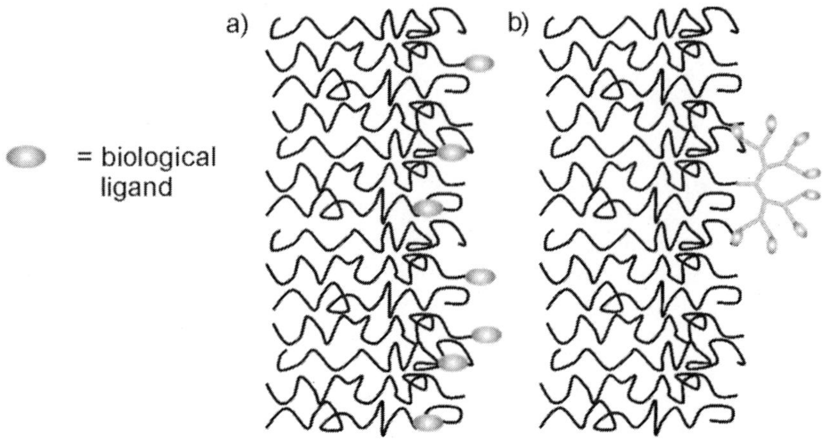

Figure 5. Schematic illustration of preformed polymersome membranes functionalized with (**a**) nondendritic and (**b**) dendritic biological ligand displays. Reproduced by permission of the American Chemical Society [31].

Polymersome-enzyme conjugates made by azide-alkyne click chemistry have been applied as nanoreactors for enzymatic cascade reactions [26]. Glucose oxidase (GOx) was encapsulated in the interior of these polymersomes and *Candida antarctica* lipase B (CalB) was incorporated into the vesicle bilayer membranes. Horseradish peroxidase (HRP) was reacted with imidazole-1-sulfonyl azide hydrochloride to introduce azide functionalities to this enzyme [57,58]. The HRP was then conjugated to the outer surface of the polymersomes via click-chemistry. The resulting construct had the three different enzymes partitioned in three different, well-defined local environments and was able to catalyze a three-step enzymatic reaction cascade, thus mimicking the confinement of individual reaction steps in compartments of living cells. Monoacetylated glucose was deprotected by CalB while diffusing from the bulk solution through the membrane into the cavity of the polymersome. In the cavity, the glucose was oxidized by GOx to gluconolactone and hydrogen peroxide. This small molecule was able to diffuse out of the nanoreactor through the membrane and was utilized by HRP to convert 2,2′-azinobis(3-ethylbenzothiazoline-6-sulfonic acid) (ABTS) to ABTS•+. In order to determine the enzyme encapsulation and conjugation efficiencies, the enzymes were decorated with a ruthenium complex and the biohybrid polymersomes were analyzed by inductively coupled plasma mass spectrometry. More than 90% (a remarkably high percentage) of available alkyne functionalities on the polymersome surface were thus found to be occupied by HRP enzymes. This example again shows the high efficiency of the azide-alkyne click reaction in aqueous media.

Reactions well-known in the field of protein conjugation chemistry, such as the reactions of maleimide functions with thiol groups [59,60], or N-hydroxy succinimidyl esters with the amine groups of lysins [61], can be applied to modify polymersomes bearing suitable functional groups on their surfaces. An example is the conjugation of the monoclonal antibody OX26 to poly(caprolactone)-*block*-poly(ethylene glycol) (PCL-*b*-PEG) polymersomes by reacting cysteine residues of the antibody with maleimido end groups of the PEG block present on the polymersome surface [41]. These decorated polymersomes were investigated as possible vessels for the delivery of peptides to the brain, as the antibody binds to a receptor that can initiate transcytosis of particles across the blood-brain barrier. A quantity of 34 OX26 antibodies on the surface of a polymersome was found to result in the greatest blood-brain barrier permeability.

Another example of a mild conjugation reaction known in biochemistry is the addition of thiols to vinyl groups. This reaction was exploited to link cysteine-containing peptides to vesicles with a reactive surface [42].

Poly(ethylene oxide)-*block*-poly(γ-methyl-ε-caprolactone) (PEG-*b*-PMCL) was functionalized at its hydrophilic terminus with a vinyl sulfone group. These block copolymers formed vesicles in aqueous solution that displayed vinyl sulfone groups on their surface. Short targeting peptides containing thiol groups did bind to the vesicles, while similar peptides lacking cysteine did not adhere. As this work was only recently published, future experiments are needed to show the efficiency of these peptide-labeled polymersomes in targeted drug-delivery.

The attachment of succinimidyl ester-activated Alexa Fluor 633 to polymersomes consisting of PDMS-*b*-PMOXA diblock copolymers comprising hydroxyl and amino end groups has been investigated very recently by Egli *et al.* [43]. The number of fluorophores per vesicle was controlled by adjusting the ratio of amine-terminated- to hydroxyl-terminated block copolymers and was measured by FCS. The fluorescence intensity for polymersomes consisting of 0.3 to 10 mol% AB-NH was determined and plotted, resulting in a linear function that quantitatively describes the number of fluorophores per polymersome. At higher ratios of AB-NH, from 10- to 100 mol%, saturation of the fluorophore attachment to the polymersome was observed, that can be explained by steric hindrance to further dye conjugation.

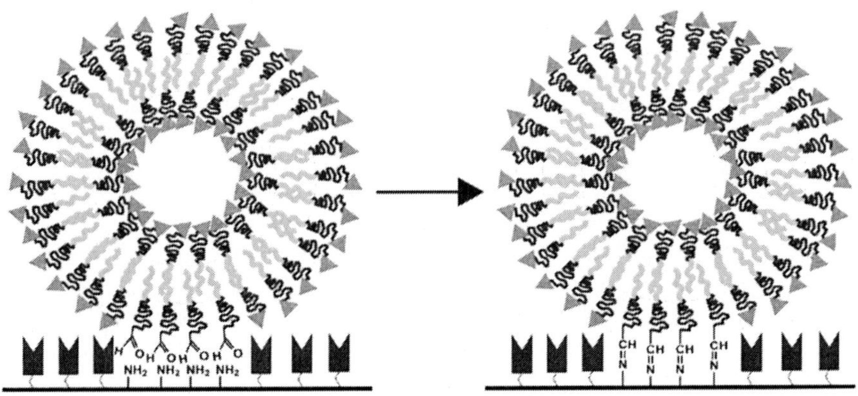

Figure 6. Covalent immobilization of aldehyde modified polymersomes on an aminated glass surface. Reproduced by permission of the American Chemical Society [19].

Foster and co-workers reported the immobilization of polymersomes composed of PCL-*b*-PEG, poly(lactide)-*block*-poly(ethylene glycol) (PLA-*b*-PEG), and poly(isoprene)-*block*-poly(ethylene glycol) (PI-*b*-PEG), each polymer with aldehyde end groups, to aminated glass surfaces, resulting in a covalent imine bond (Figure 6) [19]. Depending on the different bilayer stiffnesses, the three different species of polymersomes showed different footprint areas upon immobilization.

An extension of this concept would involve the conjugation of proteins to the aldehyde decorated polymersome surface, as proteins generally display a multitude of amine groups on their surfaces. However, this approach is unfavorable because cross-linking of polymersomes by proteins could occur, which would lead to large aggregates. Moreover, this approach would not be suitable for the formation of permanent covalent bonds in aqueous environments, since aliphatic imines are known to be unstable and to hydrolyze into their original amine and aldehyde functionalities. In contrast, aromatic imines based on a bis-arylhydrazone bond show no hydrolysis and are stable in aqueous solution in a pH range from 2–10 and up to 95 °C [62], due to their large delocalized electron system. Conjugation chemistry that forms such a bis-aryl hydrazone bond is based on two complementary hetero-bifunctional linkers, comprising an aromatic hydrazine (protected as acetone hydrazone to increase its shelf-life) and an aromatic aldehyde (Figure 7). Using two complementary hetero-bifunctional linkers confers the further advantage that cross-linking cannot occur. Such a system was used to specifically and quantitatively conjugate oligonucleotides to antibodies [63] and to bind proteins to other proteins [64] in aqueous solution. Because this conjugation reaction is carried out under mild conditions in aqueous buffer, without any additional catalyzing agents (e.g., in contrast to the copper catalyzed azide-alkyne click reaction), the resulting conjugates are applicable for pharmaceutical and therapeutic purposes. Recently, Egli *et al.* used this conjugation strategy to form covalent and stable polymersome-protein (enhanced yellow fluorescent protein, eYFP) and polymersome-antibody conjugates [43]. Polymersomes consisting of PDMS-*b*-PMOXA diblock copolymers comprising either hydroxyl or amino end groups were functionalized with an excess of succinimidyl ester-activated 4-formyl benzoic acid (S-4FB). In parallel, eYFP and antibodies were modified with succinimidyl ester-activated 6-hydrazinonicotinic acid (S-HyNic) in order to introduce the aromatic acetone hydrazone functionality. The functionalization of polymersomes and ligands with 4FB and HyNic was quantified by colorimetric reactions using 2-hydrazinopyridine (specific absorption of the formed linker at 350 nm) and 4-nitrobenzaldehyde (specific absorption of the formed linker at 390 nm), respectively. Then, solutions of both conjugation partners were mixed and incubated. The successful attachment of approximately 5 eYFP molecules per polymersome was determined by the time-dependent measurement of intensity fluctuations by FCS. Polymersome-antibody conjugates prepared by this strategy were applied to target biotinylated surfaces (using anti-biotin IgG) and Her2 receptors expressed by breast cancer cells SKBR3 (using the antibody trastuzumab). Polymersomes without antibodies attached to their surfaces showed no toxicity and only low inhibition of proliferation, whereas polymersome-trastuzumab conjungates showed specific attachment and clear reduction in proliferation of the cells.

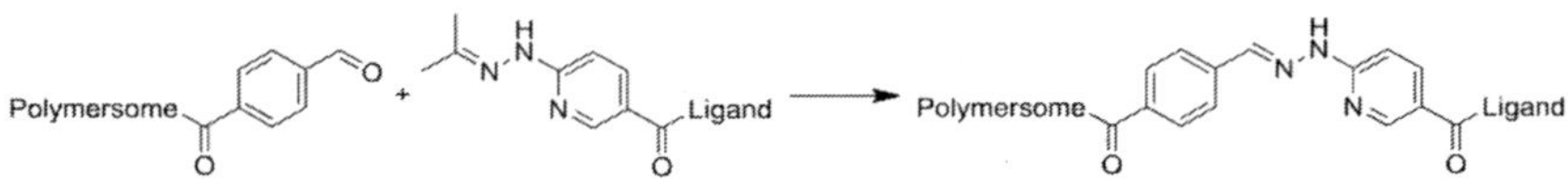

Figure 7. Specific reaction between 4FB modified polymersomes and HyNic modified ligands (e.g., proteins) forms a stable bis-aryl hydrazone bond that links the ligand to the polymersome.

In this section we have reviewed a variety of synthetic routes to modify polymersome surfaces with moieties such as biomolecules. These methods differ by their binding specificities and efficiencies, by their biocompatibilities, as well as by their binding strengths. Non-covalent (e.g., biotin-streptavidin) and covalent binding strategies (e.g., Cu-catalyzed alkyne-azide click chemistry or the reaction of aromatic acetone hydrazones with aromatic aldehydes) are possible. An advantage of non-covalent binding may be that a higher density of functional ligands can be achieved, due to rearrangement of ligands and functional anchor groups. Benefits of covalent binding strategies include improved site-specificity and reproducibility, as well as increased binding stability. The copper catalyzed azide-alkyne click chemistry has proven itself to be applicable to the conjugation of different species such as proteins, enzymes or dendrons to the polymersome surface. The use of copper as a catalyst, as well as the redox conditions while reducing Cu(II) to Cu(I), are unfavorable for biological and pharmaceutical applications. In contrast, the formation of a stable bis-arylhydrazone bond between 4FB modified polymersomes and HyNic modified ligands, such as proteins and antibodies, features the same advantages as the click chemistry, but does not require additional metal-based catalyst species.

Specific reactions to form bonds between block copolymer membranes and biological ligands, as well as the stability, and biocompatibility of the resulting bonds and linkers, are not the only points of interest when surface modified polymersomes are produced. Further investigations on the control of binding site availability, as well as membrane stability following a high degree of modification, will have to be carried out in the future.

2.2. Formation of Polymersomes from End-Functionalized Block Copolymers

In the previous chapter we showed that the surface of an existing block copolymer vesicle can be functionalized by targeting functional groups on the surface with reactive linkers or conjugation reagents that bind a ligand. A different strategy, however, involves functionalization of block copolymers at their hydrophilic chain end(s) with ligands such as carbohydrates, peptides and proteins before self-assembly into vesicles (Figure 1(b)). The advantages of this strategy are that the functionalized copolymer can be purified and characterized as a non-aggregated species and that it can be mixed with the non-functionalized copolymer prior to vesicle formation, thus allowing better control of the surface-density of exposed ligands simply by adjusting the ratio of functionalized to non-functionalized block copolymer. Furthermore, vesicles with a highly functionalized corona are obtained when the ligand-bearing copolymer is allowed to self-assemble on its own. However, the assembly behavior of the block copolymers might be affected by the end-functionalization, so that their ability to form vesicles has to be confirmed by experimental data.

Carbohydrates play an important role in cell-cell recognition and signal transduction in living organisms. Thus, decorating polymer vesicles with sugars can lead to bioactive nanostructures. Lee and co-workers synthesized rod-coil polymer amphiphiles terminated by a carbohydrate, tetra(*p*-phenylene)-*block*-PEG12-α-D-mannopyranoside. This glycopolymer formed small vesicles with a diameter of 40 nm, in which the amphiphilic bilayer membrane was covered with mannose residues [65,66]. Indeed, the surface can be regarded as supramolecular, multivalent ligand, as is shown by a hemagglutination inhibition assay with lectin concanavalin A (Con A), which recognizes R-D-mannopyranoside. The vesicle showed 800-fold higher inhibitory potency compared to free methyl mannose [66]. Moreover, the mannose-covered vesicles specifically bind to the pili of the ORN 178 *E. coli*strain, as imaged by transmission electron microscopy (TEM) (Figure 8) [65]. Bacterial pili are proteinacious fibers that bind to mannose and protrude from the surface of the cell. In control experiments, galactose-terminated amphiphilic polymers were synthesized, but the corresponding self-assembled nanostructures did not bind to the pili [66]. In addition, an *E.coli* strain that lacks the selective binding protein for mannose was tested. With these bacteria, no adhesion of the vesicles to the pili was observed. These experiments clearly show the significance of the carbohydrate end group in the interaction of polymersome and its surroundings.

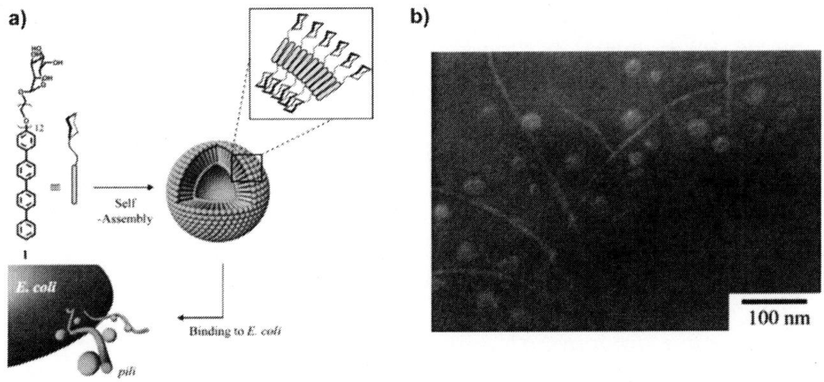

Figure 8. (**a**) Schematic representation of self-assembly of mannose-endfunctionalized polymeric rod coil amphiphiles into vesicles and their binding to the pili of *E. coli*. (**b**) TEM image of vesicles bound to the pili of an *E. coli* ORN 178 cell. Reproduced by permission of the Royal Society of Chemistry [65].

Yonese *et al.* studied the aggregation behavior of lactose-terminated PEG-*b*-poly(γ-methyl L-glutamate) in water [67]. Dynamic light scattering (DLS) indicated the formation of large aggregates, presumed to be vesicles, with a hydrodynamic radius of about 250 nm. Key to the aggregation behavior might be the association of α-helical poly(γ -methyl L-glutamate) segments, promoting the formation of planar bilayers, which then close into vesicles. The lactose units were recognized by the lectin RCA_{120}, indicating that they are located on the surface of the glycopolymer aggregate.

Spherical and worm-like polymersomes decorated with folate as a targeting ligand were investigated as nanocarrier system for a combined targeted delivery of an anticancer drug and a magnetic resonance imaging contrast agent [68,69]. To this end, heterobifunctional asymmetric triblock copolymers folate-PEG_{5000}-*b*-poly(glutamate hydrazone doxorubicin)-*b*-PEG_{2000}-acrylate, and FA-PEG_{114}-*b*-PLA-*b*-PEG_{46}-acrylate were synthesized and allowed to self-assemble into vesicles. The long PEG segments were mostly segregated in the outer hydrophilic leaflet of the membrane, whereas the shorter PEG block was mostly present in the inner hydrophilic layer. Thus, these vesicles presented folate on their surface and could be stabilized by crosslinking the polymers in the inner leaflet via the acrylate functional groups to enhance circulation time. These polymersomes showed higher cellular uptake by HeLa cells than folate-free polymersomes, due to the folate receptor-mediated endocytosis process. Accordingly, doxorubicin loaded polymersomes exhibited higher cytotoxicity than the folate free controls.

Polymersome surfaces can also be functionalized with more sophisticated biological ligands by the strategy of synthesizing end group-functionalized block copolymers and allowing them to self-assemble into polymersomes. Very recently, the group of van Hest expanded their work on functionalized polymersomes by conjugating the cell penetrating peptide Tat to PS-*b*-PEG-oxanorbornadien block copolymers via a tandem cycloaddition/retro Diels-Alder "click" reaction [70]. These polymers were admixed with 90 wt% PS-*b*-PIAT to form vesicles that presented Tat on their surfaces. The vesicles were able to encapsulate guest proteins such as green fluorescent protein and HRP, and traffic their cargo into cells with high phagocytic activity (HeLa, Jurkat, and HEK 293). It was clearly shown that the cell penetrating peptide was responsible for the cellular uptake.

In another example, Hammer and co-workers reported PBD-*b*-PEG block copolymers that were functionalized with Tat or its fluorescently-labeled version, FITC-Tat, by activating the hydroxyl end group of the block copolymers with succinimidyl carbonate, followed by conjugation with the peptide [71]. When 10 mol% of these functionalized block copolymers were mixed with the non-functionalized PBD-*b*-PEG, vesicles formed despite the presence of the rather large functional end groups (1,760, or 2,260 Da, respectively). These polymersomes enabled intracellular delivery of a hydrophobic, near-infrared emissive fluorophore embedded in the polymersome membrane (Figure 9). A future application of such a system might be the tracking of dendritic cells by NIR fluorescence based imaging.

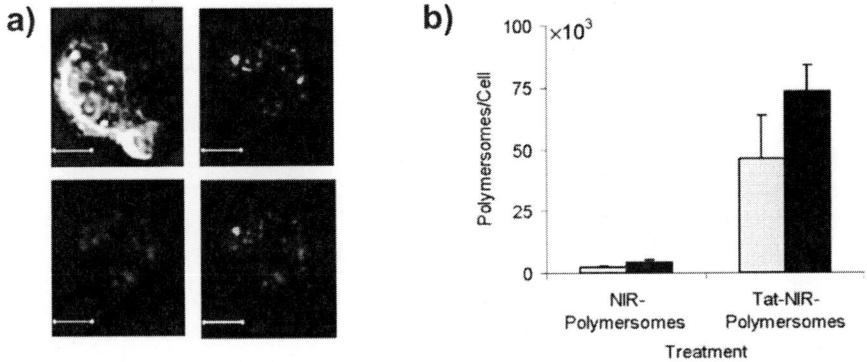

Figure 9. PBD-*b*-PEG polymersomes, surface-functionalized with the cell penetrating peptide Tat and carrying near infrared fluorophores, were used to label immature dendritic cells. (**a**) Confocal laser scanning microscopy image (transmission, FITC-Tat fluorescence (green), near-infra red (NIR) fluorescence (red), and overlay of fluorescence channels). (**b**) Comparison of intracellular uptake between non-functionalized polymersomes and Tat-functionalized polymersomes. Reproduced and adapted by permission of the American Chemical Society [71].

The biocompatibility of PBD-*b*-PEG makes this block copolymer a popular choice for cell targeting studies and potential biomedical applications [71]. Kokkoli and co-workers prepared PBD-*b*-PEG-based polymersomes that were decorated with a targeting peptide that mimics the cell adhesion site in fibronectin [72]. The peptide was conjugated at its N-terminus with a block copolymer bearing an *N*-hydroxysuccinimide end group. These functionalized polymers were mixed in 1:1 and 1:99 ratios with non-functionalized PBD-*b*-PEG and formed vesicles using the film rehydration method. The vesicles were found to be internalized within prostate cancer cells that expressed integrins on their surfaces and were able to deliver a model protein therapeutic (tumor necrosis factor-α) to the cells, dramatically enhancing its cytotoxic potential.

Using the same type of polymer and a similar strategy, Ugaz and coworkers incorporated a small fluorescent label on the surface of vesicles [73]. The coumarins were used to probe the accessibility of functional end groups on the surface of polymersomes. To this end, coumarin-functionalized PBD-*b*-PEG block copolymers were synthesized by oxidation of the PEG`s hydroxy endgroup into carboxylic acid, followed by N-hydroxysuccinimidyl-ester mediated amidation with 7-amino-4-(trifluoromethyl) coumarin. 10 wt% of coumarin-functionalized block copolymers were admixed with two different non-functionalized block copolymers, one in which the hydrophilic block was shorter than the PEG block of the functionalized block copolymer, and one that had the same length as its functionalized counterpart. The fluorescence quenching of coumarin by iodide ions was greatly reduced on the surface of the vesicles compared to free fluorophores and was smallest when the fluorophore-bearing PEG blocks were the same length as the non-functionalized hydrophilic blocks, due to retarded ion mobility in the presence of PEG.

PBD-*b*-PEG was also functionalized with another fluorophore, fluorescein, by reacting the terminal hydroxyl group of the PEG chain with fluorescein-5-carbonyl azide [74]. These block copolymers assembled into fluorescent polymersomes and were used to visualize phase separation in mixed polymersomes made of PBD-*b*-PEG and PBD-*b*-PAA.

A modular way to tailor the surface properties of polymersomes was recently introduced by Kros and co-workers [75]. They synthesized polypeptides that consisted of a hydrophobic poly(γ-benzyl L-glutamate) (PBLG) block and a hydrophilic coiled-coil forming peptide as the hydrophilic block. The latter binds noncovalently with peptides of complementary amino acid sequence, thus allowing the attachment of ligands to the peptidic block copolymers via the formation of coiled coils prior to vesicle formation. In this first study the proof-of-concept was shown using a complementary peptide and a peptide-PEG conjugate, but it can be easily envisioned that this concept would be useful to attach biological active ligands, thus creating vesicles with a functional corona.

2.3. Formation of Polymersomes from Polymers Comprising Biofunctional Hydrophilic Blocks or Side Groups

Amphiphilic block copolymers can be designed to comprise biomolecules-containing hydrophilic blocks. These blocks can be linear carbohydrates, glycopolymers, that is, synthetic polymers with carbohydrate side groups, peptides and proteins, or oligonucleotides [76,77]. Furthermore, some of these biomolecules can be attached as side chain groups to a homopolymer backbone, resulting in amphiphilic polymers as well. Many of the biomolecules exhibit biological activity, for example, they are recognized as binding sites or signaling moieties by enzymes, proteins and cells. Some of the hybrid block copolymers self-assemble in diluted aqueous solutions into vesicles (Figure 1(c)). The resulting membranes display the hydrophilic blocks towards the water-filled lumen of the vesicles and towards the bulk solution. Therefore, polymersomes with a highly functionalized inner and outer surface are obtained. Moreover, the biomolecules can also be present in deeper layers of the hydrophilic leaflet, which can result in enhanced interaction with, for example, cells, compared to end-functionalized block copolymers discussed above.

As carbohydrates exhibit specific interactions with, or recognition of cell membranes or lectins [78], polymersomes formed from glycopolymers have been extensively investigated. Schlaad *et al.* prepared a series of well-defined glycopolymers by radical photoaddition of 1-thio-β-D-glucopyranose (Glc) onto PBD homopolymers [79,80] and block copolymers, PBD-*b*-PS and PBD-*b*-PEG [81,82]. Glucosylated PBD and PBD-*b*-PEG exhibited hydrophilic weight fractions $w_{\mathrm{hydro}} > 0.5$ and can directly be dispersed in water. The glucosylated PBD-*b*-PS ($w_{\mathrm{hydro}} = 0.17$) had to be dispersed in tetrahydrofuran first, followed by a slow exchange of the solvent against water. PBD-Glc formed large, unilamellar vesicles having hydrodynamic diameters of about 500–600 nm and a membrane less than 4 nm thick, as indicated by light and X-ray scattering analysis and TEM (Figure 10(a)) [79]. The membrane appears to be a bilayer structure. The thickness of the membrane, which is similar to that of a liposome, is not affected by the length of the polymer backbone, suggesting that chains are oriented parallel to the interface. Another consequence of the parallel orientation of chains in the membrane is that vesicles are formed despite the high hydrophilic weight fraction; phospholipids are usually less hydrophilic, and block copolymers of such composition would assemble into spherical or worm-like micelles. The glucosylated PBD-*b*-PS was found to self-assemble into vesicles in a tetrahydrofuran solution as well as in water (Figure 10(b)) [81]. The apparent hydrodynamic radius of these vesicles in water was about 120 nm. They have a polystyrene bilayered wall, measuring about 20 nm across (TEM), and a Glc corona. Amphiphilic block copolymer samples of glucosylated PBD-*b*-PEG spontaneously formed very large vesicles of >500 nm in diameter [82]. The existence of unilamellar vesicles was confirmed by TEM (Figure 10(c)) and scattering analyses. Chains consisting of one hydrophobic and two hydrophilic chains can only be packed in a monolayer. These vesicles should therefore have an asymmetric membrane with glucose outside and PEG inside, as evidenced by 2D-NOESY-NMR and surface-enhanced Raman spectroscopy.

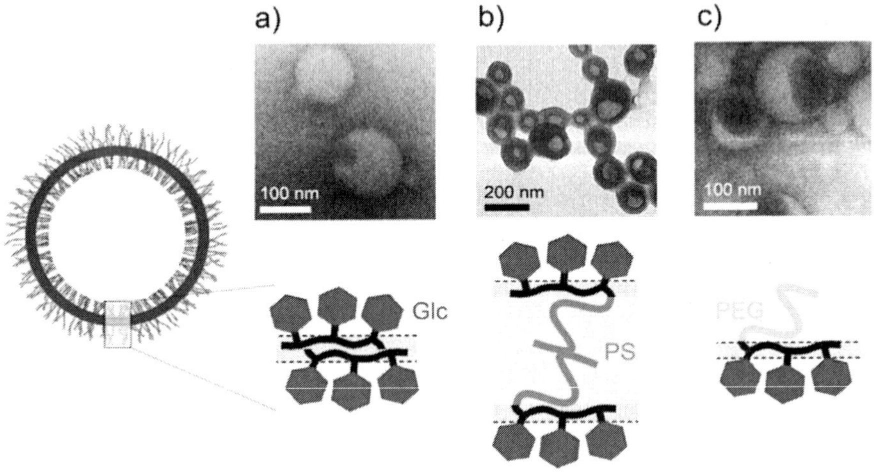

Figure 10. Transmission electron micrographs and schematic illustrations of the membrane morphology of vesicles formed by (**a**) glucosylated PBD, (**b**) glucosylated PBD-*b*-PS, and (**c**) glucosylated PBD-*b*-PEG in dilute aqueous solution.

To date, these glycopolymer vesicles have not been investigated in terms of their drug release behavior or specific interactions with biological cell membranes or lectins.

The same group synthesized glucopolyoxazoline homopolymers by cationic polymerization and thiol-ene "click" chemistry, which in pure water spontaneously assembled into spherical vesicles and hollow nanotubes [83]. However, aggregates were not observed in HEPES buffered saline solution. Adding Glc-binding lectin Con A to an aqueous glucopolyoxazoline solution led to an increase in the optical density (turbidity), indicating the formation of large aggregates due to multivalent interactions between protein and polymer ("cluster glycoside effect" [84]).

Amphiphilic block copolymers that include sugar moieties in one of the blocks are a further example of vesicle-forming glycopolymers. Li *et al.* synthesized a series of amphiphilic block copolymers consisting of hydrophobic PS and hydrophilic poly(2-(β-D-glucopyranosyloxy)ethyl acrylate) (PGEA) blocks using atom transfer radical polymerization (ATRP) [85,86]. Copolymers were first dissolved in a common solvent (*N,N*-dimethylformamide, tetrahydrofuran, dioxane, and their mixtures), followed by the addition of water to induce self-assembly. The observed morphologies included micelles, vesicles, hollow tubules, porous spheres, and large compound vesicles, depending on the composition of the copolymer and the solvent mixture. Vesicles usually exhibited polydisperse size distributions, diameters ranging from 70 to 350 nm, and the membranes measured about 20–25 nm across. The structures are envisioned to find applications as models for cell-specific drug delivery systems; to date, experimental studies have not been reported.

Dong *et al.* investigated a series of biodegradable and biomimetic star-shaped block copolymers based on PCL and either poly(lactobionamidoethyl methacrylate) (PLAMA) [87] or poly(gluconamidoethyl methacrylate) (PGAMA) [88], synthesized via ATRP (Figure 11). Biohybrid amphiphiles with hydrophobic PCL mole fractions of 62% and 76%, respectively, were found to self-assemble into vesicles measuring about 150–250 nm, eventually 750 nm, in diameter in dilute aqueous solution. The PCL-PLAMA and PCL-PGAMA copolymers showed specific recognition for RCA120 lectin and Con A, respectively.

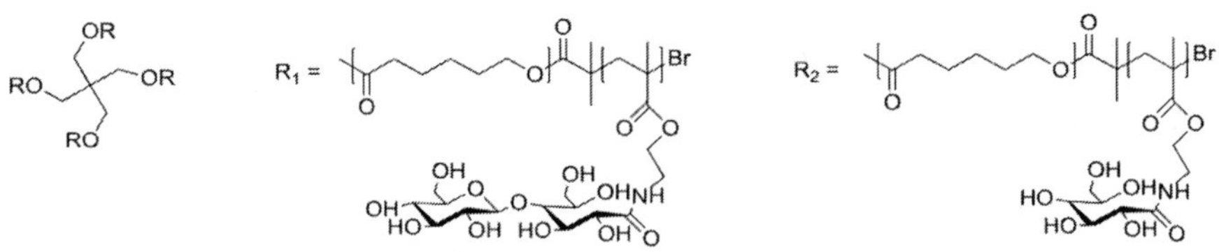

Figure 11. Biomimetic block copolymers based on poly(ε-caprolactone): Star-shaped PCL-PLAMA (R1), and star-shaped SPCL-PGAMA (R_2).

Pasparakis and Alexander synthesized block copolymers based on hydrophilic poly(2-glucosyl-oxyethyl methacrylate) (PGEMA) and poly(diethyleneglycol methacrylate) (PDEGMA) by ATRP and reversible addition-fragmentation chain-transfer (RAFT) polymerization [89]. In dilute aqueous solution, the polymers assembled into vesicles with mean diameters of approximately 250 nm (polymer containing 15 wt% Glc) and 500 nm (34 wt% Glc) at 20 °C, as indicated by dynamic light scattering and scanning electron microscopy. At 37 °C, above the lower critical solution temperature (LCST) of PDEGMA, the size of the vesicles decreased to 180 nm and 300 nm, respectively, which was attributed to the increased hydrophobicity and collapse of the PDEGMA blocks. It was demonstrated that the glycopolymer vesicles were able to accommodate the protein lecitin at their surfaces. Further studies involved carbohydrate interaction of the vesicles with a mutant *E. coli* strain that is both green fluorescent and expresses the *fim*H protein, which has binding specificity for glucose and mannose. The smaller vesicles formed large aggregates with the bacteria, the clusters comprising about 100–150 bacteria and 60–90 vesicles; the larger vesicles were bound to *E. coli* but did not form clusters. Hence, the size of vesicles controls the vesicle-cell interactions from bulk aggregation to individual associations. Furthermore, it was demonstrated that a molecular "information" load, that is a fluorescent dye, could be specifically transferred from the vesicle membrane to the cytoplasm of the bacteria (Figure 12). The vesicles can thus be considered to represent primitive mimics of natural cells with their associated glycocalix, with potential applications in cell sensing, therapeutics, and synthetic biology.

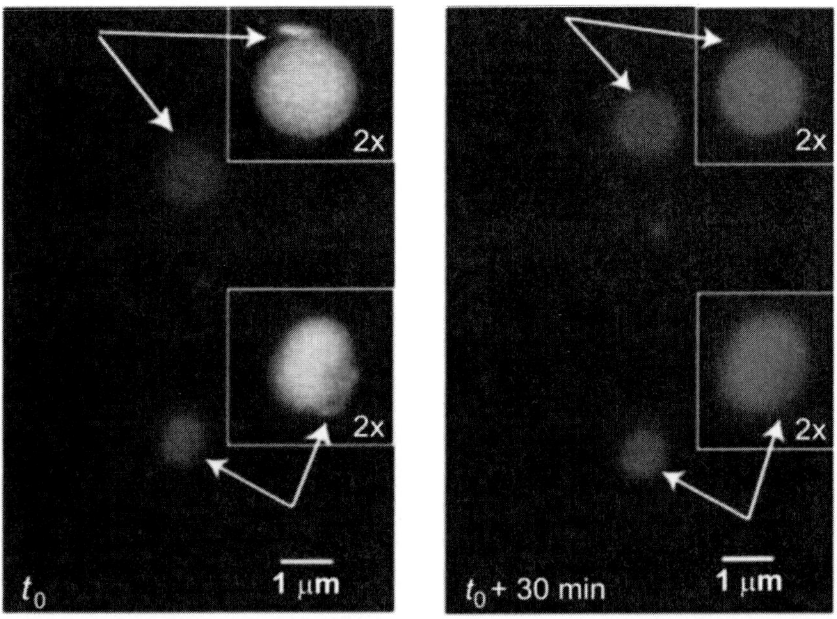

Figure 12. Information transfer at interfaces between PGEMA-*b*-PDEGMA block copolymer vesicles and *E.coli* cells. Fluorescence microscopy image (left) of orange-red fluorescing vesicles containing ethidium bromide, and green fluorescing bacteria. The right image shows the same vesicle-cell partners after 30 min, with the bacteria fluorescing in orange-red owing to transfer of the dye. Insets show the objects at higher image contrast and 2× magnification. Reproduced by permission of Wiley [89].

Hydrophilic polysaccharides can be converted to amphiphilic polymers by conjugation of hydrophobic side chains to the sugar moieties, as described, for example, by Mély and Babak *et al.* [90] and Li *et al.* [91]. In neutral water *N,N*-dialkyl chitosans (alkyl = octyl, decyl, and dodecyl) formed stable bilayered vesicles having hydrodynamic diameters in the range of 100–200 nm as determined by dynamic light scattering [91]. The size of vesicles was found to increase with the increasing molecular weight of the hydrophilic chitosan backbone and/or the increasing length of the hydrophobic alkyl side chains. This effect was attributed to a more compact structure of the membrane. The drug (vitamin B_{12}) release behavior exhibited by *N,N*-dialkyl chitosan vesicles are characterized by biphasic drug release kinetics, including an initial burst phase and a following plateau phase. Studies on specific interactions of the polysaccharide surface have not been reported.

Mishra and Lecommandoux *et al.* synthesized poly(γ-benzyl L-glutamate)-*block*-hyaluronan (PBLG-*b*-HYA) by "click" coupling of PBLG-azide ($M_n \sim 5$ kDa) and HYA-alkyne (Mw ~ 5 kDa), which self-assembled into polymersomes in dilute aqueous solution [22]. The intracellular delivery of doxorubicin loaded polymersomes (PolyDOX) was investigated in high (MCF-7) and low (U87) CD44 expressing cancer cell models. HYA is a major ligand for CD44 and can be used to target breast cancer cells having over-expressed CD44 glycoproteins. In fact, flow cytometry data suggested successful uptake of PolyDOX in cells and a higher accumulation in MCF-7 than U87 cells. PolyDOX significantly enhanced reactive oxygen species level in both cells. PolyDOX suppressed growth of breast tumor on female Sprague-Dawley rats as compared to phosphate buffer saline pH 7.4 (PBS) control group and exhibited lesser cardiotoxicity.

Polypeptides and their larger equivalents—fully folded proteins or enzymes—are macromolecules that expose hydrophilic functional groups towards an aqueous medium. When these biomolecules are linked to hydrophobic polymers, amphiphilic molecules that can assemble into vesicular structures are obtained. Block copolymers comprising a polypeptide block, and the polymersomes that result, have been extensively reviewed elsewhere [77,92,93]. Simple polypeptides (glutamate, lysine, *etc.*) can be considered as polyelectrolytes without special biological or other specific functions. Functionality is absent in the corona of such vesicles and therefore they are beyond the scope of this review.

Functional enzymes can form the surfaces of vesicles if the protein is incorporated in giant amphiphiles as macromolecular hydrophilic head groups of hydrophobic polymers. Thus, these vesicles possess a functional surface. Nolte and co-workers introduced a route to giant amphiphiles by cofactor reconstitution (Figure 13(a)). A synthetic polymer, PS-*b*-PEG, was first functionalized with a heme cofactor, protoporphyrin IX, and subsequently reconstituted with an apoprotein, myoglobin (Mb) or HRP [94,95]. The PS-*b*-PEG precursor formed small spherical micelles in water; the solution was prepared by injection of the polymer in THF into water. After reconstitution with the hydrophilic apo-HRP, the triblock biohybrid sample PEG-*b*-PS-*b*-HRP assembled into submicron vesicles with broad size distribution (Figure 13(b)). The hydrophobic membrane is built by the polystyrene middle block and is stabilized by PEG chains and protein. Whether the membrane has a mono- or a bilayered structure, and whether the PEG and HRP are segregated or mixed in the corona, is not yet known. The enzyme-polymer hybrid vesicles were further investigated according to their enzymatic activity [95,96]. The enzymatic activity decreased for HRP and the stability of the oxy complex was reduced for Mb. This may be explained by the introduction of a polystyrene chain, which seems to result in a somewhat modified binding of the heme in the reconstituted protein. Several factors could be responsible for this behavior, such as an unfavorable interaction of the protein with the PS chain or a disturbance of the three-dimensional structure of the protein, due to the aggregation of the hybrid molecules. In addition, both proteins have their substrate access channels located near the site of attachment of the polystyrene chains, which means that these channels might be partly shielded from the aqueous solution. However, the residual activity also might have arisen from smaller (micellar) aggregates having a more dynamic structure than the larger vesicles.

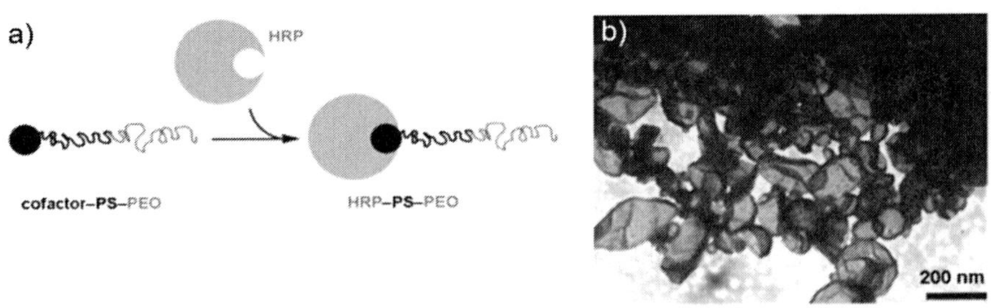

Figure 13. (**a**) Schematic representation of the formation of giant amphiphiles with an enzyme head group by cofactor reconstitution. (**b**) Transmission electron micrograph of PEG-*b*-PS-*b*-HRP vesicles. Reproduced by permission of the American Chemical Society [94].

Another route to giant amphiphiles is to graft a hydrophobic polymer chain from a protein. Such amphiphiles were realized by Le Droumaguet and Velonia [97]. They initiated the ATRP of styrene with an ATRP-initiator conjugated to serum albumins and reduced human calcitonin. The resulting amphiphiles spontaneously formed vesicles with diameters between 20 and 100 nm. The giant amphiphiles were able to encapsulate active enzymes, so that nanoreactors were formed. However, a specific interaction of the protein corona with other biomolecules was not investigated.

Vesicles with an oligonucleotide corona have been produced by the self-assembly of block copolymers with a hydrophobic PBD block and a hydrophilic oligonucleotide block [98,99]. The diblock copolymers were synthesized by linking an amine-terminated PBD to a carboxy-modified, 12-mer oligonucleotide sequence on the solid support on which the oligonucleotide had been synthesized. After cleavage from the solid support and purification, the block copolymer formed vesicles in dilute aqueous solution. These vesicles were stabilized further by photo-crosslinking the PBD blocks [99]. The oligonucleotide shell of such vesicles was able to bind to complementary oligonucleotide sequences on a glass substrate, thus allowing immobilization of the vesicles. Furthermore, the oligonucleotide sequences on the substrate and/or on the vesicles was recognized by microorganisms, as shown by enhanced expression of bacterial organelles of adhesion (curli) in *E. coli* cells during biofilm growth [99]. This is the first example of oligonucleotide-coated vesicles being recognized by microorganisms and cells. Such vesicles potentially influence cell-signaling, internalization and bacterial adhesion. The surface of vesicles might be further modified with ligands through the binding of a complementary oligonucleotide sequence.

2.4. Reconstitution of Membrane Proteins in Polymersomes

A very specific way to introduce recognition sites on vesicle surfaces is the reconstitution of membrane proteins in amphiphilic block copolymer membranes during the self-assembly process. Meier and co-workers pioneered the field of membrane-protein equipped polymersomes. They created polymer vesicles with defined, tunable permeability that resulted from the presence of channel proteins and porins in the membrane [5,10,13,23,100,101]. However, channel proteins provide not only pores, but can also act as binding sites for viral phages through their surface-exposed regions. Indeed, LamB channel proteins reconstituted in polymersomes from PMOXA-*b*-PDMS-*b*-PMOXA triblock copolymers were recognized by bacteriophage λ [7]. The phage underwent binding to the poylmersome and injected its DNA into the vesicle (Figure 14). This work clearly showed that membrane proteins incorporated in polymersomal membranes can provide specific binding sites on the polymersome surface. However, the concept has not yet been applied to other examples.

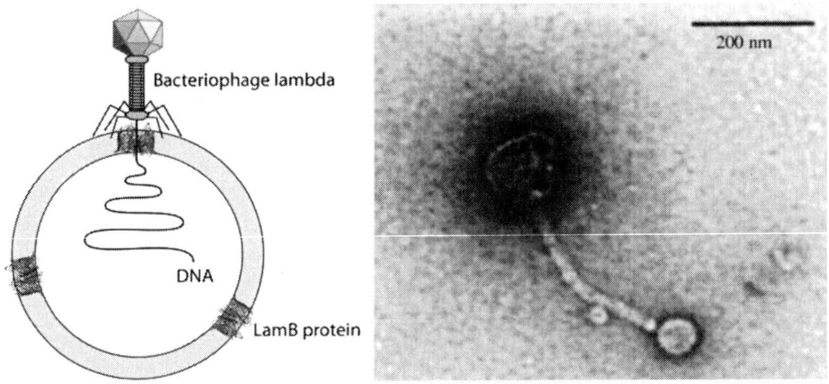

Figure 14. Introduction of surface-exposed binding sites into block copolymer vesicles by reconstituting membrane proteins in the vesicle membrane. The channel protein LamB in PMOXA-*b*-PDMS-*b*-PMOXA vesicles was recognized by a λ phage and triggered the injection of DNA from the phage into the vesicle. (**a**) Schematic representation, (**b**) TEM micrograph showing a phage bound to a vesicle. Reproduced and adapted by permission of the National Academy of Sciences of the United States of America [7].

3. CONCLUSIONS

Cells and cellular compartments interact with their environment through a multitude of functional moieties that include antigens and receptors present on their surfaces. These functional ligands are often anchored within the lipid bilayer membrane that forms the boundary shell of cells and organelles.

In the search for synthetic analogs of vesicles, organelles, and even cells, structures formed by amphiphilic block copolymers have attained a great deal of attention and research activity. Amphiphilic block copolymers can self-assemble into membranes comprising a central hydrophobic layer between two hydrophilic leaflets. This structural similarity to lipid-based membranes makes them mimics of natural membranes. Moreover, block copolymer membranes tend to form vesicles, *i.e.*, closed nano-to microsized compartments that engulf a central pool of water. Such polymersomes are more stable than their lipid-based counterparts and thus show promising applications in biomedical settings such as drug-delivery. However, the surface of a conventional polymer vesicle does not display any functional moieties that could orchestrate interaction with the environment. Therefore, a variety of methods have been developed to functionalize the surface of polymersomes with active ligands, binding sites, or signaling residues. These methods can either target fully assembled polymer vesicles by non-covalent or covalent conjugation, or the polymers that form vesicles can be designed to encompass functional groups prior to vesicle formation. Both strategies have proven successful in yielding highly functionalized polymersomes. Thus, a new generation of multifunctional polymer vesicles is emerging, capable of being designed to specifically bind, adhere to, or target surfaces, cell lines, or biomolecules while still retaining the ability to encapsulate therapeutic and diagnostic cargo, or active (bio)catalysts.

ACKNOWLEDGMENTS

This work was financially supported by the European Science Foundation through their collaborative research project Biofunctional Self-Organized Nanostructures of ionic/non-ionic amphiphilic copolymers, biopolymers-biomacromoleculres and nanoparticles (BIOSONS), by the Swiss National Science Foundation, and the NCCR Nanosciences. Nico Bruns gratefully acknowledges the Marie Curie Actions of the European Commission for an Intra-European Fellowship. We thank Mark Inglin for editing the manuscript.

REFERENCES

1. Singer, S.J.; Nicolson, G.L. Fluid mosaic model of the structure of cell membranes. *Science* **1972**,*175*, 720–731.

2. Kita-Tokarczyk, K.; Grumelard, J.; Haefele, T.; Meier, W. Block copolymer vesicles-using concepts from polymer chemistry to mimic biomembranes. *Polymer* **2005**, *46*, 3540–3563.

3. Antonietti, M.; Förster, S. Vesicles and liposomes: A self-assembly principle beyond lipids. *Adv. Mater.* **2003**, *15*, 1323–1333.

4. Discher, D.E.; Eisenberg, A. Materials science: Soft surfaces: Polymer vesicles. *Science* **2002**, *297*, 967–973.

5. Meier, W.; Nardin, C.; Winterhalter, M. Reconstitution of channel proteins in (polymerized) ABA triblock copolymer membranes. *Angew. Chem. Int. Ed.* **2000**, *39*, 4599–4602.

6. Luo, L.; Eisenberg, A. Thermodynamic size control of block copolymer vesicles in solution.*Langmuir* **2001**, *17*, 6804–6811.

7. Graff, A.; Sauer, M.; Van Gelder, P.; Meier, W. Virus-assisted loading of polymer nanocontainer.*Proc. Natl. Acad. Sci. USA* **2002**, *99*, 5064–5068.

8. Ahmed, F.; Pakunlu, R.I.; Brannan, A.; Bates, F.; Minko, T.; Discher, D.E. Biodegradable polymersomes loaded with both paclitaxel and doxorubicin permeate and shrink tumors, inducing apoptosis in proportion to accumulated drug. *J. Control. Release* **2006**, *116*, 150–158.

9. Meng, F.; Zhong, Z.; Feijen, J. Stimuli-responsive polymersomes for programmed drug delivery.*Biomacromolecules* **2009**, *10*, 197–209.

10. Onaca, O.; Enea, R.; Hughes, D.W.; Meier, W. Stimuli-responsive polymersomes as nanocarriers for drug and gene delivery. *Macromol. Biosci.* **2009**, *9*, 129–139.

11. Li, M.-H.; Keller, P. Stimuli-responsive polymer vesicles. *Soft Matter* **2009**, *5*, 927–937.

12. Du, J.; O'Reilly, R.K. Advances and challenges in smart and functional polymer vesicles. *Soft Matter* **2009**, *5*, 3544–3561.

13. Renggli, K.; Baumann, P.; Langowska, K.; Onaca, O.; Bruns, N.; Meier, W. Selective and responsive nanoreactors. *Adv. Funct. Mater.* **2010**. in press.

14. Boerner, H.G.; Schlaad, H. Bioinspired functional block copolymers. *Soft Matter* **2007**, *3*, 394–408.

15. Duncan, R. The dawning era of polymer therapeutics. *Nat. Rev. Drug Discovery* **2003**, *2*, 347–360.

16. McCarron, P.A.; Marouf, W.M.; Donnelly, R.F.; Scott, C. Enhanced surface attachment of protein-type targeting ligands to poly(lactide-*co*-glycolide) nanoparticles using variable expression of polymeric acid functionality. *J. Biomed. Mater. Res. A* **2008**, *87A*, 873–884.

17. Grzelakowski, M.; Onaca, O.; Rigler, P.; Kumar, M.; Meier, W. Immobilized Protein-polymer nanoreactors. *Small* **2009**, *5*, 2545–2548.

18. Li, F.; Ketelaar, T.; Cohen Stuart, M.A.; Sudhoelter, E.J.R.; Leermakers, F.A.M.; Marcelis, A.T.M. Gentle immobilization of nonionic polymersomes on solid substrates. *Langmuir* **2008**, *24*, 76–82.

19. Domes, S.; Filiz, V.; Nitsche, J.; Fromsdorf, A.; Forster, S. Covalent attachment of polymersomes to surfaces. *Langmuir* **2010**, *26*, 6927–6931.

20. Broz, P.; Benito, S.M.; Saw, C.; Burger, P.; Heider, H.; Pfisterer, M.; Marsch, S.; Meier, W.; Hunziker, P. Cell targeting by a generic receptor-targeted polymer nanocontainer platform. *J. Control. Release***2005**, *102*, 475–488.

21. Uchegbu, I.F. Pharmaceutical nanotechnology: Polymeric vesicles for drug and gene delivery.*Expert Opin. Drug Delivery* **2006**, *3*, 629–640.

22. Upadhyay, K.K.; Bhatt, A.N.; Mishra, A.K.; Dwarakanath, B.S.; Jain, S.; Schatz, C.; Le Meins, J.-F.; Farooque, A.; Chandraiah, G.; Jain, A.K.; Misra, A.; Lecommandoux, S. The intracellular drug delivery and anti tumor activity of doxorubicin loaded poly(gamma - benzyl -glutamate)-*b*-hyaluronan polymersomes. *Biomaterials* **2010**, *31*, 2882–2892.

23. Nardin, C.; Thoeni, S.; Widmer, J.; Winterhalter, M.; Meier, W. Nanoreactors based on (polymerized) ABA-triblock copolymer vesicles. *Chem. Commun.* **2000**, *36*, 1433–1434.

24. Axthelm, F.; Casse, O.; Koppenol, W.H.; Nauser, T.; Meier, W.; Palivan, C.G. Antioxidant nanoreactor based on superoxide dismutase encapsulated in superoxide-permeable vesicles. *J. Phys. Chem. B* **2008**, *112*, 8211–8217.

25. De Vocht, C.; Ranquin, A.; Willaert, R.; Van Ginderachter, J.A.; Vanhaecke, T.; Rogiers, V.; Versees, W.; van Gelder, P.; Steyaert, J. Assessment of stability, toxicity and immunogenicity of new polymeric nanoreactors for use in enzyme replacement therapy of mitochondrial neurogastrointestinal encephalomyopathy. *J. Control. Release* **2009**, *137*, 246–254.

26. van Dongen, S.F.M.; Nallani, M.; Cornelissen, J.J.L.M.; Nolte, R.J.M.; van Hest, J.C.M. A three-enzyme cascade reaction through positional assembly of enzymes in a polymersome nanoreactor.*Chem.-Eur. J.* **2009**, *15*, 1107–1114.

27. Onaca, O.; Hughes, D.W.; Balasubramanian, V.; Grzelakowski, M.; Meier, W.; Palivan, C.G. SOD Antioxidant nanoreactors: Influence

of Block copolymer composition on the nanoreactor efficiency. *Macromol. Biosci.* **2010**, *10*, 531–538.

28. Broz, P.; Driamov, S.; Ziegler, J.; Ben-Haim, N.; Marsch, S.; Meier, W.; Hunziker, P. Toward intelligent nanosize bioreactors: A ph-switchable, channel-equipped, functional polymer nanocontainer. *Nano Lett.* **2006**, *6*, 2349–2353.

29. van Dongen, S.F.M.; Nallani, M.; Schoffelen, S.; Cornelissen, J.J.L.M.; Nolte, R.J.M.; van Hest, J.C.M. A block copolymer for functionalisation of polymersome surfaces. *Macromol. Rapid Commun.* **2008**, *29*, 321–325.

30. Li, B.; Martin, A.L.; Gillies, E.R. Multivalent polymer vesicles via surface functionalization. *Chem. Commun.* **2007**, *43*, 5217–5219.

31. Martin, A.L.; Li, B.; Gillies, E.R. Surface functionalization of nanomaterials with dendritic groups: Toward enhanced binding to biological targets. *J. Am. Chem. Soc.* **2009**, *131*, 734–741.

32. Opsteen, J.A.; Brinkhuis, R.P.; Teeuwen, R.L.M.; Loewik, D.W.P.M.; van Hest, J.C.M. "Clickable" polymersomes. *Chem. Commun.* **2007**, *43*, 3136–3138.

33. Lin, J.J.; Silas, J.A.; Bermudez, H.; Milam, V.T.; Bates, F.S.; Hammer, D.A. The effect of polymer chain length and surface density on the adhesiveness of functionalized polymersomes. *Langmuir***2004**, *20*, 5493–5500.

34. Broz, P.; Ben-Haim, N.; Grzelakowski, M.; Marsch, S.; Meier, W.; Hunziker, P. Inhibition of macrophage phagocytotic activity by a receptor-targeted polymer vesicle-based drug delivery formulation of pravastatin. *J. Cardiovasc. Pharmacol.* **2008**, *51*, 246–252.

35. Rigler, P.; Meier, W. Encapsulation of fluorescent molecules by functionalized polymeric nanocontainers: Investigation by confocal fluorescence imaging and fluorescence correlation spectroscopy. *J. Am. Chem. Soc.* **2006**, *128*, 367–373.

36. Lin, J.J.; Ghoroghchian, P.P.; Zhang, Y.; Hammer, D.A. Adhesion of antibody-functionalized polymersomes. *Langmuir* **2006**, *22*, 3975–3979.

37. Nehring, R.; Palivan, C.G.; Casse, O.; Tanner, P.; Tuxen, J.; Meier, W. Amphiphilic diblock copolymers for molecular recognition: Metal-nitrilotriacetic acid functionalized vesicles. *Langmuir***2009**, *25*, 1122–1130.

38. Nehring, R.; Palivan, C.G.; Moreno-Flores, S.; Mantion, A.; Tanner, P.; Toca-Herrera, J.L.; Thuenemann, A.; Meier, W. Protein decorated membranes by specific molecular interactions. *Soft Matter* **2010**, *6*, 2815–2824.

39. Felici, M.; Marza-Perez, M.; Hatzakis, N.S.; Nolte, R.J.M.; Feiters, M.C. P-Cyclodextrin-appended giant amphiphile: Aggregation to vesicle polymersomes and immobilisation of enzymes. *Chem. Eur. J.* **2008**, *14*, 9914–9920.

40. Guo, M.; Jiang, M.; Zhang, G. Surface modification of polymeric vesicles via host-guest inclusion complexation. *Langmuir* **2008**, *24*, 10583–10586.

41. Pang, Z.; Lu, W.; Gao, H.; Hu, K.; Chen, J.; Zhang, C.; Gao, X.; Jiang, X.; Zhu, C. Preparation and brain delivery property of biodegradable polymersomes conjugated with OX26. *J. Control. Release***2008**, *128*, 120–127.

42. Petersen, M.A.; Yin, L.; Kokkoli, E.; Hillmyer, M.A. Synthesis and characterization of reactive PEO-PMCL polymersomes. *Polym. Chem.* **2010**, *1*, 1281–1290.

43. Egli, S.; Nussbaumer, M.G.; Balasubramanian, V.; Chami, M.; Bruns, N.; Palivan, C.G.; Meier, W. Biocompatible functionalization of polymersome surfaces: New approach for surface immobilization and cell targeting using polymersomes. *J. Am. Chem. Soc.* **2010**. submitted.

44. Nilsson, K.; Mosbach, K. Immobilization of ligands with organic sulfonyl chlorides. *Methods Enzymol.* **1984**, *104*, 56–69.

45. Hammer, D.A.; Robbins, G.P.; Haun, J.B.; Lin, J.J.; Qi, W.; Smith, L.A.; Peter Ghoroghchian, P.; Therien, M.J.; Bates, F.S. Leuko-polymersomes. *Faraday Discuss.* **2008**, *139*, 129–141.

46. Weber, P.C.; Ohlendorf, D.H.; Wendoloski, J.J.; Salemme, F.R. Structural origins of high-affinity biotin binding to streptavidin. *Science* **1989**, *243*, 85–88.

47. Holmberg, A.; Blomstergren, A.; Nord, O.; Lukacs, M.; Lundeberg, J.; Uhlen, M. The biotin-streptavidin interaction can be reversibly broken using water at elevated temperatures.*Electrophoresis* **2005**, *26*, 501–510.

48. Lin, J.J.; Bates, F.S.; Hammer, D.A.; Silas, J.A. Adhesion of polymer vesicles. *Phys. Rev. Lett.* **2005**,*95*, 026101.

49. Robbins, G.P.; Saunders, R.L.; Haun, J.B.; Rawson, J.; Therien, M.J.; Hammer, D.A. Tunable Leuko-polymersomes That Adhere

Specifically to Inflammatory Markers. *Langmuir* **2010**, *26*, 14089–14096.

50. Alon, R.; Hershkoviz, R.; Bayer, E.A.; Wilchek, M.; Lider, O. Streptavidin blocks immune reactions mediated by fibronectin-VLA-5 recognition through an Arg-Gly-Asp mimicking site. *Eur. J. Immunol.* **1993**, *23*, 893–898.

51. Dorn, I.T.; Neumaier, K.R.; Tampe, R. Molecular recognition of histidine-tagged molecules by metal-chelating lipids monitored by fluorescence energy transfer and correlation spectroscopy. *J. Am. Chem. Soc.* **1998**, *120*, 2753–2763.

52. Rahimi, Y.; Shrestha, S.; Banerjee, T.; Deo, S.K. Copper sensing based on the far-red fluorescent protein, HcRed, from Heteractis crispa. *Anal. Biochem.* **2007**, *370*, 60–67.

53. Kornberg, R.D.; Darst, S.A. Two-dimensional crystals of proteins on lipid layers. *Curr. Opin. Struct. Biol.* **1991**, *1*, 642–646.

54. Rostovtsev, V.V.; Green, L.G.; Fokin, V.V.; Sharpless, K.B. A stepwise Huisgen cycloaddition process: Copper(I)-catalyzed regioselective "ligation" of azides and terminal alkynes. *Angew. Chem. Int. Ed.* **2002**, *41*, 2596–2599.

55. Binder, W.H.; Sachsenhofer, R. 'Click' chemistry in polymer and materials science. *Macromol. Rapid Commun.* **2007**, *28*, 15–54.

56. Evans, C.E.; Lovell, P.A. Click chemistry as a route to surface functionalization of polymer particles dispersed in aqueous media. *Chem. Commun.* **2009**, *45*, 2305–2307.

57. Goddard-Borger, E.D.; Stick, R.V. An efficient, inexpensive, and shelf-stable diazotransfer reagent: Imidazole-1-sulfonyl Azide hydrochloride. *Org. Lett.* **2007**, *9*, 3797–3800.

58. van Dongen, S.F.M.; Teeuwen, R.L.M.; Nallani, M.; van Berkel, S.S.; Cornelissen, J.J.L.M.; Nolte, R.J.M.; van Hest, J.C.M. Single-step azide introduction in proteins via an aqueous diazo transfer.*Bioconjugate Chem.* **2009**, *20*, 20–23.

59. Ghosh, S.S.; Kao, P.M.; McCue, A.W.; Chappelle, H.L. Use of maleimide-thiol coupling chemistry for efficient syntheses of oligonucleotide-enzyme conjugate hybridization probes. *Bioconjugate Chem.* **1990**, *1*, 71–76.

60. Velonia, K.; Rowan, A.E.; Nolte, R.J.M. Lipase polystyrene giant amphiphiles. *J. Am. Chem. Soc.***2002**, *124*, 4224–4225.

61. Xu, H.; Kaar, J.L.; Russell, A.J.; Wagner, W.R. Characterizing the modification of surface proteins with poly(ethylene glycol) to interrupt platelet adhesion. *Biomaterials* **2006**, *27*, 3125–3135.

62. Hermanson, G.T. *Bioconjugate Techniques*, 2nd ed.; Academic Press: London, UK, 2008.

63. Kozlov, I.A.; Melnyk, P.C.; Stromsborg, K.E.; Chee, M.S.; Barker, D.L.; Zhao, C. Efficient strategies for the conjugation of oligonucleotides to antibodies enabling highly sensitive protein detection.*Biopolymers* **2004**, *73*, 621–630.

64. Bruns, N.; Pustelny, K.; Bergeron, L.M.; Whitehead, T.A.; Clark, D.S. Mechanical nanosensor based on fret within a thermosome: Damage-reporting polymeric materials. *Angew. Chem., Int. Ed.***2009**, *48*, 5666–5669.

65. Kim, B.-S.; Yang, W.-Y.; Ryu, J.-H.; Yoo, Y.-S.; Lee, M. Carbohydrate-coated nanocapsules from amphiphilic rod-coil molecule: Binding to bacterial type 1 pili. *Chem. Commun.* **2005**, *41*, 2035–2037.

66. Kim, B.-S.; Hong, D.-J.; Bae, J.; Lee, M. Controlled self-assembly of carbohydrate conjugate rod-coil amphiphiles for supramolecular multivalent ligands. *J. Am. Chem. Soc.* **2005**, *127*, 16333–16337.

67. Toyotama, A.; Kugimiya, S.-I.; Yamanaka, J.; Yonese, M. Preparation a novel aggregate like sugar-ball micelle composed of poly(methylglutamate) and poly(ethyleneglycol) modified by lactose and its molecular recognition by lectin. *Chem. Pharm. Bull.* **2001**, *49*, 169–172.

68. Yang, X.; Grailer, J.J.; Rowland, I.J.; Javadi, A.; Hurley, S.A.; Matson, V.Z.; Steeber, D.A.; Gong, S. Multifunctional stable and ph-responsive polymer vesicles formed by heterofunctional triblock copolymer for targeted anticancer drug delivery and ultrasensitive MR imaging. *ACS Nano* **2010**,*4*, 6805–6817.

69. Yang, X.; Grailer, J.J.; Rowland, I.J.; Javadi, A.; Hurley, S.A.; Steeber, D.A.; Gong, S. Multifunctional SPIO/DOX-loaded wormlike polymer vesicles for cancer therapy and MR imaging. *Biomaterials* **2010**, *31*, 9065–9073.

70. van Dongen, S.F.M.; Verdurmen, W.P.R.; Peters, R.J.R.W.; Nolte, R.J.M.; Brock, R.; van Hest, J.C.M. Cellular integration of an enzyme-loaded polymersome nanoreactor. *Angew. Chem. Int. Ed.* **2010**,*49*, 7213–7216.

71. Christian, N.A.; Milone, M.C.; Ranka, S.S.; Li, G.; Frail, P.R.; Davis, K.P.; Bates, F.S.; Therien, M.J.; Ghoroghchian, P.P.; June, C.H.; Hammer, D.A. Tat-functionalized near-infrared emissive polymersomes for dendritic cell labeling. *Bioconjugate Chem.* **2007**, *18*, 31–40.

72. Demirgoz, D.; Pangburn, T.O.; Davis, K.P.; Lee, S.; Bates, F.S.; Kokkoli, E. PR_b-targeted delivery of tumor necrosis factor-α by polymersomes for the treatment of prostate cancer. *Soft Matter* **2009**,*5*, 2011–2019.

73. Chang, Y.-W.; Silas, J.A.; Ugaz, V.M. A direct probe of the interplay between bilayer morphology and surface reactivity in polymersomes. *Langmuir* **2010**, *26*, 12132–12139.

74. Rajagopal, K.; Christian, D.A.; Harada, T.; Tian, A.; Discher, D.E. Polymersomes and wormlike micelles made fluorescent by direct modifications of block copolymer amphiphiles. *Int. J. Polym. Sci.* **2010**.

75. Marsden, H.R.; Handgraaf, J.-W.; Nudelman, F.; Sommerdijk, N.A.J.M.; Kros, A. Uniting polypeptides with sequence-designed peptides: Synthesis and assembly of poly(γ-benzyl l-glutamate)-*b*-coiled-coil peptide copolymers. *J. Am. Chem. Soc.* **2010**, *132*, 2370–2377.

76. Schatz, C.; Lecommandoux, S. Polysaccharide-containing block copolymers: Synthesis, properties and applications of an emerging family of glycoconjugates. *Macromol. Rapid Commun.* **2010**, *31*, 1664–1684.

77. van Dongen, S.F.M.; de Hoog, H.P.M.; Peters, R.; Nallani, M.; Nolte, R.J.M.; van Hest, J.C.M. Biohybrid polymer capsules. *Chem. Rev.* **2009**, *109*, 6212–6274.

78. Wu, A.; Lisowska, E.; Duk, M.; Yang, Z. Lectins as tools in glycoconjugate research. *Glycoconjugate J.* **2009**, *26*, 899–913.

79. Hordyjewicz-Baran, Z.; You, L.; Smarsly, B.; Sigel, R.; Schlaad, H. Bioinspired polymer vesicles based on hydrophilically modified polybutadienes. *Macromolecules* **2007**, *40*, 3901–3903.

80. ten Brummelhuis, N.; Diehl, C.; Schlaad, H. Thiol-ene modification of 1,2-polybutadiene using UV light or sunlight. *Macromolecules* **2008**, *41*, 9946–9947.

81. You, L.; Schlaad, H. An easy way to sugar-containing polymer vesicles or glycosomes. *J. Am. Chem. Soc.* **2006**, *128*, 13336–13337.

82. Schlaad, H.; You, L.; Sigel, R.; Smarsly, B.; Heydenreich, M.; Mantion, A.; Mašić, A. Glycopolymer vesicles with an asymmetric membrane. *Chem. Commun.* **2009**, *45*, 1478–1480.

83. Gress, A.; Smarsly, B.; Schlaad, H. Formation of glycopolyamide nanofibers. *Macromol. Rapid Commun.* **2008**, *29*, 304–308.

84. Lundquist, J.J.; Toone, E.J. The cluster glycoside effect. *Chem. Rev.* **2002**, *102*, 555–578.

85. Li, Z.-C.; Liang, Y.-Z.; Li, F.-M. Multiple morphologies of aggregates from block copolymers containing glycopolymer segments. *Chem. Commun.* **1999**, *35*, 1557–1558.

86. Liang, Y.-Z.; Li, Z.-C.; Li, F.-M. Multiple morphologies of molecular assemblies formed by polystyrene-*block*-poly[2-(β-D-glucopyranosyloxy)ethyl acrylate] in water. *New J. Chem.* **2000**, *24*, 323–328.

87. Zhou, W.; Dai, X.-H.; Dong, C.-M. Biodegradable and biomimetic poly(ε-caprolactone)/ poly(lactobionamidoethyl methacrylate) biohybrids: Synthesis, lactose-installed nanoparticles and recognition properties. *Macromol. Biosci.* **2008**, *8*, 268–278.

88. Dai, X.-H.; Dong, C.-M. Synthesis, self-assembly and recognition properties of biomimetic star-shaped poly(ε-caprolactone)-b-glycopolymer block copolymers. *J. Polym. Sci. A Polym. Chem.* **2008**, *46*, 817–829.

89. Pasparakis, G.; Alexander, C. Sweet talking double hydrophilic block copolymer vesicles. *Angew. Chem. Int. Ed.* **2008**, *47*, 4847–4850.

90. Ercelen, S.; Zhang, X.; Duportail, G.; Grandfils, C.; Desbrières, J.; Karaeva, S.; Tikhonov, V.; Mély, Y.; Babak, V. Physicochemical properties of low molecular weight alkylated chitosans: A new class of potential nonviral vectors for gene delivery. *Colloids Surf. B* **2006**, *51*, 140–148.

91. Li, M.; Su, S.; Xin, M.; Liao, Y. Relationship between N,N-dialkyl chitosan monolayer and corresponding vesicle. *J. Colloid Interf. Sci.* **2007**, *311*, 285–288.

92. Bertin, A.; Hermes, F.; Schlaad, H. Biohybrid and peptide-based polymer vesicles. *Adv. Polym. Sci.* **2010**, *224*, 167–195.

93. Carlsen, A.; Lecommandoux, S. Self-assembly of polypeptide-based block copolymer amphiphiles. *Curr. Opin. Colloid Interface Sci.* **2009**, *14*, 329–339.

94. Reynhout, I.C.; Cornelissen, J.J.L.M.; Nolte, R.J.M. Self-assembled architectures from biohybrid triblock copolymers. *J. Am. Chem. Soc.* **2007**, *129*, 2327–2332.

95. Boerakker, M.J.; Botterhuis, N.E.; Bomans, P.H. H.; Frederik, P.M.; Meijer, E.M.; Nolte, R.J.M.; Sommerdijk, N.A.J.M. Aggregation behavior of giant amphiphiles prepared by cofactor reconstitution. *Chem.-Eur. J.* **2006**, *12*, 6071–6080.

96. Boerakker, M.J.; Hannink, J.M.; Bomans, P.H. H.; Frederik, P.M.; Nolte, R.J.M.; Meijer, E.M.; Sommerdijk, N.A.J.M. Giant Amphiphiles by cofactor reconstitution. *Angew. Chem. Int. Ed.* **2002**,*41*, 4239–4241.

97. Le Droumaguet, B.; Velonia, K. *In situ* ATRP-Mediated hierarchical formation of giant amphiphile bionanoreactors. *Angew. Chem., Int. Ed.* **2008**, *47*, 6263–6266.

98. Teixeira, F.; Rigler, P.; Vebert-Nardin, C. Nucleo-copolymers: Oligonucleotide-based amphiphilic diblock copolymers. *Chem. Commun.* **2007**, *43*, 1130–1132.

99. Cottenye, N.; Teixeira, F.; Ponche, A.; Reiter, G.; Anselme, K.; Meier, W.; Ploux, L.; Vebert-Nardin, C. Oligonucleotide Nanostructured surfaces: Effect on escherichia coli curli expression. *Macromol. Biosci.* **2008**, *8*, 1161–1172.

100. Nardin, C.; Widmer, J.; Winterhalter, M.; Meier, W. Amphiphilic block copolymer nanocontainers as bioreactors. *Eur. Phys. J. E Soft Matter Biol. Phys.* **2001**, *4*, 403–410.

101. Kumar, M.; Grzelakowski, M.; Zilles, J.; Clark, M.; Meier, W. Highly permeable polymeric membranes based on the incorporation of the functional water channel protein Aquaporin Z.*Proc. Natl. Acad. Sci. USA* **2007**, *104*, 20719–20724.

CHAPTER 4

Spontaneous Enhancement of Packing Regularity of Spherical Microdomains in the Body-Centered Cubic Lattice upon Uniaxial Stretching of Elastomeric Triblock Copolymers

Takuya Kota [1], Kimio Imaizumi [2], Sono Sasaki [1] and Shinichi Sakurai [1,3,*]

[1]*Department of Biobased Materials Science, Graduate School of Science and Technology, Kyoto Institute of Technology, Matsugasaki, Sakyo-ku, Kyoto 606-8585, Japan*

[2]*Asahi-kasei Corporation, 1-1-2 Yurakucho, Chiyoda-ku, Tokyo 100-8440, Japan*

[3]*Center for Fiber and Textile Science, Kyoto Institute of Technology, Matsugasaki, Sakyo-ku, Kyoto 606-8585, Japan*

ABSTRACT

Block copolymers forming glassy spheres in the matrix of rubbery chains can exhibit elastomeric properties. It is well known that the spherical microdomains are arranged in the body-center cubic (bcc) lattice. However, recently, we have found packing in the face-centered cubic (fcc) lattice, which is easily transformed into the bcc lattice upon uniaxial stretching. In the same time, the packing regularity of the spheres in the bcc lattice was found to be enhanced for samples completely recovered from the stretched state. This reminds us that a cycle of stretching-and-releasing plays an important role from analogy of densification of the packing in granules upon shaking. In the current paper, we quantify the enhancement of packing regularity of spherical microdomains in the bcc lattice upon uniaxial stretching of the same elastomeric triblock copolymer as used in our previous work by conducting small-angle X-ray scattering (SAXS) measurements using high brilliant synchrotron radiation. Isotropically circular rings of the lattice peaks observed for the unstretched sample turned into deformed ellipsoidal rings upon the uniaxial stretching, with sharpening of the peaks in the direction parallel to the stretching direction and almost disappearing of the peaks in the perpendicular direction. By quantitatively analyzing the SAXS results, it was found that the packing regularity of the spherical microdomains was enhanced in the parallel direction while it was spoiled in the perpendicular direction under the stretched state. The enhanced regularity of packing was unchanged even if the stretching load was completely removed.

Keywords: block copolymers; thermoplastic elastomer; small-angle X-ray scattering; synchrotron; spherical microdomain; uniaxial stretching; body-centered cubic lattice; face-centered cubic lattice; packing regularity; paracrystal model

1. INTRODUCTION

In block copolymers comprising rubbery and glassy chains with the glassy minor component, spherical microdomains are formed due to microphase separation, which are dispersed in the rubbery matrix. Such a microphase-separated structure imparts elastomeric properties to the material since the glassy spherical microdomains play the role of a crosslinker. Meanwhile, they can be melt-processed above the glass transition temperature of the glassy component. Thus, they are referred to as a thermoplastic elastomer [1]. The microphase-separated structures in block copolymers have been studied for many decades and much has been uncovered [2–10], such as that spherical microdomains regularly order in the body-centered cubic (bcc) lattice [6]. We also have studied why the block copolymers favor the bcc over the face-centered cubic (fcc) lattice, and have been able to demonstrate a very rare case of spherical microdomains exhibiting the fcc lattice [9]. In the course of the study, we have accidentally discovered the transition from fcc to bcc upon uniaxial stretching of the block copolymer sample. Furthermore, it seems that the resultant bcc lattice (after releasing the tensile force from the stretched state) achieved enhanced packing regularity in the direction parallel to the uniaxial stretching. This fact seems to imply that mechanical stretching can play the role of an external field. However, the ability to enhance the packing regularity of spherical microdomains is exclusive to the mechanical stretching, bearing in mind that other external fields are generally effective for alignment or orientation of microdomains [7,8,10]. It should be further mentioned that a cycle of stretching-and-releasing plays an important role from analogy of densification of packing in granules upon shaking. In the current paper, we quantify the enhancement of packing regularity of spherical microdomains in the bcc lattice upon uniaxial stretching of the same elastomeric triblock copolymer as used in our previous work [9] by conducting small-angle X-ray scattering (SAXS) measurements using high brilliant synchrotron radiation.

2. EXPERIMENTAL SECTION

The material used is a poly(1,3-cyclohexadiene)-*block*-poly(ethylene-*co*-but-1-ene)-*block*-poly(1,3-cyclohexadiene) (PCHD-PEB-PCHD) triblock copolymer, having the volume fraction of PCHD (ϕ_{PCHD}) of 0.293 (determined by ^{1}H-NMR), Mn = 5.73 × 10^4, and Mw/Mn = 1.09 (determined by gel permeation chromatography, GPC) where Mn and Mw are the number- and weight-average molecular weight, respectively. Note here that the GPC column was calibrated with the polystyrene standard (Polymer Laboratories, Inc.) so that the molecular weight is polystyrene equivalent. It should be also noted that the glass transition temperatures (Tg), which were determined from peak positions of the mechanical tan δ for PCHD and PEB are 150 °C and −58 °C, respectively, indicating that the PCHD domains play a role of physical crosslinking points in the rubbery PEB matrix. We prepared a film of this sample by compression molding at 230 °C (as-mold sample).

The SAXS measurements were performed at SAXS beamlines BL-10C and BL-15A in the Photon Factory of the High Energy Accelerator Research Organization, Tsukuba, Japan [11,12], and at SAXS beamline BL45XU of SPring-8 in the Japan Synchrotron Radiation Research Institute [13]. The wavelength of the X-rays was tuned to λ = 0.149 nm for BL-10C, λ = 0.150 nm for BL-15A, and λ = 0.099 nm for BL45XU. Imaging plates of 250 × 200 mm size and 100 × 100 μm resolution were used as detectors at BL-10C and BL-15A. The measured two-dimensional SAXS pattern was converted to the one-dimensional profile by conducting sector averaging

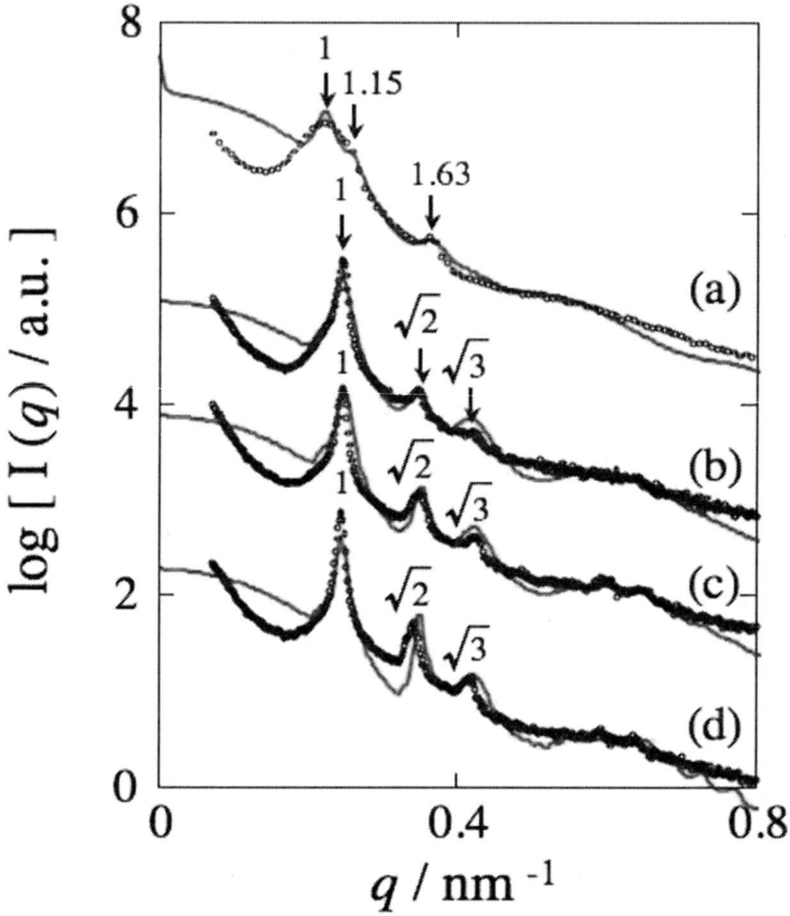

Figure 1. SAXS profiles measured at room temperature (a) in the unstretched state, and for the samples suffering the uniaxial stretching at room temperature up to the stretching ratio of (b) $\alpha = 4.0$, (c) $\alpha = 7.5$, and (d) $\alpha = 8.0$. Note that SAXS profiles (b-d) were obtained for the samples after complete removal of the mechanical stretching, *i.e.*, under the state being relaxed from the stretched state. Here, the sample subjected to the measurement (a) was prepared as follows. The as-mold sample was further treated by imposing the planar shear flow (flow-imposed sample) at 220 °C for 2 h, followed by further annealing at 210 °C for 2 h. It should be also noted that all SAXS profiles were obtained by conducting sector average in the direction parallel to the original stretching direction ($\mu = 0° \pm 5°$). The solid curves are results of the calculated profiles based on the fcc and bcc models.

in the direction where diffraction peaks were most prominent. The annealed film was trimmed into a rectangular shape (5 mm wide, 20 mm long with a thickness of 0.5 mm) and it was cramped with a pair of jaws on the homemade stretching device (seeFigure 1 of [14] for the schematic illustration of the stretching device) with the gauge length of 15 mm, which enabled 2d-SAXS measurements under the stretched state at room temperature [14].

Transmission electron microscopy (TEM) was performed for ultrathin sections, each with a thickness of *ca.* 80 nm. The sample was sliced at 183 K using a Reichert-Jung Ultracut T with a cryo-sectioning system (Reichert-Nissei FC-S). The thin section was picked up on 400-mesh copper grids and exposed to ruthenium tetraoxide (RuO_4) vapor for about 12 h to stain selectively the PCHD domain. TEM was performed at 100 kV on a JEM-2010 instrument (Nippon Denshi Co., Ltd.).

3. RESULTS AND DISCUSSION

Before going into the details, we will mention our previous SAXS results, shown in Figure 1, where the logarithm of the scattering intensity, $I(q)$, is plotted as a function of the magnitude of the scattering vector, $q = (4\pi/\lambda) \sin(\theta/2)$, with λ and θ being the wavelength of X-ray and the scattering angle, respectively [9]. Here, the sample subjected to the measurement was prepared as follows. The as-mold sample was further treated by imposing the planar shear flow (flow-imposed sample) at 220 °C for two hours, followed by further annealing at 210 °C for two hours without the flow in order to eliminate any effects of the flow. The 1d-SAXS profiles shown in Figure 1 were obtained by conducting sector average in the direction parallel to the original stretching direction ($\mu = 0° \pm 5°$ where μ is the azimuthal angle defined as increasing clockwise with respect to the meridional direction pointing upward ($\varphi = 0°$; parallel to the stretching direction). For the as-annealed sample, prior to being subjected to the uniaxial stretching, the scattering peaks were observed at the relative q-positions of 1:1.15 ($\sqrt{(4/3)}$:1.63 ($\sqrt{(8/3)}$). This fact reminds us that the scattering originates from the fcc lattice [(111) and (200) reflections, respectively]. Therefore, we conducted the fcc model calculation of the scattering curve based on the same procedure [9]. See the Appendix for details of the formulation. The solid curve is the result of the model calculation, which shows a good agreement with the experimental result (open circles), although we found here that better fitting was attained with a new set of structural parameters as compared to those reported before [9]. The structural parameters include d, the Bragg spacing of the (111) plane giving rise to the first-order peak, the g-

factor, the so-called paracrystalline distortion factor, R, the average radius of spherical microdomains, σ_R the standard deviation in R, and t_I, the characteristic thickness of the interfacial region of the spherical microdomains. The values of the structural parameters used for the model calculation are summarized in Table 1. Here, the values of the average radius of spheres R are considerably larger than observed by TEM (Figure 2), except for the values listed in Table 1(b) under the stretched state in the perpendicular direction, which are around 4.2 ~ 5 nm, showing good agreement with the TEM results. The fact that the radius is larger in the parallel direction than in the perpendicular direction does not represent deformation of the spheres, because the spheres are composed of PCHD, such that they were in the glassy state at room temperature where the *in situ*SAXS experiments were conducted. The overestimation of the radius might be an artifact, though the reason is not clear at present. There may be limitation of the applicability of the model calculation based on the isotropic orientation (random orientation) of the cubic lattice to the deformed cubic lattice, as well. For the purpose of the current work, we just recognize the structural values evaluated from the model calculation for the case of the *in situ* SAXS profiles as values equivalent to the undeformed cubic lattice.

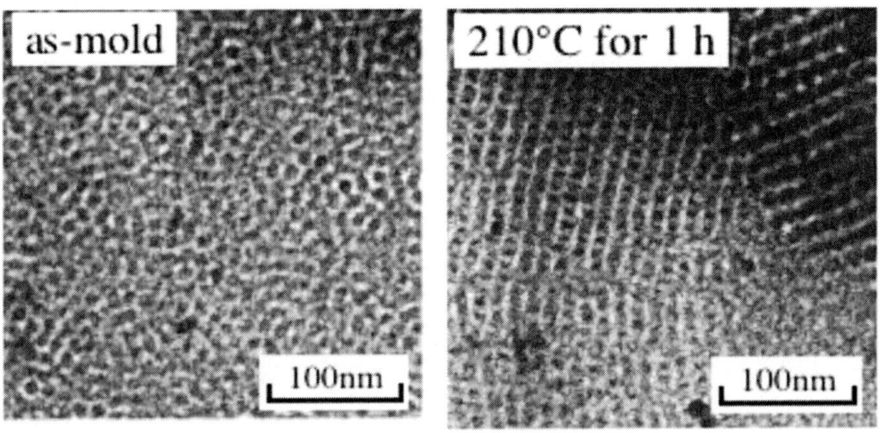

Figure 2. Transmission electron microscopic photographs for the as-mold sample and the sample further annealed at 210 °C for 1 h. The PCHD phase is selectively stained with RuO_4 and appears dark.

Table 1. Values of structural parameters used for the SAXS profiles based on the fcc and bcc paracrystalline models.

	Stretching Ratio	Direction	Type of Lattice	g	d (nm)	R (nm)	σ_R (nm)	t_I (nm)
(a) shown in Figure 1	1.0	//	fcc	0.085	28.0	8.8	2.5	1.5
	4.0	//		0.070	25.5	8.5	1.5	1.0
	7.5	//	bcc	0.060	25.5	8.5	1.5	1.0
	8.0	//		0.050	25.5	8.5	1.5	1.0
(b) shown in Figures 5 and 6	1.0	//	bcc	0.090	26.0	8.3	1.0	1.0
		⊥		0.080	24.5	8.5	1.0	1.0
	4.5	//		0.070	63.7	8.7	1.0	0.5
		⊥		0.090	16.6	5.0	1.2	1.0
	5.4	//		0.065	69.0	8.7	0.8	0.5
		⊥		0.090	16.2	4.2	2.0	1.0
	6.8	//		0.065	89.0	8.7	1.0	0.5
		⊥		0.150	15.5	4.2	2.0	1.0
	7.9	//		0.065	105	9.0	1.0	0.5
		⊥		0.120	10.5	4.2	1.5	1.0
	after releasing the load	//		0.067	31.0	8.7	1.0	1.0
		⊥		0.080	22.5	8.0	1.0	1.0

Figure 1 also contains the SAXS results of the samples suffering the uniaxial stretching at room temperature up to the stretching ratio of (b) α = 4.0, (c) α = 7.5, and (d) α = 8.0. The stretching ratio α is defined by $\alpha = \ell / \ell_0$ with ℓ_0 and ℓ being the original length of the sample before stretching and the length of the sample at the stretched state, respectively. Note that these SAXS profiles were obtained for the samples after complete removal of the mechanical stretching, *i.e.*, under the state being relaxed from the stretched state. It was found that all SAXS profiles other than that for the unstretched sample show the scattering peaks at the relative q-positions of $1{:}\sqrt{2}{:}\sqrt{3}$ which are ascribed to the bcc lattice, clearly demonstrating that the mechanical stretching induces transformation of the packing manner from fcc to bcc. It can be further recognized that the peaks are sharper for the sample suffering the higher extent of uniaxial stretching. The peak sharpness correlates with regularity of the packing of spherical microdomains. To quantify the degree of the packing regularity, we calculated the model scattering function based on the bcc lattice by following a similar procedure [6,15,16]. The solid curves are the results of the calculation by appropriately adjusting the structural parameters as mentioned above where d denotes, in this particular case for the bcc lattice, the Bragg spacing of the (110) plane giving rise to the first-order peak of the bcc lattice. Among them, the g-factor rationalizes the degree of the packing regularity as $g = \Delta a / a$ where Δa and a denote, respectively, the standard deviation in a due to the paracrystalline distortion and the edge length of the unit cell. Note that the smaller value of the g-factor indicates better packing regularity. As expected, the g-factor exhibits decreasing tendency (*i.e.*, the packing regularity is being enhanced) as a function α of the sample suffering prior to the SAXS measurement (see Figure 3). Thus, it is quantitatively confirmed that the regularity of the bcc lattice is better when the sample suffers a higher extent of uniaxial stretching. This result, furthermore, reminds us that a cycle of stretching-and-releasing plays an important role from analogy of densification of packing in granules upon shaking.

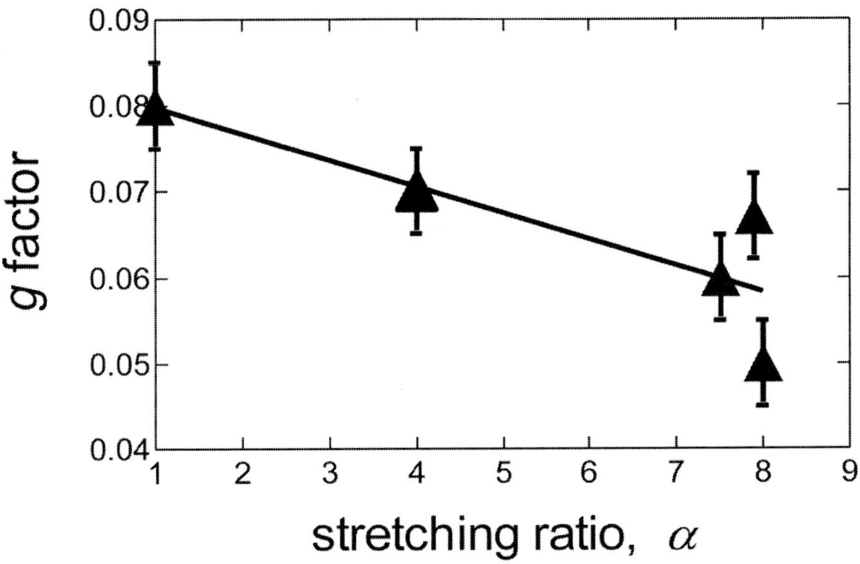

Figure 3. Variation of the g-factor as a function of the stretching ratio (α), evaluated from the model fitting of the SAXS profiles shown in Figure 1.

Now, we confirm whether the enhancement of packing regularity of spherical microdomains in the bcc lattice takes place in the sample under the stretched state or not. For this purpose, we conducted *in situ* 2d-SAXS measurements of the sample under the uniaxial stretching using the homemade stretching device [14] with the initial gauge length of 15 mm and the initial thickness and width of the film being 0.5 mm and 5 mm, respectively. Here, the sample subjected to the measurement was prepared as follows. The as-mold sample was further annealed at 210 °C for one hour, without imposing the planar shear flow. Figure 2 shows TEM photographs for the as-mold sample and the sample further annealed at 210 °C for one hour. The PCHD phase is selectively stained with RuO_4 and appears dark. Although for both samples, PCHD spheres were identified, the packing regularity is very different; the regularity is much better for the annealed sample. This annealed sample was used for the *in situ* 2d-SAXS measurements. It is noticed here that the average radius was found to be approximately 5 nm from the TEM photos for both samples.

Figure 4 displays 2d-SAXS patterns as a function of the stretching ratio. From the isotropic circular peaks observed for the unstretched state to the elliptic patterns under stretched states, the 2d-SAXS pattern changed. In the direction parallel to the stretching direction, the peaks moved towards the smaller q-position while the peaks moved towards the higher q-positions in the direction perpendicular to the stretching direction, both of which give rise qualitatively to the elliptic shape of the 2d-SAXS pattern. Furthermore, the scattering peaks appeared much clearer for the samples under higher stretching in the direction parallel to the stretching direction. These results qualitatively remind us that the bcc regularity already becomes better even in the samples under the stretched state. To quantify the enhancement of the packing regularity, the model SAXS function was calculated and was fitted to the experimental results in the direction parallel to the stretching direction.

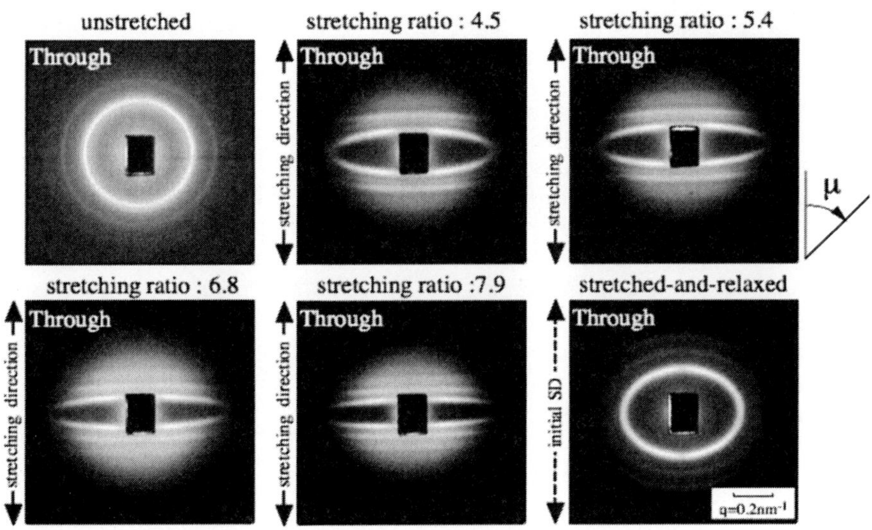

Figure 4. 2d-SAXS patterns obtained from *in situ* measurements of the sample under the uniaxial stretching using the homemade stretching device. The stretching direction is vertical.

Figure 5 displays the fitting, where the open circles denote the experimental result and the solid curve the result of the model calculation. Meanwhile, for the direction perpendicular to the stretching direction, similar plots are shown in Figure 6. The evaluated g-factor is shown in Figure 7 as a function of α. It is clearly found that the plots exhibit a monotonic decrease in the g-factor with an increase in α. To compare with the previous results plotted in Figure 3, the approximated line for these data is included in Figure 7. The fact that the line expresses firmly the tendency of data points indicates definitely that the enhancement of the packing regularity is already attained in the course of the uniaxial stretching and the better regularity attained is not spoiled in the following releasing process. In other words, the releasing process after the stretching is not necessarily required for the enhancement of the packing regularity. As for the g-factor in the direction perpendicular to the stretching direction, the results are shown together in Figure 7, and the g-factor is observed to considerably increase with an increase in α. Thus, in the perpendicular direction, the packing regularity becomes worse in the course of the uniaxial stretching. Although this result seems to be reasonable because it compensates the enhancement of the regularity in the parallel direction, the regularity in the perpendicular direction is much worse! In other words, the better regularity in the parallel direction is a bit independent. Another important aspect is that the enhanced regularity in the parallel direction attained in the course of the uniaxial stretching was able to be maintained after releasing of the tensile load, as shown in Figure 7.

These two facts imply there is a special mechanism behind. To uncover the mechanism, it is worthwhile examining the change in the domain spacing d along with the uniaxial stretching as shown in Figure 8. The solid curves indicate behaviors when assuming that the affine deformation takes place. Note that the affine deformation stands for the fact that macroscopic deformation can be transmitted to the microscopic level, and for the particular case of the block copolymers the degree of deformation in the microdomain repeating distance can be identical to the macroscopic deformation ratio. The solid curves shown in Figure 8 are based on the following equations:

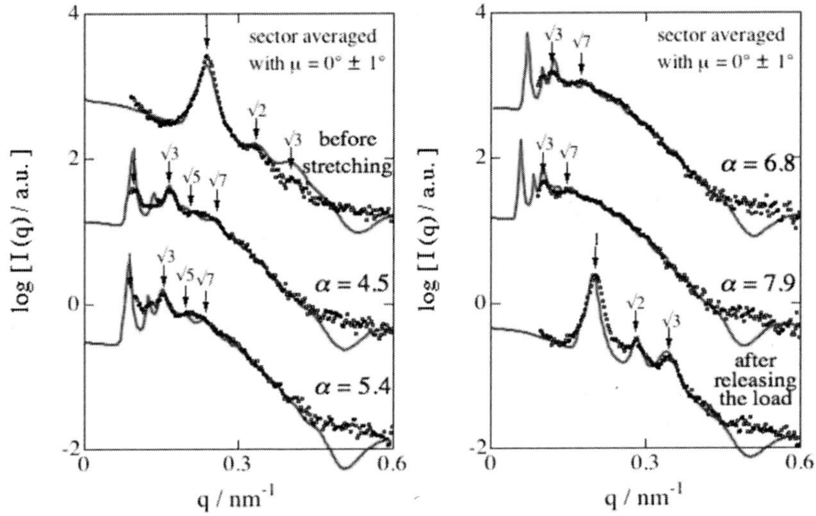

Figure 5. 1d-SAXS profiles parallel to the stretching direction converted from the 2d-SAXS patterns shown in Figure 4 by conducting the sector averaging over the azimuthal angle range with μ = 0° ± 1°. The solid curves are the results of the calculated profiles based on the bcc model.

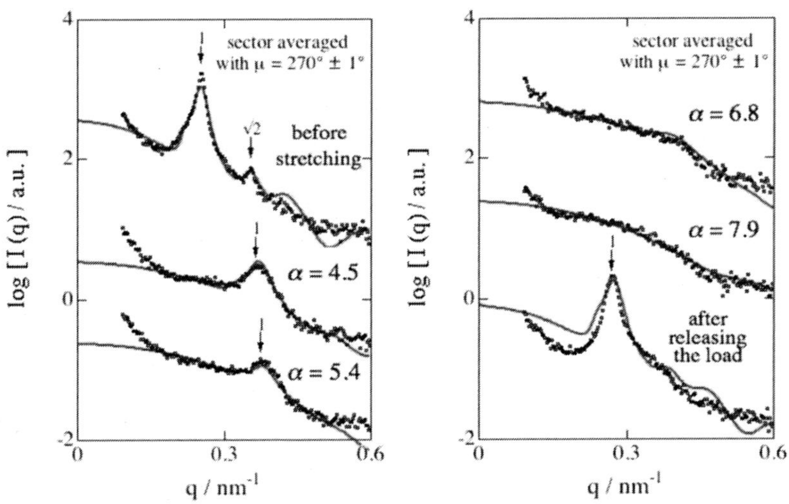

Figure 6. 1d-SAXS profiles perpendicular to the stretching direction converted from 2d-SAXS patterns shown in Figure 4 by conducting the sector averaging over the azimuthal angle range with μ = 270° ± 1°. The solid curves are results of the calculated profiles based on the bcc model.

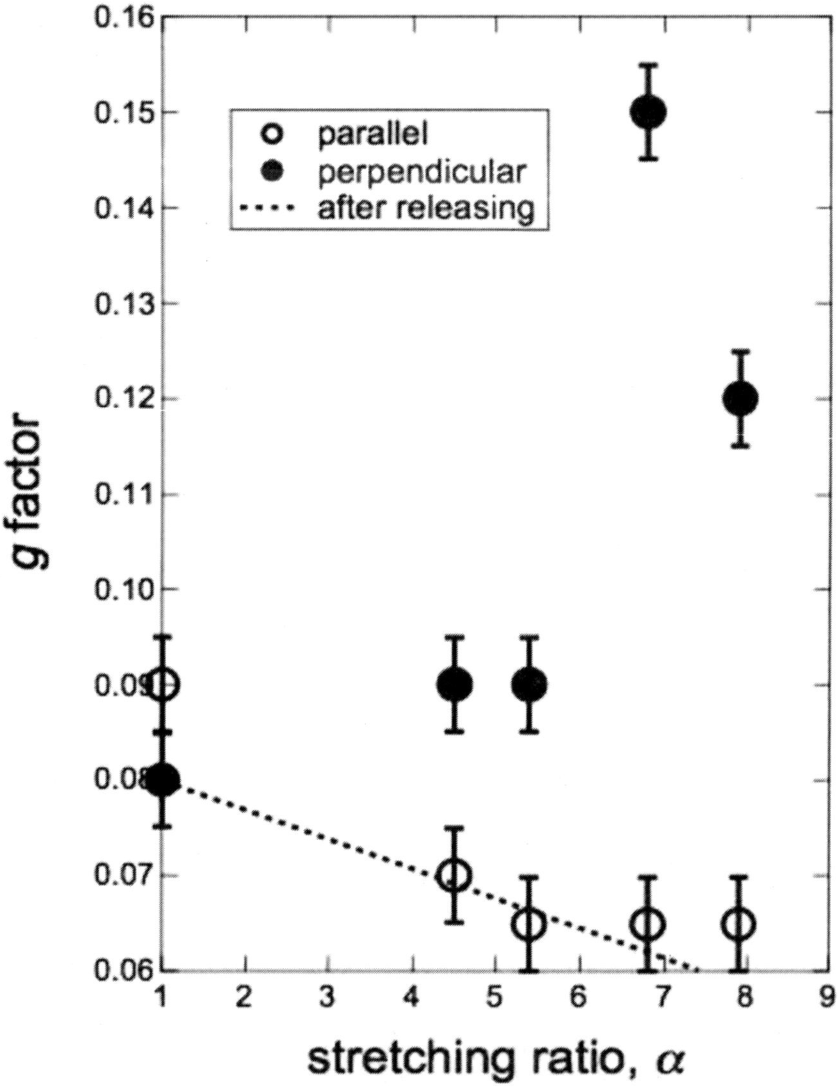

Figure 7. Variation of the g-factor as a function of the stretching ratio (α), evaluated from the model fitting of the SAXS profiles shown in Figures 5 and 6. To compare with the previous results plotted in Figure 3, the approximated line for those data is shown by the dotted line.

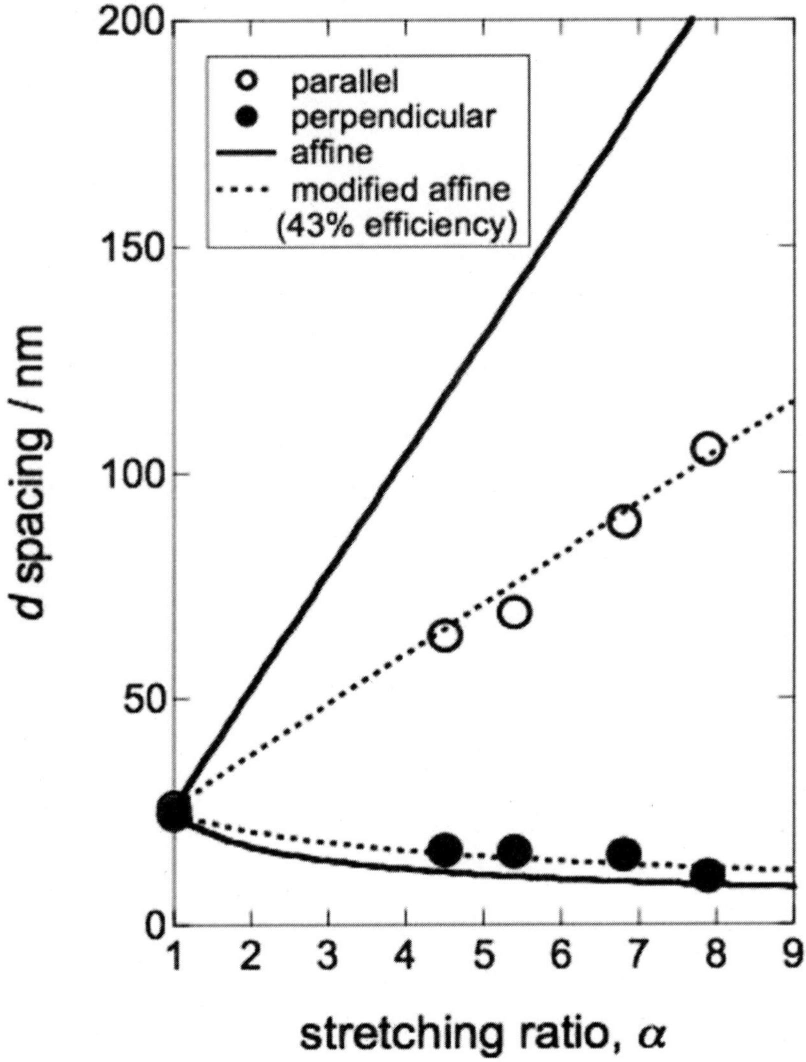

Figure 8. Change in the domain spacing *d* along with the uniaxial stretching. The solid curves indicate behaviors by assuming the affine deformation and that the total volume of the sample does not change during the stretching, based on Equations (1) and (2). The dotted curves are the best fits to the data expressed by Equations (3)and (4), the modified affine with $k = 0.43$.

$$d_{/\!/}(\alpha) = d_{/\!/}(1)\,\alpha \qquad (1)$$

$$d_{\perp}(\alpha) = d_{\perp}(1)\,\alpha^{-0.5} \qquad (2)$$

where $d_{/\!/}(\alpha)$ and $d_{\perp}(\alpha)$ are the d spacing as a function of α in the direction parallel and perpendicular to the stretching direction, respectively. And $d_{/\!/}(1)$ and $d_{\perp}(1)$ are the values of the d spacing in the unstretched state for the samples used for the SAXS measurements along with the uniaxial stretching in the direction parallel and perpendicular to the stretching direction, respectively. In Equation (2), it is assumed that the total volume of the sample does not change during the stretching.

As clearly seen in Figure 8, the data points do not follow the affine deformation, which suggests slip during the stretching. To quantify the apparent efficiency of the stretching, the above equation is modified by introducing a parameter k in the degree of elongation ($\alpha-1$), giving the modified affine deformation as:

$$d_{/\!/}(\alpha) = d_{/\!/}(1) \times \{1 + k(\alpha - 1)\} \qquad (3)$$

$$d_{\perp}(\alpha) = d_{\perp}(1) \times \{1 + k(\alpha - 1)\}^{-0.5} \qquad (4)$$

By adjusting the parameter k, good agreement between the data points and the modified affine behaviors can be confirmed in Figure 8, where the broken curves in Figure 8 show the modified affine with $k = 0.43$. Macroscopic slippage of the film sample at the jaws of the stretching device cannot explain such a tremendous deficiency of the mechanical stretching.

Then the possible explanation for the deviation from the affine deformation exists in the sample, which is referred to as inner slippage, and void formation during the stretching may be responsible. Inoue *et al.* reported a peculiar light scattering pattern (so-called butterfly pattern) for sphere-forming elastomeric block copolymer samples (comprising polystyrene and polyisoprene blocks) under stretched state at room temperature [17]. They ascribed the results to microvoid formation for which such a mechanism as that shown schematically in Figure 9 was also incorporated [17]. This model can be explained as follows: In the initial stage before stretching, there is some extent of number density fluctuation of the hard spheres, which might be due to paracrystalline distortion, and the fluctuation is enhanced upon the stretching. The resultant regions where the number density is lower turn into microvoids due to stress concentration. Interestingly, such microvoid regions are elongated in the direction perpendicular to the stretching direction. This two-region model can explain the failure of the affine deformation. Namely, the microvoid regions are heavily deformed while the other (microdomain) regions are less deformed than the macroscopic deformation. It is needless to note that the microdomain regions are also deformed, but what we like to say here is that the degree of the deformation is less than the macroscopic one. Due to the heavy deformation in the microvoid regions, the effective strain in the microdomain regions is 43% of the macroscopic one, considering k = 0.43 evaluated above.

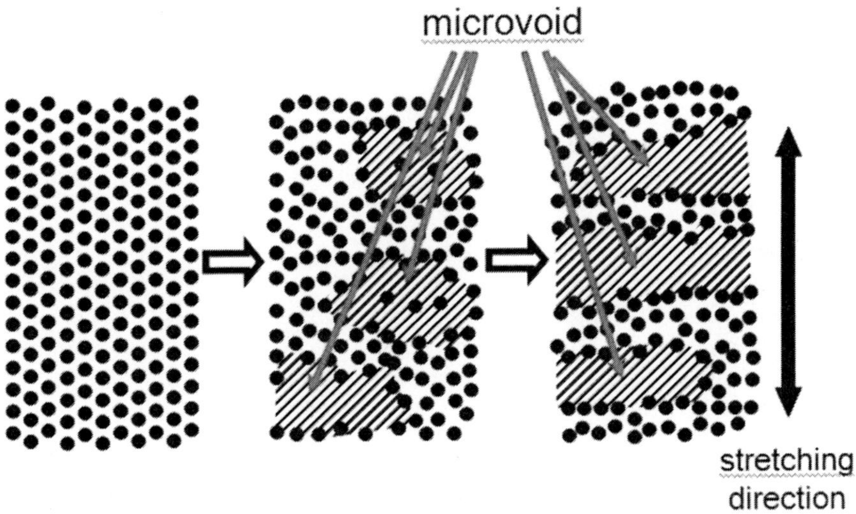

Figure 9. Schematic illustration of the two-region model with microvoids. The stretching direction is vertical.

The microvoid formation results in non-recovery eternal strain after releasing the tensile load, which is so-called remaining strain. Using the values of the d spacing, the remaining strain can be estimated.

$$\alpha_{//,\mathrm{off}} = d_{//,\mathrm{off}} \,/\, d_{//}(1) \tag{5}$$

$$\alpha_{\perp,\mathrm{off}} = \{ d_{\perp,\mathrm{off}} \,/\, d_{\perp}(1) \}^{-0.5} \tag{6}$$

where $\alpha_{//,\mathrm{off}}$ and $\alpha_{\perp,\mathrm{off}}$ denote, respectively, the remaining strain calculated independently from the values of $d_{//,\mathrm{off}}$ and $d_{\perp,\mathrm{off}}$ which are, respectively, values of the d spacing for the samples used for the SAXS measurements in the direction parallel and perpendicular to the stretching direction after releasing the tensile load. Substituting $d_{/\!/,\mathrm{off}}$ = 31.0 nm, $d_{/\!/}(1)$ = 26.0 nm, $d_{\perp,\mathrm{off}}$ = 22.5 nm, and $d_{\perp,\mathrm{off}}(1)$ = 24.5 nm from the d values tabulated in Table 1, $\alpha_{/\!/,\mathrm{off}} = \alpha_{\perp,\mathrm{off}}$ = 0.19 are evaluated. It is quite reasonable to obtain the identical values even if the apparently independent dual Equations (5) and (6) were used. Such a small value of the remaining strain ensures almost complete recovery of the deformation in the microdomain lattice upon the release of the tensile load.

At a first glance, the two-region model proposed in Figure 9 cannot explain our result of the small remaining strain. However, if we consider that the microvoid region does not really contain voids but is filled with heavily stretched rubbery chains, then this region also can exhibit complete recovery upon removal of the stretching load. Therefore, such an elastic region, which is heavily deformed under the stretched state, is responsible for both the non-affine deformation and the small remaining strain.

Finally, it is worth mentioning that the g factor did not change upon the release of the stretching load for the direction parallel to the stretching direction, which is recognized when comparing the values of the g factor for α = 7.9 and the unstretched states in Table 1. In other words, not only the enhancement of the regularity in the stretching direction remained as it was upon the release of the stretching load. It is characteristic of block copolymers that the anisotropy of the degree of the packing regularity of the spherical microdomains remains in the sample recovered from the stretched state without being homogenized.

4. CONCLUSIONS

We conducted *in situ* 2d-SAXS measurements at room temperature during the uniaxial stretching of the elastomeric sphere-forming triblock copolymer (PCHD-PEB-PCHD) using synchrotron radiation as an X-ray source. Isotropically circular rings of the lattice peaks observed for the unstretched sample turned into deformed elliptic rings upon the uniaxial stretching, with sharpening of the peaks in the direction parallel to the stretching direction and an almost disappearing of the peaks in the perpendicular direction. By quantitatively analyzing the SAXS results, it was found that the packing regularity of the spherical microdomains was enhanced in the parallel direction while it was spoiled in the perpendicular direction under the stretched state. The enhanced regularity of packing was unchanged even if the stretching load was completely removed.

Although our result seems to indicate that the packing regularity in the parallel direction is enhanced by extensional deformation, it is unclear whether compression in the perpendicular direction plays a role or not. 2d-SAXS studies on the sample under simple compression or simultaneous biaxial stretching deserves future work. Another interesting aspect is that the state attained in the stretched state can be maintained even after the complete removal of the stretching load. Why does homogenization not take place? Considering that this result is obtained right after the release of the stretching, gradual homogenization may occur. In this case, how fast the decay rate is and what determines the decay rate deserves also future work. It is needless to say that future work should be carried out to reveal the mechanism of the enhancement of the packing regularity upon the uniaxial stretching.

ACKNOWLEDGMENTS

SAXS experiments were performed at the Photon Factory of the High Energy Accelerator Research Organization (approval numbers 2003G228 and 2007G546), and at SPring-8 (approval number 2006B1125 and 2009A1573).

APPENDIX

Here we present the model calculation of the SAXS profiles for spheres ordered in bcc and fcc lattices based on the Hosemann's paracrystal theory. According to the Hosemann's paracrystal theory, the scattering intensity $I(q)$ from microphase-separated block copolymers can be formulated as follows [15,16]:

$$I(q) = \langle |f^2(q)| \rangle - |\langle f(q) \rangle|^2 + |\langle f(q) \rangle|^2 Z(q) \quad \text{(A1)}$$

where $f(q)$ and $Z(q)$ denote the scattering amplitude due to the intraparticle interference and the scattering intensity due to the interparticle interference, respectively. Hereafter, $f(q)$ and $Z(q)$ are referred to as the particle and the lattice factor, respectively. In Equation A1, $< x >$ is the average of the quantity of x. $f(q)$ for a spherical particle with its radius, R, can be given by the Fourier tranformation of the variation of the electron density in the real space and reads

$$f(q) = 3A_e \Delta\rho V \frac{\sin(qR) - qR\cos(qR)}{(qR)^3} \exp\left(-\frac{\sigma_s^2 q^2}{2}\right) \quad \text{(A2)}$$

where A_e is the scattering amplitude of Thomson scattering, $\Delta\rho$ is the difference in the electron density between sphere and matrix, V is the volume of the spheres, and σ_s is a parameter characterizing the variation of the (electron) density in the interfacial region (σ_s determines the characteristic interfacial thickness t_I, which is mentioned in the text, through the relationship $t_I = (2\pi)^{1/2}\sigma_s$). We use the Gaussian distribution for the radius of sphere upon averaging with σ_R being the standard deviation for the distribution of the sphere radius. On the other hand, the lattice factor $Z(q)$ is given by Equation A3 with Eulerian angles, θ and ϕ, which define orientation of the unit cell of a given grain with respect to the experimental Cartesian coordinates:

$$Z(q,\theta,\phi) = \frac{(1-F_k^2)^3}{\left[1 - 2F_k\cos\left(\frac{\sqrt{6}}{4}qd\alpha\right) + F_k^2\right]\left[1 - 2F_k\cos\left(\frac{\sqrt{6}}{4}qd\beta\right) + F_k^2\right]\left[1 - 2F_k\cos\left(\frac{\sqrt{6}}{4}qd\gamma\right) + F_k^2\right]} \quad \text{(A3)}$$

with

$$F_k = \exp\left[-\frac{3}{16}g^2d^2q^2(\alpha^2 + \beta^2 + \gamma^2)\right] \quad \text{(A4)}$$

in which

$$\alpha = \sin\theta\cos\phi + \sin\theta\sin\phi + \cos\theta \quad \text{(A5)}$$

$$\beta = -\sin\theta\cos\phi - \sin\theta\sin\phi + \cos\theta \quad \text{(A6)}$$

$$\gamma = -\sin\theta\cos\phi + \sin\theta\sin\phi - \cos\theta \quad \text{(A7)}$$

for the bcc lattice, and

$$\alpha = \sin\theta\,\sin\phi + \cos\theta \quad \text{(A8)}$$

$$\beta = -\sin\theta\,\cos\phi + \cos\theta \quad \text{(A9)}$$

$$\gamma = -\sin\theta\,\cos\phi + \sin\theta\,\sin\phi \quad \text{(A10)}$$

for the fcc lattice. In Equations A3? and A4, d denotes the Bragg spacing, which gives rise to the first-order peak and corresponds to the spacing for {110} and {111} planes for the bcc and fcc lattices, respectively. For randomly oriented polygrains in actual samples, the scattering is isotropic. Therefore, $Z(q, \theta, \phi)$ is averaged with respect to θ and ϕ to obtain isotropic $Z(q)$:

$$Z(q) = \frac{1}{4\pi}\int_0^{2\pi} d\phi \int_0^{\pi} Z(q,\theta,\phi) \sin\theta \, d\theta \quad \text{(A11)}$$

REFERENCES AND NOTES

1. Holden, G.; Legge, N.R.; Quirk, R.; Schroeder, H.E.; Hanser, M. *Thermoplastic Elastomers*, 2nd ed.; Hanser: New York, NY, USA, 1996.

2. Hashimoto, T. Order-Disorder Transition in Block Polymers. In *Thermoplastic Elastomers*, 2nd ed.; Holden, G., Legge, N.R., Quirk, R., Schroeder, H.E., Hanser, M., Eds.; Hanser: New York, NY, USA, 1996; pp. 429–463.

3. Matsen, M.W.; Bates, F.S. Unifying Weak- and Strong-Segregation Block Copolymer Theories.*Macromolecules* **1996**, *29*, 1091–1998.

4. Hamley, I.W. *The Physics of Block Copolymers*; Oxford University Press: Oxford, UK, 1998.

5. Hamley, I.W. *Developments in Block Copolymer Science and Technology*; John Wiley & Sons: Lodon, UK, 2004.

6. Kim, J.K.; Lee, H.H.; Sakurai, S.; Aida, S.; Masamoto, J.; Nomura, S.; Kitagawa, Y.; Suda, Y. Lattice Disordering and Domain Dissolution Transitions in Polystyrene-*block*-poly(ethylene-*co*-but-1-ene)-*block*-polystyrene Triblock Copolymer Having a Highly Asymmetric Composition.*Macromolecules* **1999**, *32*, 6707–6717.

7. Okamoto, S.; Saijo, K.; Hashimoto, T. Dynamic SAXS Studies of Sphere-Forming Block Copolymers under Large Oscillatory Shear Deformation. *Macromolecules* **1994**, *27*, 3753–3758.

8. Shin, G.; Sakamoto, N.; Saijo, K.; Suehiro, S.; Hashimoto, T.; Ito, K.; Amemiya, Y. Time-Resolved SAXS Studies of a Sphere-Forming Block Copolymer under Large Oscillatory Shear Deformation.*Macromolecules* **2000**, *33*, 9002–9014.

9. Imaizumi, K.; Ono, T.; Kota, T.; Okamoto, S.; Sakurai, S. Transformation of Cubic Symmetry for Spherical Microdomains from Face-Centered to Body-Centered Cubic upon Uniaxial Elongation in an Elastomeric Triblock Copolymer. *J. Appl. Cryst.* **2003**, *36*, 976–981.

10. Sakurai, S. Progress in Control of Microdomain Orientation in Block Copolymers—Efficiencies of Various External Fields. *Polymer* **2008**, *49*, 2781–2796.

11. Sakurai, S.; Aida, S.; Okamoto, S.; Ono, T.; Imaizumi, K.; Nomura, S. Preferential Orientation of Lamellar Microdomains Induced by Uniaxial Stretching of Cross-Linked Polystyrene-*block*-polybutadiene-*block*-polystyrene Triblock Copolymer. *Macromolecules* **2001**, *34*, 3672.

12. Sakurai, S.; Isobe, D.; Okamoto, S.; Yao, T.; Nomura, S. Collapse of the Ia3d Cubic Symmetry by Uniaxial Stretching of a Double-Gyroid Block Copolymer. *Phys. Rev. E* **2001**, *63*, 061803:1–061803:5.

13. Fujisawa, T.; Inoko, Y.; Yagi, N. The use of a Hamamatsu X-ray image intensifier with a cooled CCD as a solution X-ray scattering detector. *J. Synchrotron Rad.* **1999**, *6*, 1106.

14. Sakurai, S.; Kota, T.; Isobe, D.; Okamoto, S.; Sakurai, K.; Ono, T.; Imaizumi, K.; Nomura, S. Synchrotron Small-Angle X-ray Scattering Studies on Flow-Induced Gyroid to Cylinder Transition in an Elastomeric SBS Triblock Copolymer. *J. Macromol. Sci. Phys.* **2004**, *43*, 1–11.

15. Matsuoka, H.; Tanaka, H.; Hashimoto, T.; Ise, N. Elastic Scattering from Cubic Lattice Systems with Paracrystalline Distortion. *Phys. Rev. B* **1987**, *36*, 1754.

16. Matsuoka, H.; Tanaka, H.; Iizuka, N.; Hashimoto, T.; Ise, N. Elastic Scattering from Cubic Lattice Systems with Paracrystalline Distortion. II. *Phys. Rev. B* **1990**, *41*, 3854.

17. Inoue, T.; Moritani, M.; Hashimoto, T.; Kawai, H. Deformation Mechanism of Elastomeric Block Copolymers Having Spherical Domains of Hard Segments under Uniaxial Tensile Stress.*Macromolecules* **1971**, *4*, 500–507.

5

CHAPTER

Development and Characterization of Biocompatible Fullerene [C60]/Amphiphilic Block Copolymer Nanocomposite

Alok Chaurasia,[1] Nanda Gopal Sahoo,[2] James T. McLeskey Jr.,[3] and Xiao Hu[1]

[1]*School of Materials Science and Engineering, Nanyang Technological University, Singapore 639798*

[2]*Department of Chemistry, DSB Campus, Kumaun University, Nainital 263002, India*

[3]*Department of Physics, Randolph-Macon College, Ashland, VA 23005, USA*

ABSTRACT

We report a supramolecular process for the synthesis of well-defined fullerene (C_{60})/polymer colloid nanocomposites in an aqueous solution via complex formation. A biocompatible triblock poly(4-vinylpyridine)-b-polyethylene-b-poly(4-vinylpyridine), $P4VP_8$-b-PEO_{105}-b-$P4VP_8$, was synthesized by atom transfer radical polymerization. The block copolymer formed complexes with C_{60} in toluene and resulted in fullerene assembly in cluster form. Nanocomposite dispersion in an aqueous solution could be obtained using an aged solution of the polymer/C_{60}/toluene solution by a solvent evaporation technique. The UV-Vis and FTIR spectroscopy confirmed the complex formation of fullerene with the polymer which plays a significant role in controlling the PDI and size of polymer/C_{60} micelles in the toluene solution. The particle size and morphology of $P4VP_8$-b-PEO_{105}-b-$P4VP_8$ and $P4VP_8$-b-PEO_{105}-b-$P4VP_8/C_{60}$ mixture were studied by dynamic light scattering (DLS) and transmission electron microscopy (TEM). In a cytotoxicity test, both pure polymer and the resulting polymer/C_{60} composite in water showed more than 90% cell viability at 1 mg/mL concentration.

1. INTRODUCTION

The organization of fullerenes (C_{60}) and their derivatives into nanostructures within polymer systems has potential applications in solar cells and biomedicine [1–4]. For example, interesting results on polymer solar cells were reported using a blend of a fullerene derivative and a block copolymer poly(4-vinylpyridine) of poly(3-hexyl thiophene) (P3HT-b-P4VP) [5]. It was also found that a fullerene in an aqueous solution can generate a singlet oxygen under photoirradiation which has implications in the fields of biomedical and environmental science [4, 6]. One of the biologically most relevant features of C_{60} is the ability to function as a "free radical sponge" and quench various free radicals more efficiently than conventional antioxidants [7]. Fullerenes also have widespread applications ranging from drug-delivery and tissue-scaffolding systems to consumer products [8], and they have been explored in the area of biological chemistry, for activities such as enzyme inhibition, antiviral activity, DNA cleavage, and photodynamic therapy [9].

However, many of C_{60}'s potential applications have been seriously hampered [10, 11] by its extremely low solubility in water. Derivatization of the fullerene molecule with various functional groups and other solubilization procedures using surfactants or long chain polymers [12, 13] through covalent interactions [14–17] may change the properties of C_{60}. It was reported that $^1O_2$ quantum yield decreases with decreasing double bonds in the fullerene molecule due to the modification [18, 19]. Generally, introduction of any substituent into the fullerene core can diminish these properties [18–20]. A more effective and promising approach is to organize the fullerenes by specific noncovalent interactions with amphiphilic block copolymers through charge transfer complexation between fullerenes and block copolymers. Laiho and coworkers fabricated functionally unmodified C_{60}-containing nanostructures via a combination of PS-b-P4VP block copolymer self-assembly and charge-transfer complexation between fullerenes and P4VP segments in an organic solvent [21]. However, synthesis of water-soluble adducts of this spherical carbon allotrope (potential biological applications) became a particularly challenging topic. Therefore, in order to enhance the processability and water solubility of the materials, self-assembly systems of pristine C_{60} with block copolymers have been explored. For example, water-soluble C_{60}/PS-b-PEO composites obtained through a tedious process exhibited multidistribution of particle sizes [22]. However, supramolecular aggregates with a more uniform size distribution could be obtained. This was done through the aggregation of C_{60} with PS-b-PDMAEMA in water using a dialysis process, where the fullerene preferably goes to PS region due to hydrophobic interaction [23]. However, in most cases, incorporation of fullerenes into water-soluble supramolecular structures using surfactants or other modifying agent results in the fullerene core being completely covered by the modifying agent. This reduces possibility of contact between water-soluble oxygen and fullerenes, thereby inhibiting some of the desired properties such as the ability to quench singlet oxygen ($^1O_2$). Clearly, the preparation of well structure-defined aggregates from C_{60} with biocompatible polymers in an aqueous medium remains a challenge to overcome.

The present study focuses on the development of a simple method which will allow the assembly of fullerene with an amphiphilic block copolymer $P4VP_8$-b-PEO_{105}-b-$P4VP_8$ in aqueous and organic media. The choice of the P4VP block is due to its more hydrophilic nature as compared to PS which will be more accessible to dissolved oxygen. The $P4VP_8$-b-PEO_{105}-b-$P4VP_8$ amphiphilic block copolymer was synthesized using atom transfer radical polymerization [24, 25].

2. MATERIALS AND METHODS

2.1. Chemicals and Materials

The hydroxy-end-capped PEO [degree of polymerization (DP): 105; PDI: 1.08] was purchased from Aldrich. 4-Vinylpyridine (VP) (99%) was purified by stirring over CaH_2 and distilled under vacuum. 2-Bromoisobutyryl bromide, triethylamine, CuCl (99.99%), N,N,N′,N″,N″-pentamethyldiethylenetriamine (PMDETA 95%), and fullerene (99.5%) were purchased from Aldrich and used without further purification.

2.2. Charge Transfer Complex Formation

The triblock copolymer $P4VP_8$-b-PEO_{105}-b-$P4VP_8$ was synthesized by using the ATRP method described elsewhere [26]. The polymer was dissolved in toluene (1 mg/mL) and kept stirring for 24 hrs. The fullerenes were separately dissolved in toluene with concentration of 1 mg/mL. The color of the solution was purple. The fullerene solution (1 mg/mL) was then added to polymer stock solution, keeping polymer and fullerene ratio of 95 : 5 (w/w). The initial color of solution was purple but with time it gradually turned into brown, indicates complex formation. The formation of the charge transfer complex as a function of time was monitored by UV-Vis spectrophotometry.

2.3. Preparation of Aqueous Colloidal Solution of Fullerene/Polymer Nanocomposite

The aqueous colloidal solution of fullerene/polymer nanocomposite was obtained by pouring 10 mL of a two-month aged solution in toluene into 100 mL of water. The toluene was allowed to evaporate at 80°C for 6 hrs and the final volume of fullerene/polymer composite in aqueous solution was kept at 10 mL. No precipitation was observed.

2.4. In Vitro Cytotoxicity Test

An in vitro cytotoxicity test was performed to study the biocompatibility of the polymer, polymer/C_{60}nanocomposite, in aqueous solution. For the cytotoxicity test, cell culture experiments were carried out using L929 mouse fibroblasts which were obtained from American Type Culture Collection (ATCC). Cells were grown in Dulbecco's modified Eagle's Medium (DMEM) supplemented with 10% heat inactivated Fetal Bovine Serum (FBS) (Lonza) and 1% Penicillin-Streptomycin-Glutamine (Antibiotic) (Gibco) maintained at 37°C in 5% CO_2 and 95% relative humidity. Cell cytotoxicity test was performed using the Promega CellTiter 96 One Solution Cell Proliferation Assay (MTS). This is a colorimetric method and the solution contains a tetrazolium compound [3-(4,5-dimethylthiazol-2-yl)-5-(3-carboxymethoxyphenyl)-2-(4-sulfophenyl)-2H tetrazolium] (MTS) and an electron coupling reagent (phenazine ethosulfate, PES). The compound is bioreduced by living cells into a colored formazan product that is soluble in tissue culture medium. The quantity of formazan product is measured at 490 nm at which the intensity of absorbance is directly proportional to the number of living cells in culture [27–29]. Cells were seeded at a density of 2×10^4 cells/well in 96-well plates. After 24 hrs, culture medium was replaced by 200 µL serial dilutions of the polymer solutions and the cells were incubated for 24 hrs. Polymer solutions were prepared in serum supplemented DMEM and sterilized by filtration (0.2 µm). An MTS assay was done for four different concentrations, 0.1 mg/mL, 0.25 mg/mL, 0.5 mg/mL, and 1 mg/mL. At the end of incubation time period (24 hrs), polymer solutions were aspirated and replaced with 100 µL fresh DMEM, to which 20 µL CellTiter 96 One Solution was added. The cells were then incubated for 4 hrs at 37°C in 5% CO_2 incubator. Then the absorbance was measured at 490 nm in a microplate reader (TECAN, Tecan Group Ltd., Switzerland). The cell viability (%) related to control wells containing cell culture medium without polymer was calculated by equation. Each sample was done in triplicate.

2.5. Characterization

The Fourier Transform Infrared (FTIR) spectra were recorded on a Perkin Elmer system 2000. UV Spectrophotometer (UV2501PC, Pharmaspec, Shimadzu) was used to monitor the complex formation. Dynamic light scattering (DLS) [Nano Zetasize ZS90 (Malvern)] and transmission electron microscopy (TEM) (JEOL JEM-2010) were used to monitor the aggregation behavior.

3. RESULTS AND DISCUSSION

The block copolymer $P4VP_8$-b-PEO_{105}-b-$P4VP_8$ was prepared via atom transfer radical polymerization (ATRP) and its complex with fullerene as outlined in Figure 1.

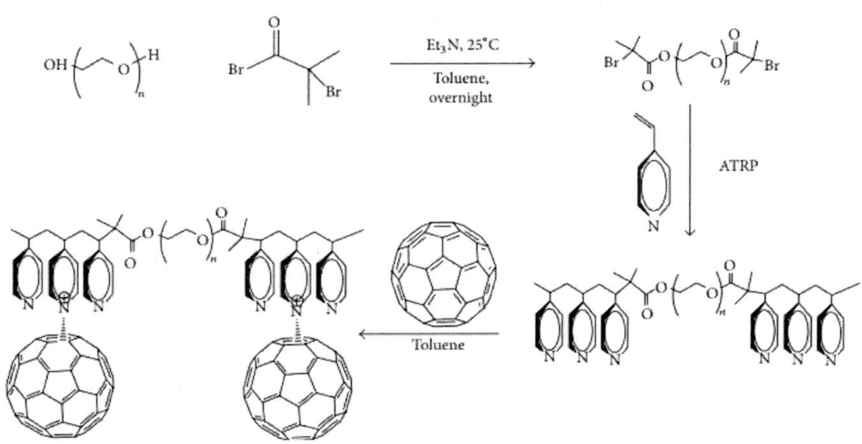

Figure 1: Scheme for synthesis of triblock copolymer and charge-transfer complex formation between $P4VP_8$-b-PEO_{105}-b-$P4VP_8$ and C_{60}.

Fullerenes (C_{60}) dissolve readily in toluene, which leads to a purple color solution. A fullerene solution in toluene (1 mg/mL) was then added to polymer stock solution (1 mg/mL), keeping polymer fullerene ratio of 95 : 5 by weight. The initial color of solution was purple but with time it gradually turned into brown indicating complex formation as shown in Figure 2.

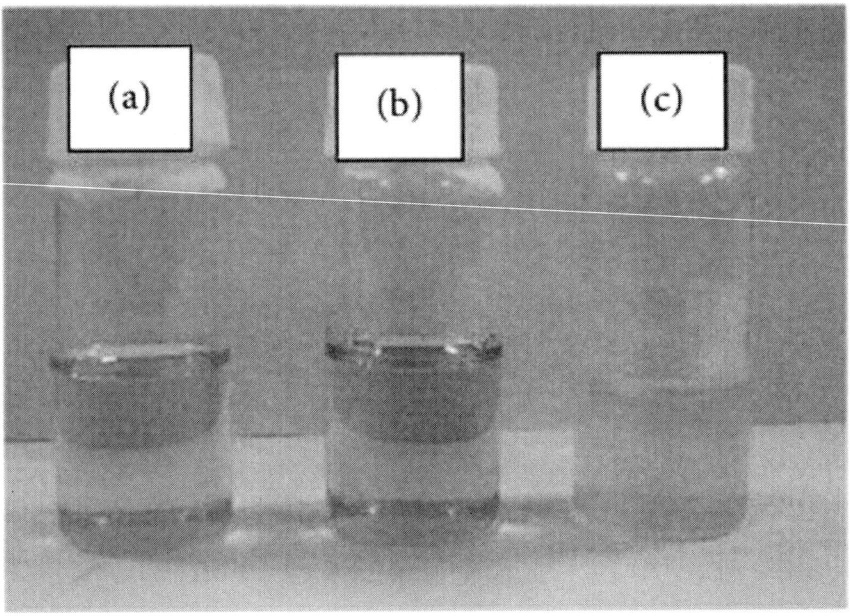

Figure 2: Colors of the solution for (a) freshly prepared $P4VP_8$-b-PEO_{105}-b-$P4VP_8$/C_{60}(95/5 w/w) in toluene, (b) two-month aged $P4VP_8$-b-PEO_{105}-b-$P4VP_8$/C_{60} (95/5 w/w), and (c) $P4VP_8$-b-PEO_{105}-b-$P4VP_8$/C_{60} (95/5 w/w) in water prepared from two-month aged solution by a solvent exchange process.

The color change with time was similar to that for the solution of C_{60} in pyridine. This agrees with the formation of charge-transfer complexes between the electron-donating pyridine solvent and the electron-accepting fullerenes [30, 31]. The changes in the visible spectrum are shown in Figure 3. The solution shows a broad and weak absorption band ranging from 410 to 650 nm. The absorbance in the region increases with aging. The spectral change (Figure 3(iii)) is attributed to the formation of charge transfer complex [30, 31].

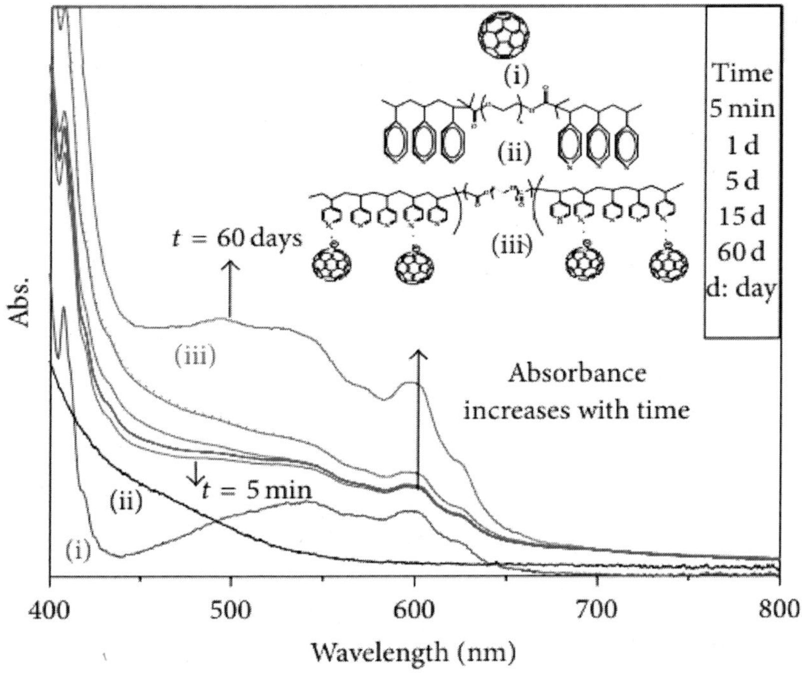

Figure 3: UV-Vis spectra of (i) C_{60} in toluene, purple line, (ii) $P4VP_8$-b-PEO_{105}-b-$P4VP_8$in toluene, black line, and (iii) $P4VP_8$-b-PEO_{105}-b-$P4VP_8/C_{60}$ (95/5 w/w) dissolved in toluene at different aging times (up to two months), brown line.

In the solution of $P4VP_8$-b-PEO_{105}-b-$P4VP_8/C_{60}$ (95/5 w/w) the most characteristic absorption peak of fullerenes at ~332 nm was not observed. It is because, due to higher concentration of polymer/C_{60} in solution, UV-Vis spectroscopy crosses the limit and was not able to pinpoint the most characteristic peak at ~332 nm. But the most characteristic peak at ~332 nm is clearly observed when the polymer/C_{60} solution concentration was reduced by 50%, as it is shown in Figure 4.

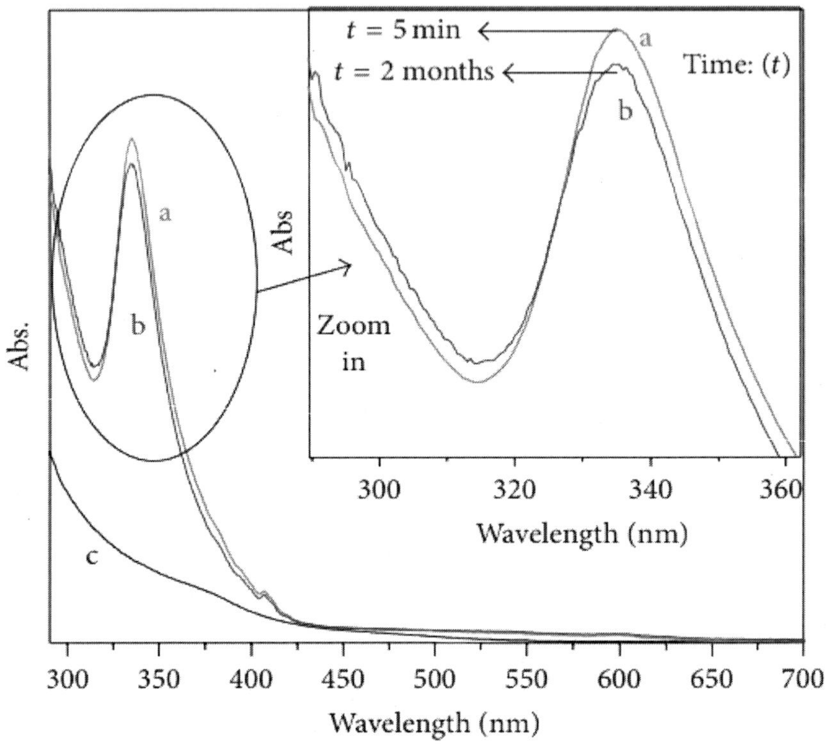

Figure 4: "a" line indicated UV-Vis spectrum of $P4VP_8$-b-PEO_{105}-b-$P4VP_8/C_{60}$ (95/5 w/w) dissolved in toluene (0.5 mg/mL) after 5 min aging; "b" line indicated UV-Vis spectrum of $P4VP_8$-b-PEO_{105}-b-$P4VP_8/C_{60}$ (95/5 w/w) dissolved in toluene (0.5 mg/mL) after 2-month aging; "c" line indicated UV-Vis spectrum of $P4VP_8$-b-PEO_{105}-b-$P4VP_8$dissolved in toluene.

FTIR spectroscopy provides further evidence for the complex formation. Upon aging of $P4VP_8$-b-PEO_{105}-b-$P4VP_8/C_{60}$ (95/5 w/w) in toluene, the pyridine peak at 1603 cm^{-1} is slowly shifted to 1618 cm^{-1} as shown in Figure 5. The shifting of the most characteristic peak due to interaction of the P4VP near 1600 cm^{-1} for C=N stretching indicates that a slight electron transfer took place from the 4-vinylpyridine to C_{60}.

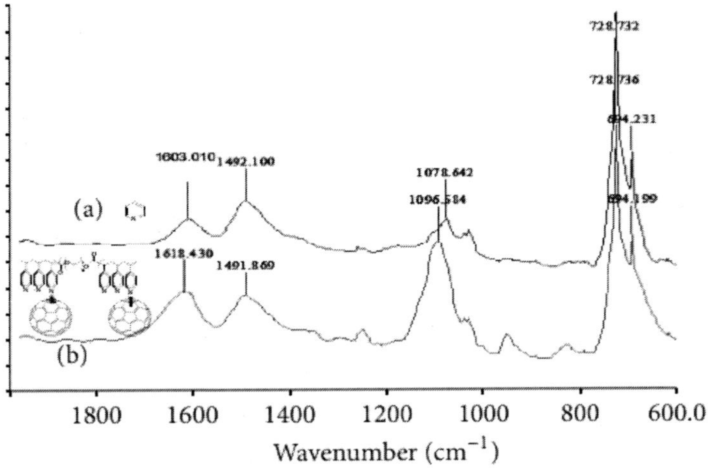

Figure 5: FT-IR spectra of (a) P4VP$_8$-b-PEO$_{105}$-b-P4VP$_8$/toluene and (b) two-month aged solution of P4VP$_8$-b-PEO$_{105}$-b-P4VP$_8$/C_{60} (95/5 w/w) in toluene.

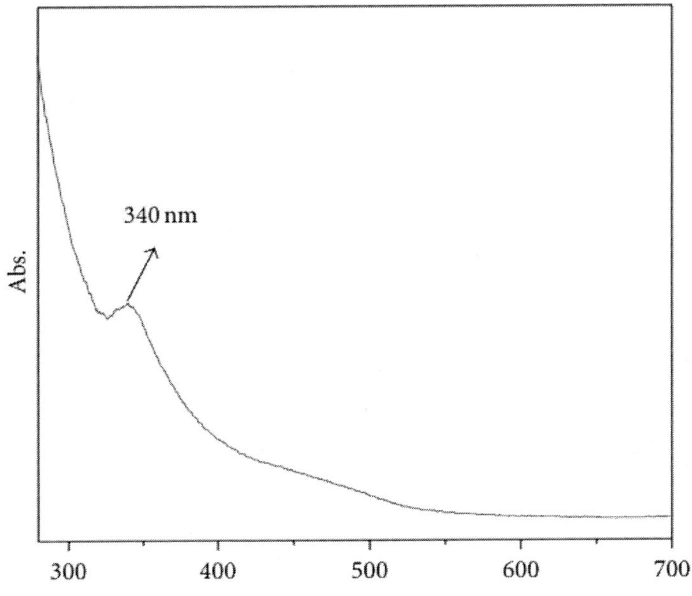

Figure 6: UV-Vis spectrum of polymer/C_{60} nanocomposite in aqueous solution.

The UV-Vis spectrum of polymer/C_{60} nanocomposite in an aqueous solution shows a broad peak from the fullerenes at about 340 nm. This peak clearly indicates the presence of fullerene in aqueous solution as shown in Figure 6.

The aggregation behavior of $P4VP_8$-b-PEO_{105}-b-$P4VP_8$ block copolymers was studied by dynamic light scattering using Zetasizer Nano ZS and ZS90 (Malvern). The apparent hydrodynamic radius (R_h) can be calculated using the Stokes-Einstein equation:

$$R_h = \frac{kT}{6\pi\eta D}, \tag{1}$$

where is the Boltzmann constant, is the absolute temperature, is the solvent viscosity, and is the diffusion coefficient. The DLS was conducted at a scattering angle of 90°. The polymer in toluene solution (1 mg/mL) showed bimodal peaks with mean diameters of ~24.4 nm and ~199 nm, respectively, and the PDI was 0.534 (Figure 7(a)). The particle size ranges from 10 to 400 nm. However, in the presence of C_{60} the polymer solution aggregates into one type of particle with mean diameter of 28.2 nm (Figure 7(b)) with PDI of 0.205. The particle size varies from 10 to 100 nm. The change in the aggregation behavior is clearly due to the C_{60}complex formation.

TEM micrographs are shown in Figure 8. Figure 8(a) shows the micelle consisting of a compact P4VP core and PEO chains as the corona. These samples were stained with iodine. On the other hand, the polymer/fullerene composites in toluene and water are shown in Figures 8(b) and 8(c), respectively, without iodine staining. The lack of iodine staining was to avoid confusion between P4VP and fullerene which shows up clearly in the fullerene aggregates. The particle size from TEM studies is in good agreement with the light scattering data for the $P4VP_8$-b-PEO_{105}-b-$P4VP_8$ polymer and C_{60} complexes.

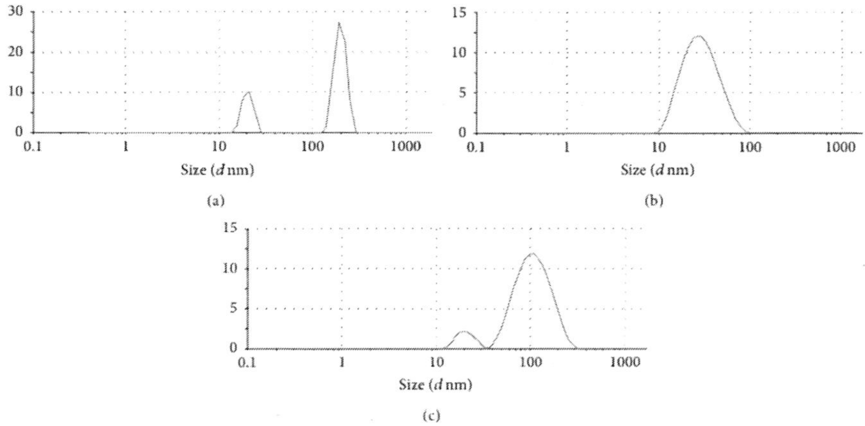

Figure 7: Dynamic light scattering of (a) $P4VP_8$-b-PEO_{105}-b-$P4VP_8$ in toluene, (b) two-month aged $P4VP_8$-b-PEO_{105}-b-$P4VP_8/C_{60}$ (95/5 w/w) in toluene, and (c) $P4VP_8$-b-PEO_{105}-b-$P4VP_8/C_{60}$ nanocomposite in aqueous solution.

From the TEM, it was determined that the C_{60} clusters were aggregated with a size of ~30 nm. In light scattering studies (Figure 7(c)), the polymer/C_{60} nanocomposite micelle in aqueous solution shows the bimodal distribution with mean diameters of ~30 nm and ~110 nm. This agrees reasonably well with the TEM images. The TEM images confirm the formation of fullerene clusters in both processes. The results confirm that the micelle in solution consists of P4VP in the core and PEO in the corona. The fullerenes are located in the core of the polymer aggregates. A plausible mechanism of the micelle formation of $P4VP_8$-b-PEO_{105}-b-$P4VP_8$and $P4VP_8$-b-PEO_{105}-b-$P4VP_8/C_{60}$ (95/5 w/w) is shown in Scheme 1.

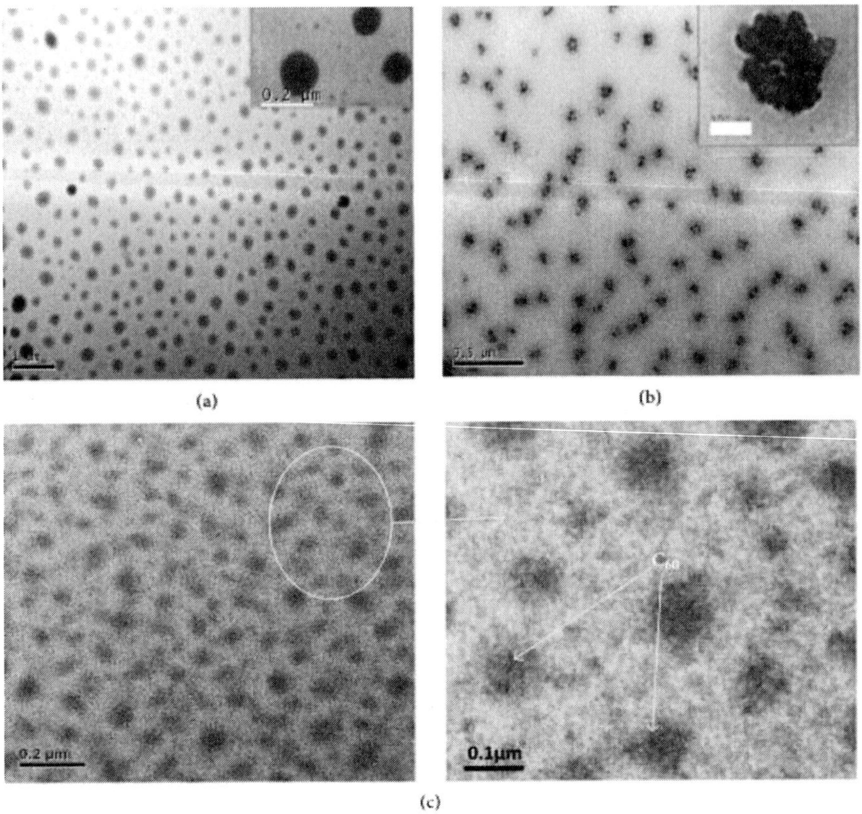

Figure 8: TEM images of (a) $P4VP_8$-b-PEO_{105}-b-$P4VP_8$ in toluene with iodine stain, (b) aged $P4VP_8$-b-PEO_{105}-b-$P4VP_8/C_{60}$ (95/5 w/w) in toluene solution (scale bar: 500 nm) without iodine staining, (c) fullerene cluster from aqueous solution (200 nm), and zoom in form of "c" where fullerene clusters are visible (100 nm), images were taken without iodine staining.

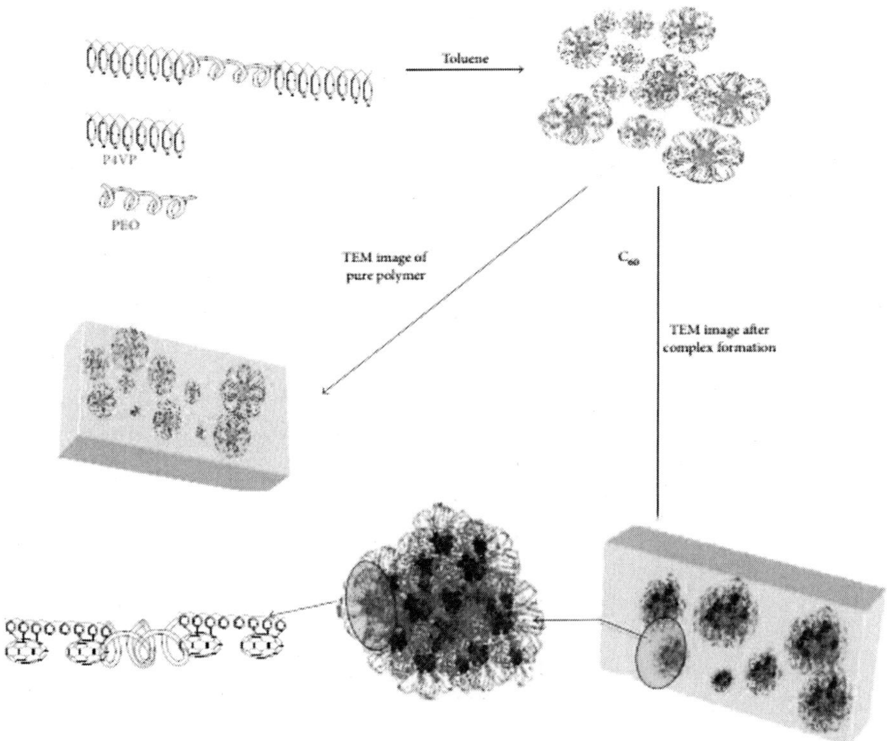

Scheme 1: Plausible mechanism of the micelle formation in $P4VP_8$-b-PEO_{105}-b-$P4VP_8$and $P4VP_8$-b-PEO_{105}-b-$P4VP_8$/C_{60} (95/5 w/w).

To evaluate the possibility of using the pure polymer and the polymer/C_{60} nanocomposite in aqueous solution for bioapplications, in vitro cytotoxicity studies were performed by MTS assay. This is a colorimetric test, based on the selective ability of viable cells to reduce the tetrazolium component of MTS into purple colored formazan crystals. MTS assay was done for four different concentrations, namely, 0.1 mg/mL, 0.25 mg/mL, 0.5 mg/mL, and 1 mg/mL. The results (Figure 9) show that there is no significant cytotoxicity of the four different concentrations of the pure polymer and the polymer/C_{60} nanocomposite. It was found that more than 90% of the cells are viable in all the different concentrations for both cases.

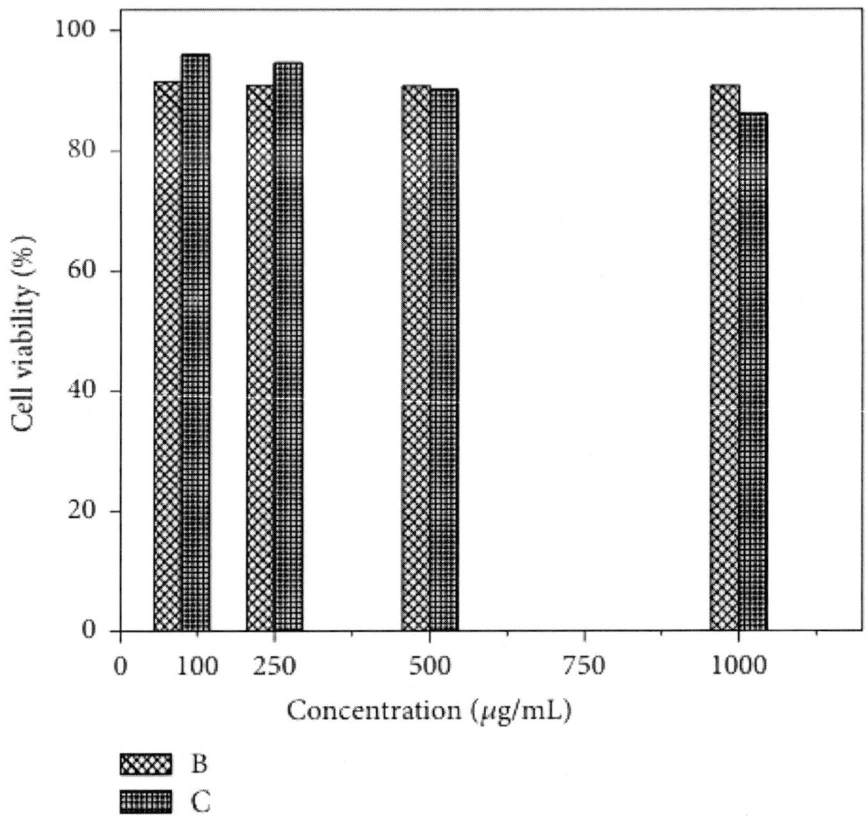

Figure 9: In vitro cytotoxicity studies of (B) polymer in water and (C) polymer/C_{60}nanocomposite in water.

4. CONCLUSIONS

This study shows that $P4VP_8$-b-PEO_{105}-b-$P4VP_8$/C_{60} systems can lead to the formation of self-assembled structures. The C_{60} complexation process has significant effect on the aggregation behaviour. Upon aging of the solution, a charge transfer complex between vinylpyridine unit and C_{60} was formed, which leads to the formation of self-assembled fullerene clusters. These results were supported by UV-Vis, TEM, and DLS studies. The formation of biocompatible $P4VP_8$-b-PEO_{105}-b-$P4VP_8$/C_{60} nanocomposites in water may have potential application in biomedicine, cosmetic (as singlet oxygen generator) or other useful applications. In summary, organic-soluble and water-soluble, structurally well-defined C_{60}/polymer nanocomposites were achieved through the self-assembly of pristine C_{60} with polymer. Though the rate of complex formation presented in this study was found to be slow, the work has opened the door and enlightened how noncovalent interaction can be used to obtain water-soluble, biocompatible, polymer-fullerene nanoparticles.

CONFLICT OF INTERESTS

The authors declare that there is no conflict of interests regarding the publication of this paper.

ACKNOWLEDGMENTS

The authors acknowledge funding from National Research Foundation of Singapore under a competitive research program (NRF-G-CRP-2007-01). One of the authors (Alok Chaurasia) would like thank Nanyang Technological University for a graduate scholarship.

REFERENCES

1. L.-M. Chen, Z. Hong, G. Li, and Y. Yang, "Recent progress in polymer solar cells: manipulation of polymer: fullerene morphology and the formation of efficient inverted polymer solar cells," Advanced Materials, vol. 21, no. 14-15, pp. 1434–1449, 2009.
2. R. Po, M. Maggini, and N. Camaioni, "Polymer solar cells: recent approaches and achievements," The Journal of Physical Chemistry C, vol. 114, no. 2, pp. 695–706, 2010.
3. N. S. Sariciftci, D. Braun, C. Zhang et al., "Semiconducting polymer-buckminsterfullerene heterojunctions: diodes, photodiodes, and photovoltaic cells," Applied Physics Letters, vol. 62, no. 6, pp. 585–587, 1993.
4. M. Orfanopoulos and S. Kambourakis, "Chemical evidence of singlet oxygen production from C60 and C70 in aqueous and other polar media," Tetrahedron Letters, vol. 36, no. 3, pp. 435–438, 1995.
5. N. Sary, F. Richard, C. Brochon et al., "A new supramolecular route for using rod-coil block copolymers in photovoltaic

applications," Advanced Materials, vol. 22, no. 6, pp. 763–768, 2010.

6. J. L. Anderson, Y.-Z. An, Y. Rubin, and C. S. Foote, "Photophysical characterization and singlet oxygen yield of a dihydrofullerene," Journal of the American Chemical Society, vol. 116, no. 21, pp. 9763–9764, 1994.

7. P. J. Krusic, E. Wasserman, P. N. Keizer, J. R. Morton, and K. F. Preston, "Radical reactions of C60,"Science, vol. 254, no. 5035, pp. 1183–1185, 1991.

8. Z. Markovic and V. Trajkovic, "Biomedical potential of the reactive oxygen species generation and quenching by fullerenes (C60)," Biomaterials, vol. 29, no. 26, pp. 3561–3573, 2008.

9. A. S. Boutorine, H. Tokuyama, M. Takasugi, H. Isobe, E. Nakamura, and C. Hélène, "Fullerene-oligonucleotide conjugates: photoinduced sequence-specific DNA cleavage," Angewandte Chemie, vol. 33, no. 23-24, pp. 2462–2465, 1995.

10. C. C. Zhu, Y. Xu, Y. Q. Liu, and D. B. Zhu, "Preparation and characterization of novel amphiphilic C_{60}derivatives," Journal of Organic Chemistry, vol. 62, no. 7, pp. 1996–2000, 1997.

11. P. Ravi, S. Dai, K. M. Hong, K. C. Tam, and L. H. Gan, "Self-assembly of C60 containing poly(methyl methacrylate) in ethyl acetate/decalin mixtures solvent," Polymer, vol. 46, no. 13, pp. 4714–4721, 2005.

12. W. T. Ford, T. Nishioka, S. C. McCleskey, T. H. Mourey, and P. Kahol, "Structure and radical mechanism of formation of copolymers of C_{60} with styrene and with methyl methacrylate," Macromolecules, vol. 33, no. 7, pp. 2413–2423, 2000.

13. S. Mehrotra, A. Nigam, and R. Malhotra, "Effect of [60]fullerene on the radical polymerization of alkenes," Chemical Communications, no. 5, pp. 463–464, 1997.

14. H. Yu, L. H. Gan, X. Hu, S. S. Venkatraman, K. C. Tam, and Y. Y. Gan, "A novel amphiphilic double-[60]fullerene-capped triblock copolymer," Macromolecules, vol. 38, no. 23, pp. 9889–9893, 2005.

15. J.-F. Nierengarten, M. Guttiérez-Nava, S. Zhang et al., "Fullerene-containing macromolecules for materials science applications," Carbon, vol. 42, no. 5-6, pp. 1077–1083, 2004.

16. Z. T. Ball, K. Sivula, and J. M. J. Fréchet, "Well-defined fullerene-containing homopolymers and diblock copolymers with high fullerene content and their use for solution-phase and bulk organization,"Macromolecules, vol. 39, no. 1, pp. 70–72, 2006.

17. K. Sivula, Z. T. Ball, N. Watanabe, and J. M. J. Fréchet, "Amphiphilic diblock copolymer compatibilizers and their effect on the morphology and performance of polythiophene: fullerene solar cells," Advanced Materials, vol. 18, no. 2, pp. 206–210, 2006.

18. T. Hamano, K. Okuda, T. Mashino et al., "Singlet oxygen production from fullerene derivatives: effect of sequential functionalization of the fullerene core," Chemical Communications, no. 1, pp. 21–22, 1997.

19. D. M. Guldi and K.-D. Asmus, "Photophysical properties of mono- and multiply-functionalized fullerene derivatives," Journal of Physical Chemistry A, vol. 101, no. 8, pp. 1472–1481, 1997.

20. F. Prat, R. Stackow, R. Bernstein, W. Qian, Y. Rubin, and C. S. Foote, "Triplet-state properties and singlet oxygen generation in a homologous series of functionalized fullerene derivatives," The Journal of Physical Chemistry A, vol. 103, no. 36, pp. 7230–7235, 1999.

21. A. Laiho, R. H. A. Ras, S. Valkama, J. Ruokolainen, R. Österbacka, and O. Ikkala, "Control of self-assembly by charge-transfer complexation between C_{60} fullerene and electron donating units of block copolymers," Macromolecules, vol. 39, no. 22, pp. 7648–7653, 2006.

22. X. L. Chen and S. A. Jenekhe, "Solubilization and encapsulation of fullerenes by amphiphilic block copolymers," Langmuir, vol. 15, no. 23, pp. 8007–8017, 1999.

23. X.-S. Wang, T. Metanawin, X.-Y. Zheng, P.-Y. Wang, M. Ali, and D. Vernon, "Structure-defined C_{60}polymer colloids supramolecular nanocomposites in water," Langmuir, vol. 24, no. 17, pp. 9230–9232, 2008.

24. S. N. Sidorov, L. M. Bronstein, Y. A. Kabachii et al., "Influence of metalation on the morphologies of poly(ethylene oxide)-block-poly(4-vinylpyridine) block copolymer micelles," Langmuir, vol. 20, no. 9, pp. 3543–3550, 2004.

25. H. Yu, L. H. Gan, X. Hu, and Y. Y. Gan, "A pH-sensitive double [60]fullerene-end-capped polymers via ATRP: synthesis and aggregation behavior," Polymer, vol. 48, no. 8, pp. 2312–2321, 2007.

26. C. Alok, L. H. Gan, and X. Hu, "Dramatic solvent effect on interaction kinetics and self-organization of phenyl-C61 butyric acid methyl ester in a triblock copolymer," Macromolecular Chemistry and Physics, vol. 212, no. 8, pp. 813–820, 2011.

27. T. M. Buttke, J. A. McCubrey, and T. C. Owen, "Use of an aqueous soluble tetrazolium/formazan assay to measure viability and proliferation of lymphokine-dependent cell lines," Journal of Immunological Methods, vol. 157, no. 1-2, pp. 233–240, 1993.

28. T. Mosmann, "Rapid colorimetric assay for cellular growth and survival: application to proliferation and cytotoxicity assays," Journal of Immunological Methods, vol. 65, no. 1-2, pp. 55–63, 1983.

29. G. Malich, B. Markovic, and C. Winder, "The sensitivity and specificity of the MTS tetrazolium assay for detecting the in vitro cytotoxicity of 20 chemicals using human cell lines," Toxicology, vol. 124, no. 3, pp. 179–192, 1997.

30. S. Bhattacharya, S. K. Nayak, S. Chattopadhyay, M. Banerjee, and A. K. Mukherjee, "Study of molecular complex formation between [60]fullerene and two series of donors by the NMR method," The Journal of Physical Chemistry A, vol. 105, no. 43, pp. 9865–9868, 2001.

31. S. Bhattacharya, S. K. Nayak, S. K. Chattopadhyay, M. Banerjee, and A. K. Mukherjee, "Absorption spectroscopic study of EDA complexes of [70] fullerene with a series of methyl benzenes," Spectrochimica Acta, Part A: Molecular and Biomolecular Spectroscopy, vol. 57, no. 2, pp. 309–313, 2001.

6

CHAPTER

Superparamagnetic Nanocomposites Templated with Pyrazole-Containing Diblock Copolymers

Sanchita Biswas [1], Kevin D. Belfield [1,*], Ritesh K. Das [2], Siddhartha Ghosh [2] and Arthur F. Hebard [2]

[1]Department of Chemistry, University of Central Florida, P.O. Box 162366, Orlando, FL 32816, USA

[2]Department of Physics, University of Florida, Gainesville, FL 32611, USA

ABSTRACT

Monodisperse maghemite nanoparticles, templated in novel, well-defined pyrazole-containing norbornene-based block copolymers, provided a superparamagnetic nanocomposite with high saturation magnetization at room temperature under an applied magnetic field. The synthesis of the polymer nanocomposites and physical, morphological, and magnetic chracaterization of the nanocomposites are reported. Micelle-encapsulated superparamagnetic nanocomposites were generated for dispersal in aqueous medium. Their stability in water in the presence of a magnetic field was investigated as was their morphology and cell viability, strongly suggesting the potential of these superparamagnetic polymer-based nanocomposites in certain biomedical imaging and associated applications.

Keywords: block copolymer; polymeric nanocomposite; maghemite; ROMP; chelating ligand

1. INTRODUCTION

There has been great interest in magnetic nanoparticles for a number of biomedical applications, including medical diagnosis and imaging, magnetic separation, magnetic hyperthermia, and controlled drug delivery [1,2,3,4,5,6,7] as well as for biomimetic applications [8,9,10], such as magnetically controlled locomotion of artificial bacterial flagellar motors [11] and artificial muscles [12]. Magnetic resonance imaging (MRI) is a powerful 3D technique for noninvasive imaging (deeper penetration) of anatomic and physicochemical details of tissue and blood flow and diagnosis of a number of diseases [13,14]. To improve MRI resolution and image quality several paramagnetic iron oxide nanoparticles, based on Fe_2O_3 and Fe_3O_4 [15,16], have been employed as negative contrast agents in clinical trials [1,17,18,19,20,21]. Among these, newly developed ultrasmall superparamagnetic iron oxide (USPIO) is, perhaps, the most promising because of its absence of magnetic hysteresis at ambient temperature, high magnetization change under applied magnetic field, tunability of sizes of the nanoparticles, non-toxicity, and rapid clearance from the organs [22,23,24,25]. USPIO enhances MRI images by decreasing the T_1- and T_2- relaxation behavior of the water molecules through interaction with the superparamagnetic iron oxide core, inducing signal increases in T_1-weighted images (T_1-w) and signal reduction in T_2-weighted-gradient-echo (T^*_2-w) images [26,27,28]. Control of uniformity and stability of magnetic nanoparticles is often challenging due to their inherent nature to form clusters, particularly for nano-sized particles given the tendency to reduce surface energy. Thus, strategies to prepare stable magnetic nanoparticle dispersions are of significant interest, e.g., via coating the nanoparticle surface with ligands that act to prevent agglomeration.

Ligands grafted to a polymer backbone offer a means to control stability as well as spatial behavior [29]. An iron-binding, anchoring ligand (siderophore) is present in one block of the diblock copolymer to chelate iron-oxide nanoparticles while the other block contains a steric stabilizing group which helps isolate the magnetic nanoparticles from each other and facilitate dispersion. Our previous success in formation of stable and well-dispersed maghemite (γ-Fe_2O_3)-polymer nanocomposites led to the design of more effective ligand systems through exploiting the chemistry of the oxirane group [30,31]. There are several natural iron-protein complexes that play an important role in a number of physiological processes, e.g., transferring [32,33], ferritin, and hemeproteins (e.g., hemoglobin, myoglobin, neuroglobin), where stable bonding between iron (III or II) core and protoporphrin IX and the imidazole group of the histidine protein residue forms. Here, we functionalized the oxirane with the 2,3-dimethylpyrazole group, a very close analog of imidazole, to study the chelation behavior with the iron-oxide nanoparticles for better stabilization compared to common chelating functional groups such as -COOH, -OH, $-NH_2$, phosphate. Ring opening metathesis polymerization (ROMP), a powerful and broadly applicable method compared to other controlled living polymerization techniques, was employed to synthesize highly functionalized block copolymers with low polydispersity to form superparamagnetic nanocomposite [7,34,35,36].

We present a strategy for the synthesis of a novel norbornene-based diblock copolymer, via ROMP of a functional monomer, and subsequent *in situ* generation and stabilization of maghemite nanoparticles to form superparamagnetic nanocomposites. The polynorbornene backbone was selected for its high thermal stability, optical transparency, low dielectric constant, and good mechanical properties [37]. In addition, functionalized norbornene-based monomers are readily polymerizable, which is thermodynamically favorable by releasing the strain in the monomer, by highly tolerant Grubbs catalysts into low polydispersity copolymers with readily tailored block ratios. The block copolymers design, contains "chelating groups" in one block to bind with the metal nanoparticles and "steric-stabilizing" block in another block to prevent agglomeration, was preferred for the nanocomposite generation. The anchoring group was synthesized by taking advantage of the versatility of oxirane ring-opening by the heterocycle pyrazole, creating a ligand to stabilize maghemite nanopaticles through strong coordination. The stable binding of this ligand system represents a significant advance over our previously reported materials [23,24]. Well-defined novel diblock copolymers, containing 1:1 anchoring and steric stabilizing blocks, were characterized by NMR, gel permeation chromatography (GPC), thermogravimetric analysis (TGA), and differential scanning calorimetry (DSC). The ligand-stabilized organic-inorganic hybrid nanocomposites were prepared into self-assembled 1:1 block copolymer matrix *in situ* by a non-hydrolytic procedure using $Fe(CO)_5$ [31] and morphology and particle nature were characterized with TEM, Electron diffraction pattern, SEM and XRD. Magnetic measurements on the nanocomposite were performed using SQUID magnetometer demonstrates the typical superparamagnetic properties with better magnetization in presence of magnetic field.

For biomedical applications, biocompatibility along with aqueous dispersibility and stability of the superparamagnetic nanocomposites are essential. Pluronic™ copolymers are well known in the pharmaceutical industry to improve the solubility of hydrophobic drugs [38] and to increase biocompatibility and blood circulation time [39,40]. Pluronic™ F127 is a triblock polymer of poly(ethylene oxide)-poly(propylene-oxide)-poly(ethylene oxide) (PEO-PPO-PEO), consisting of 70 wt% PEO. Stabilization of several types of nanoparticles by Pluronic F127 was reported to maintain stable suspensions in a high ionic strength environment [41]. It is reported in literature that iron oxide nanoparticles were stabilized in water by phase transfer from an organic medium while maintaining its inherent magnetic properties for drug delivery purposes [42,43]. Pluronic F127 also provides antifouling properties to prevent aggregation and protein adsorption along with recognition by reticuloendothelial system (RES) [44]. But Pluronic F127 is a stabilizer of intermediate strength for nanoparticles. Here, Pluronic F127 was chosen as a carrier to encapsulate the stable, new pyrazole block copolymer-stabilized nanocomposites, to provide an aqueous dispersibility and biocompatibility [45]. The magnetic nanocomposite was encapsulated with F127 to provide water dispersability, subsequent determination of aqueous stability was examined by TEM. To assess cytotoxicity of the resulting micelle-encapsulated magnetic nanocomposites, a cell viability assay was performed with Hela cells.

2. EXPERIMENTAL

2.1. General

2.1.1. Materials

Bis(tricyclohexylphosphine)-benzylideneruthenium dichloride Grubbs' second generation catalyst, 3,5-dimethylpyrazole, $Fe(CO)_5$ (99.9%), and trimethylamine *N*-oxide (98%), DMF were purchased from Aldrich and used as received. CH_2Cl_2 and $CHCl_3$ were dried over $CaCl_2$ and distilled. Tetrahydrofuran (THF) was distilled over sodium and benzophenone under N_2 before use. All catalyst solutions were prepared in a glovebox. Pluoronic F127 (M_w~12600) prill was purchased from BASF. Promega CellTiter 96® AQueous One Solution Reagent was purchased from Fisher. Hela cells were purchased from ATCC (America Type Culture Collection, Manssas, VA, USA). All cells were incubated in Minimum Essential Media (MEM, Invitrogen, Carlsbad, CA, USA), supplemented with 10% fetal bovine serum (FBS, Atlanta Biologicals, Lawrenceville, GA, USA), 100 units/mL penicillin-streptomycin (Atlanta Biologicals, Lawrenceville, GA, USA) and incubated at 37 °C in a 95% humidified atmosphere containing 5% CO_2.

2.1.2. Characterization

^{1}H NMR and ^{13}C NMR spectra were acquired on a Varian spectrometer at 500 and 125 MHz, respectively, using $CDCl_3$ as the solvent for all monomers and polymers. Elemental analysis was performed at Atlantic Microlab, Inc., Norcross, GA. All FTIR studies were done using a Perkin-Elmer Spectrum One FTIR spectrometer from 4000 to 500 cm^{-1}. Gel permeation chromatography (GPC) was conducted with a Waters 2414 refractive index detector, Waters 2996 photodiode array, and Waters 1525 binary HPLC pump (THF as the mobile phase, flow rate of 1 mL/min) using Waters styragel HR2 and HR5E columns, and polystyrene standards. Thermogravimetric analysis (TGA) was performed with a TA Instruments model Q5000 TGA, from room temperature to 600 °C at 20 °C/min. All samples were dried under vacuum for 2 days before measurement. Differential scanning calorimetry (DSC) was conducted with a TA Instruments Q1000 DSC, from −10 to 120 °C at a rate of 10 °C/min. Transmission electron microscopy (TEM) was accomplished using a JEOL 1011 TEM, operated at 100 kV. The samples were prepared by casting a thin film of nanocomposite on carbon-coated copper TEM grids. Particles size distribution, on the base of the profile of randomly selected quasi-spherically shaped particles, has been obtained using the ImageJ program. Selected area electron diffraction patterns were also obtained. XRD (Geigerflex Rigaku2, 2θ = 0–80°, step = 0.05, dwell (s) = 3) was used to obtain powder X-ray diffraction pattern spectra using Cu-$K\alpha$ radiation(λ = 0.154 nm). Noise corrections were made by using MDI Jade 7 software. The morphology of the block copolymers and the nanocomposite in powder form (sputter coated on sample holder) were examined by SEM (Zeiss ultra 55). Magnetic properties of the nanocomposites were measured using a superconducting quantum interference device (SQUID) magnetometer from Quantum Design. All the measurements were done in powder form of the sample after vacuum drying. The temperature dependence of the magnetization was determined by zero field-cooled (ZFC) and field-cooled (FC) measurements. The ZFC curve was obtained by cooling down to 4 K at zero magnetic fields and then measuring the magnetization under a 500 Oe applied magnetic field upto 300 K. The corresponding FC curves were similarly obtained, except that this time the sample was cooled while applying a 500 Oe magnetic field. The magnetizations as a function of applied magnetic field were also studied at 100 K above the blocking temperature.

2.1.3. Cytotoxicity (MTS) Assay

To test the cytotoxicity of the nanoparticles, 4 × 10^3 per well of Hela cells in 96-well plates were incubated in 90 μL of RPMI medium without phenol red, supplemented with 10% FBS and 100 units/mL penicillin-streptomycin for 24 h. Then the cells were incubated with various amounts of nanoparticles (1 μg/mL, 10 μg/mL, 20 μg/mL, 50 μg/mL) for additional 20 h. Subsequently, 20 μL of CellTiter 96®AQueous One Solution reagent was added into each well, followed by further incubation for 4 h at 37 °C. The relative viability of the cells incubated with nanoparticles to untreated cells was determined by measuring the MTS-formazan absorbance on a Kinetic microplate reader (Spectra Max M5, Molecular Devices, Sunnyvale, CA, USA) at 490 nm with a subtraction of the absorbance of cell-free blank volume at 490 nm. The results from three individual experiments were averaged.

2.2. Synthesis of 3-(3,5-Dimethyl-1H-Pyrazol-1-yl)-2-Hydroxypropyl Bicyclo[2.2.1] Hept-5-Ene-2-Carboxylate (2)

Monomer 1 was prepared according to our previous published procedure [46]. Epoxy monomer 1 (0.23 g, 1.2 mmol) in DMF was added dropwise to a suspension of 3,5-dimethylpyrazole (0.107 g, 1.11 mmol) and anhydrous potassium carbonate (0.166 g, 1.2 mmol) in 10 mL of anhydrous DMF. The reaction was stirred at 100 °C for 9 h until the disappearance of starting materials (followed by TLC). It was then cooled to room temperature, the salt was filtered off, and the solvent was removed under reduced pressure. A wheatish creamy solid (0.15 g, 55% yield) was obtained after column chromatography (1:1 hexane:ethyl acetate on silica) followed by vacuum drying overnight. ^{1}H NMR (500 MHz, $CDCl_3$) δ:6.14–6.03 (m, 1.5H, HC=CH), 5.87 (m, 0.5H, HC=CH), 5.74 (s, 1H-Py ring), 4.52 (b, 1H, -OH), 4.17–3.85 (m, 6H), 3.15 (s, 0.5H), 2.98–2.90 (m, 1H), 2.89–2.86 (m, 1H), 2.20–2.13 (m, 6H, 2 -CH_3in Py ring), 1.88–1.84 (s, 1H), 1.69–1.19 (m, 3H). ^{13}C NMR (125 MHz, $CDCl_3$) δ: 176.05, 174.54, (C=O exo and endo) 148.24, 139.80, 138.14, 137.96, 137.92, 135.64, 132.26, 105.12 (=C-CH=C), 69.24, 65.19, 65.17, 65.04, 64.97, 49.72, 46.67, 46.36, 45.80, 45.77, 43.26, 43.04, 42.53, 41.64, 30.45, 29.31, 13.43 (Py-CH_3), 10.95 (Py-CH_3) IR (neat): 3349.48 (b, -OH), 3025.81 (=C-H stretch), 2925.08, 1732.36 (*vs.*, C=O), 1601.25 (-C=C-stretch),1553.78 (-C=N-), 1492.89, 1452.06, 1333.81, 1272.80, 1232.39, 1180.06, 1029.02, 906.84, 756.10, 620.99 (-C=C- bend), 540.10 cm^{-1}. Anal. calcd for $C_{16}H_{22}N_2O_3$: C, 66.18, H, 7.64. Found: C, 65.96, H, 7.65.

2.3. Synthesis of Diblock Copolymer 5 by ROMP of 3-(3,5-Dimethyl-1H-Pyrazol-1-yl)-2-Hydroxypropyl Bicyclo[2.2.1]Hept-5-Ene-2-Carboxylate (5)

ROMP of monomer 2 with Grubbs second generation catalyst 3 was performed as shown in Scheme 1. The glassware was dried and purged with vacuum and N_2 in a Schlenk line several times prior to conducting the polymerization reaction. A solution (0.2 M) of monomer 2 (120 mg, 4.13×10^{-4} M, 175 eq) was prepared in dry CH_2Cl_2 under N_2. The catalyst solution was prepared by dissolving the catalyst in anhydrous CH_2Cl_2 under N_2 in a glovebox. The catalyst solution (2 mg, 2.36×10^{-6} M in 0.5 mL CH_2Cl_2, 1 eq) was added to the reaction mixture and stirred for 8 h at 30 °C. A norbornene solution (38 mg, 4.13×10^{-4} M, 175 eq) was injected and stirred for another 9 h. The polymerization reaction mixture was terminated with excess ethyl vinyl ether (300 eq relative to catalyst) and stirred for another 1 h. The reaction mixture was then poured into cold methanol and stirred, purified, and dried under vacuum, yielding a flaky white solid in 72% yield. ^1H NMR (500 MHz, $CDCl_3$) δ: 5.82 (b, CH, Py ring), 5.35–5.22 (b, -HC=CH-), 4.17–3.85, 3.18, 2.96, 2.44, 2.23–2.19, 1.87–1.27(b).

2.4. Stabilized Magnetic Dispersions

Preparation of monodisperse maghemite nanoparticles within copolymer matrices was accomplished by our previously reported method [30] using cyclohexanone as solvent or others as required.

2.5. Micelles Stabilized Magnetic Nanoparticle Dispersions in Water

120 mg of F127 was dissolved in 20 mL (the concentration is higher than critical micelle concentrations (CMC); CMC of Pluronic F127 is 0.007 g/cm^3 at room temperature) [36,38,39,47,48] of water by vigorously stirring for 3 hrs. Maghemite-polymer nanocomposites (7 mg) were dispersed in CH_2Cl_2. Equal volume of these two mixtures were added together and stirred under N_2 at 40 °C (waterbath) for until all CH_2Cl_2 evaporated in a fume hude and nanocomposite covered by the F127 coating was prepared in water. Then, it was subjected to centrifuge and finally, the light brown colored supernatant was collected which contains block copolymer-maghemite nanocomposites entrapped in pluronic micelles.

3. RESULTS AND DISCUSSION

3.1. Synthesis of Monomer and Polymers

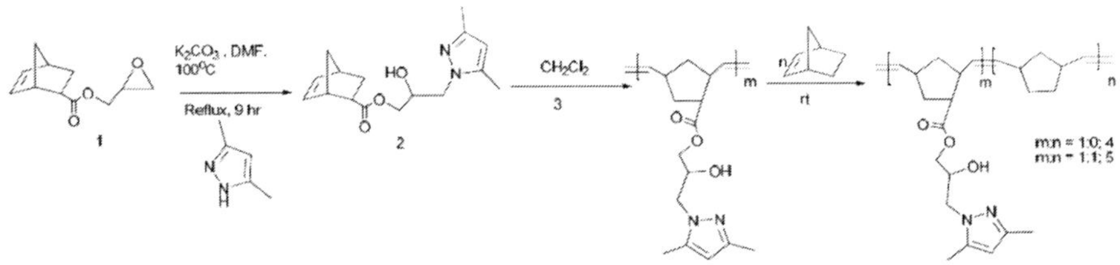

Scheme 1. Synthesis of monomers and block copolymers.

The monomer and polymers were designed and prepared by exploiting the versatility of epoxy precursor 1. In the first step, pyrazole-containing monomer 2 was synthesized from 1 [46], by opening the epoxide ring with 3,5-dimethyl pyrazole in the presence of base. Next, both a homopolymer 4 and a 1:1 diblock copolymer 5 were synthesized according to Scheme 1 via ROMP using Grubbs second generation catalyst (3, bis(tricyclohexylphosphine)-benzylideneruthenium dichloride). Copolymer 5 was prepared by sequential addition of monomer 2 and nobornene. The molecular weight and polydispersity of the polymers (Table 1) were controlled on the basis of a time-dependent study, monitoring monomer consumption by ^{1}H NMR and controlling the [M]/[I] ratio, where M is monomer and I is initiator.

The diblock copolymer template 5 contained the ligand system in one block that can chelate through both N-(from pyrazole) and O-donors (from ring-opened hydroxy and/or carboxylate groups) with the metal oxide core, while the second block contained the steric stabilizing group to disperse the nanoparticle composite and reduce interaction between nanoparticles. The pyrazole ligands played a vital role in stabilizing the maghemite nanoparticles, affording a supramolecular metal-organic hybrid. The polymers were characterized by ^{1}H NMR (Figure 1) and FTIR. In the ^{1}H NMR, distinct peaks at 6.14–5.87 ppm for -*CH*=*CH*- vinyl protons, 5.74 ppm for -*H* pyrazole ring and a broad -O*H* proton peak at 4.52 ppm were present for the monomer (Figure 1a) while the vinyl protons disappear and a new peak appears at 5.35–5.22 ppm for new alkene protons for the polymers Figure 1(b,c).The ratio between the alkene proton *vs.* pyrazole ring proton present in polymer determined the block ratio (m:n). The molecular weights of the polymers were experimentally determined by standard GPC (Figure 2) method calibrated with PS standards correspond to apparent Mn values, shown in Table1. The relatively high polydispersity index of the copolymer is possibly due to the presence of heteroatom (N, O) in the pendant group functionality.

Table 1. Properties of the polymers.

Block ratio (m:n)	M_n [a] (Theo)	m:n 1H NMR	M_n [b] (Exp)	PDI [b]	Block Ratio [c]	Tg (°C)	T [d] (°C)
200:0 (4)	58072	1:0	65085	1.42	224:0	46	170
175:175 (5)	67291	1.08:1	66331	1.33	176:163	39	217

[a]: Theoretical molecular weight calculated from[M]/[I]feed; [b]: Mn and PDI were obtained experimentally from gel permeation chromatography (GPC) in THF relative to polystyrene standards; [c]: Actual polymer block ratio was calculated from the ^{1}H NMR and GPC results; [d]: Temperature at 10% weight loss.

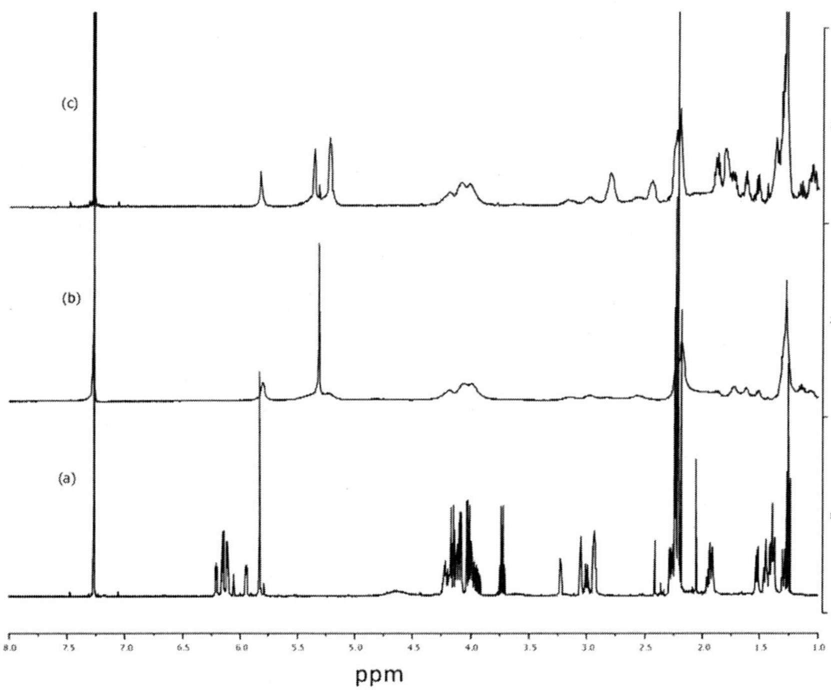

Figure 1. ^{1}H NMR analysis supporting the formation of 1:1 diblock copolymer; (**a**) monomer 3; (**b**) homopolymer 4; and (**c**) 1:1 diblock copolymer 5.

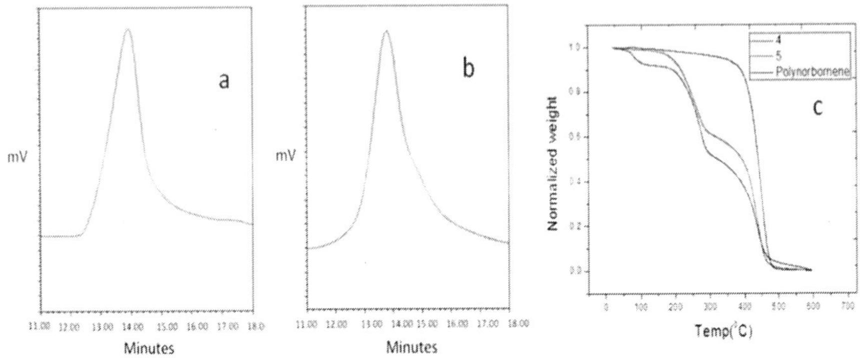

Figure 2. Gel permeation chromatography (GPC) analysis of the polymers (**a**) homopolymer 4; and (**b**) 1:1 diblock copolymer 5; (**c**) thermogravimetric analysis (TGA) of the polymers.

The glass transition temperature and thermal decomposition temperature were determined by DSC and TGA analyses, respectively (Table 1). The 1:1 block copolymer 5 was more thermally stable (decomposition at 217 °C) compare to the pyrazole group containing homopolymer 4. Because of the thermally labile bond at the side chain, it was noticed, that the thermal stability of the polymer in general decreases with respect to homopolynorbornene (450 °C) or epoxy containing homopolymer (370 °C) as reported before [30]. The glass transition temperature of the polynorbornene, homopolymer, 4, and copolymer, 5 (were approximately 31, 46, and 39 °C respectively), were in close range and broad in nature.

3.2. Synthesis and Characterization of Polymeric-Iron Oxide Nanocomposite

The second step was to generate relatively uniform, highly crystalline maghemite nanoparticles *in situ*from $Fe(CO)_5$ precursor in the presence of diblock copolymer 5 (0.68% molar wt of polymer compared to ironpentacarbonyl feed) in cyclohexanone. Supported by our previous study [31], 1:1 block copolymer ratio of the polymer was chosen to stabilize the nanopartcle stabilization at the optimum level. Trimethylamine N-oxide was used as an oxidizer to prepare the maghemite nanocomposite in a similar manner to our previously reported method [30,31]. Briefly, the block copolymer and iron-pentacarbonyl were added in cyclohexanone and refluxed under inert atmosphere until the solution turned to black. Thereafter, solution was cooled to room temperature and oxidizer was added to it, followed by reflux for another few hours, and then, purified as mentioned in literature. Finally, the nanocomposite was dispersed in solvent as required.

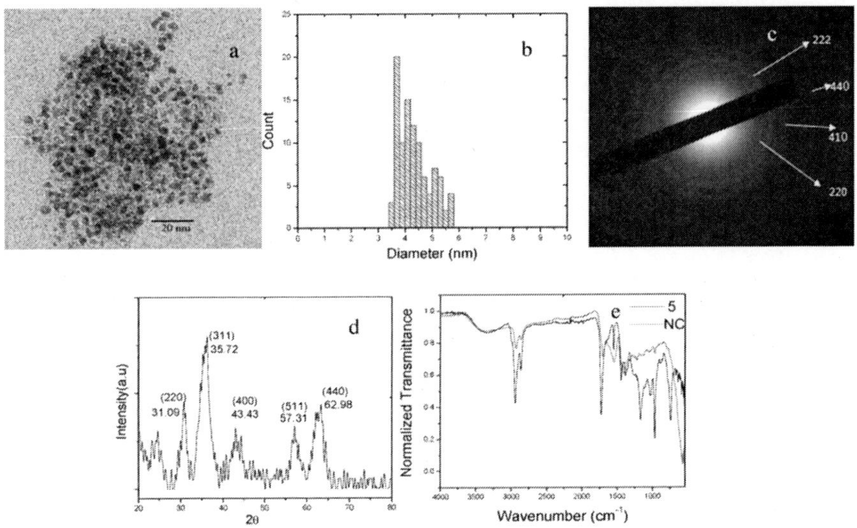

Figure 3. (**a**) TEM images of magnetic nanocomposite; (**b**) TEM nanoparticle size distribution histogram; (**c**) electron diffraction pattern of nanocomposite; (**d**) XRD analysis showing the presence of crystalline maghemite-C (NC-2); (**e**) FTIR analysis of 1:1 block copolymer (labeled 5) and magnetic nanocomposite (labeled NC).

The block copolymer-stabilized nanocomposites were characterized by TEM, SEM, XRD, and FTIR. TEM (Figure 3a) confirmed generation of relatively uniform and spherical iron oxide nanoparticles. The size distribution (Figure 3b) of the nanoparticles was in the 3–6 nm range with an average size of 4.3 nm. Evidence of formation of a crystalline structure of the Fe_2O_3 nanoparticles was obtained by SAED and XRD (Figure 3(c,d)), revealing cubic maghemite-C type structures of the nanoparticles, corresponding well with standard data [49]. The average diameter (d) of the singular γ-Fe_2O_3 nanocrystallites, estimated using the Scherrer equation [50], was 4 nm, corresponding to the strongest reflection (311) of Fe_2O_3 nanoparticles at 2θ value of 35.72°. This was in good agreement with TEM results. SEM data revealed the bulk morphology of the 1:1 block copolymer and nanocomposite. FTIR spectroscopic analysis provided compelling evidence of ligand-mediated binding of iron oxide with the pendant chelating group of the diblock copolymer (Figure 3e). The presence of a single, broad band at about 574 cm^{-1} in the FTIR spectrum of the nanocomposite is characteristic of Fe-O stretching for less than 8 nm nanoparticles [51]. Also, the shift from 1732 cm^{-1} to 1690 cm^{-1}, corresponding to C=O stretching in the polymer and nanocomposite, respectively, supported interaction between the iron oxide core and the diblock copolymer.

3.3. Magnetic Properties of Nanocomposite

Magnetization (M) measurement as a function of temperature (T) was performed on γ-Fe_2O_3-diblock copolymer nanocomposites in powder form using SQUID. Figure 4a shows the zero field cooled (ZFC) and field cooled (FC) *vs.* temperature profile, where the blocking temperature (T_B) of 9 K is denoted by the peak of the ZFC curve. The low and sharp blocking temperature of the nanocomposite is consistent with the small and uniform size of the maghemite nanoparticles, respectively. Uniform distribution of the magnetic nanoparticles within the polymer matrix is further strengthened by very close superimposition of the ZFC and FC curves after passing the T_B. The magnetic hysteresis loop at 4.2 K, below T_B, is shown in Figure 4b. The coercive field was of the order of 100 Oe. Above T_B the coercive field was zero and the data collapse onto the same curve when plotted as a function of H/T. This H/T scaling behavior indicates that the sample consists of non-interacting superparamagnetic particles. The magnetization data above T_B are fit to a Langevin function, and a saturated moment of approximately 73 emu/g is determined from the fit.

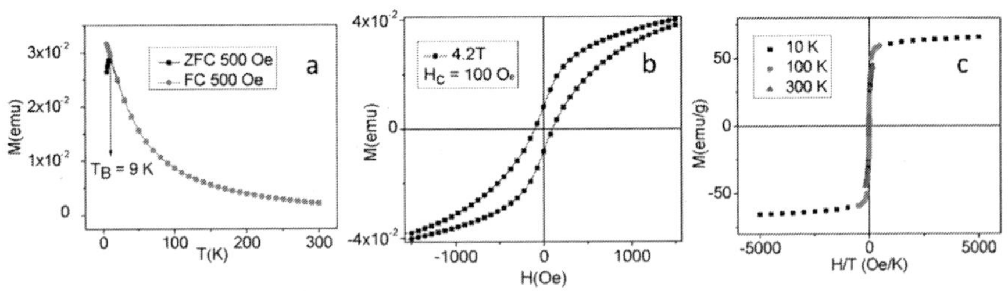

Figure 4. (**a**) magnetization (M) *vs.* temperature (T) of the nanocomposite at 500 Oe applied magnetic field (H); (**b**) M-H loop below T_B shows coercive field (H_c) of 100 Oe; (**c**) M- H loop above T_B, coercive field (H_c) is zero.

3.4. Formation of Biocompatible Nanocomposite

For potential biomedical applications, the stable dispersion in aqueous and physiological media, and biocompatibility of the pyrazole-containing block copolymer-stabilized maghemite nanocomposites was assessed. The organic soluble maghemite-polymer nanocomposites were dispersed in water using Pluoronic F127 (PEO-PPO-PEO), using to a solvent evaporation method. The Pluronic-coated, water-dispersible nanocomposite was generated by mixing an equal volume of the nanocomposite in $CH_2Cl_2$with an aqueous solution of Pluronic F127, followed by vigorous stirring and evaporation of the CH_2Cl_2. The outer hydrophilic (PEO) blocks of Pluronic F127 help to form micelles and dispersion in water while the hydrophobic block (PPO) attracts the polymer-coated maghemite nanoparticles. An aqueous dispersion of micelle-encapsulated superparamagnetic nanocomposites was stable for weeks and retained their magnetic properties.

The morphology of the Pluronic-pyrazole-maghemite nanocomposites in water was studied by TEM (Figure 5a and Figure 5b), indicating that the size of the core maghemite nanoparticles was similar before and after dispersion in water (confirming that the particles were stable and did not aggregate). The observed stability of the Pluronic-coated magnetic nanocomposites in water is important for biological use and was consistent to literature reports [45,52].

For potential biomedical applications, the cytotoxicity of Pluronic™-pyrazole block copolymer-coated superparamagnetic nanoparticles was assessed using a cell viability assay with Hela cells. Hela cells were incubated for 20 h with varying concentrations of the nanocomposites (0–50 μM), exhibiting high cell viability, 90–100%, depending on concentration. Qin *et al.* reported a novel type ferrogel (SPIONS coated with oleic acid and PF127) to release drug for magnetically controlled release of hydrophobic drug very efficiently [42]. Recently, Zhao *et al.* also demonstrated a RGD modified ferrogel, where of SPION is encapsulated in PF127, alginate and AAD, for sustain release of cells and other bioactive agents under magnetic field [53]. Thus, the Pluronic encapsulated pyrazole block copolymer stabilized iron oxide nanocomposite also provides a promising tool to encapsulate hydrophobic drug or optical contrast agent and holds the potential for dual imaging modalities or theranostic applications in biomedical field.

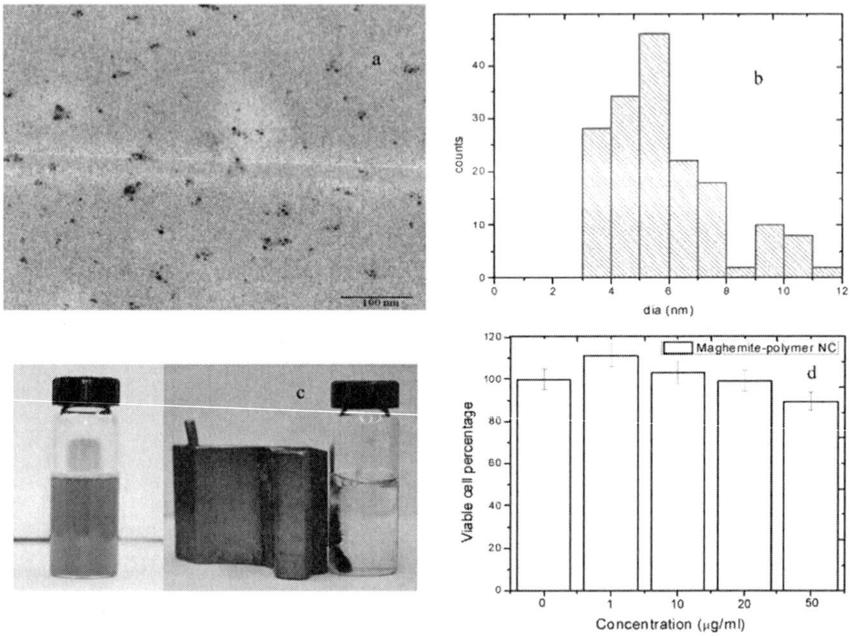

Figure 5. (**a**) TEM image of Pluronic 127-coated polymer-maghemite nanocomposites; (**b**) TEM nanoparticle size distribution histogram; (**c**) photograph of Pluronic 127-stabilized polymer-Fe_2O_3nanocomposite without (left) and with external applied magnetic field (right); and (**d**) cell viability of Pluronic 127-stabilized polymer-Fe_2O_3 nanocomposite incubated in Hela cells for 20 h as a function of concentration.

4. CONCLUSIONS

In summary, it is clear that well-defined *N*,*N*-dimethyl pyrazole-functionalized norbornene-based diblock copolymers are effective in stabilizing *in situ*-generated maghemite nanoparticles. The resulting nanocomposite exhibited strong superparamagnetism in the presence of an external magnetic field. Nanoparticle agglomeration was prevented through specific binding between the pendant ligands in the anchoring block and the iron oxide metal core. This organic block copolymer-inorganic maghemite nanocomposite underwent dynamic changes in magnetic properties and directional motion in response to an external magnetic field without destroying the integrity of the system. Pluronic127-coated magnetic nanocomposites were stable in aqueous media and possessed low cytotoxicity, making them potential candidates as contrast agents for MRI imaging, a carrier for drug delivery, or for use in magnetically-actuated biomimetic devices, subjects of future investigation.

ACKNOWLEDGEMENTS

We wish to acknowledge the National Science Foundation (CHE-0832622, and DMR-0704240) for support of this work.

REFERENCES

1. Berry, C.C.; Curtis, A.S.G. Functionalisation of magnetic nanoparticles for applications in biomedicine. *J. Phys. D Appl. Phys.* **2003**, *36*, 198–206.
2. Pankhurst, Q.A.; Connolly, J.; Jones, S.K.; Dobson, J. Applications of magnetic particles in biomedicine. *J. Phys. D Appl. Phys.* **2003**, *36*, 167–181.
3. Chertok, B.; Moffat, B.A.; David, A.E.; Yu, F.; Bergemann, C.; Ross, B.D.; Yang, V.C. Iron oxide nanoparticles as a drug delivery vehicle for MRI monitored magnetic targeting of brain tumors.*Biomaterials* **2008**, *29*, 487–496.
4. Gupta, A.K.; Naregalkar, R.R.; Vaidya, V.D.; Gupta, M. Recent advances on surface engineering of magnetic iron oxide nanoparticles and their biomedical applications. *Nanomedicine* **2007**, *2*, 23–29.
5. Sun, C.; Lee, J.S.H.; Zhang, M. Magnetic nanoparticles in MR imaging and drug delivery. *Adv. Drug Delivery Rev.* **2008**, *60*, 1252–1265.
6. Shubayev, V.I.; Pisanic, T.R.; Jin, S. Magnetic nanoparticles for theragnostics. *Adv. Drug Delivery Rev.* **2009**, *61*, 467–477.
7. Oh, J.K.; Park, J.M. Iron oxide-based superparamagnetic polymeric nanomaterials: Design, preparation, and biomedical application. *Prog. Polym. Sci.* **2011**, *36*, 168–189.
8. Xulu, P.M.; Filipcsei, G.; Zrinyi, M. Preparation and responsive properties of magnetically soft poly(N-isopropylacrylamide) gels. *Macromolecules* **2000**, *33*, 1716–1719.

9. Csetneki, I.; Filipcsei, G.; Zrínyi, M. Smart nanocomposite polymer membranes with on/off switching control. *Macromolecules* **2006**, *39*, 1939–1942.

10. Filipcsei, G.; Csetneki, I.; Szilágyi, A.; Zrínyi, M. Magnetic field-responsive smart polymer composites. *Adv. Polym. Sci.* **2007**, *206*, 137–189.

11. Zhang, L.; Abbott, J.J.; Dong, L.; Peyer, K.E.; Kratochvil, B.E.; Zhang, H.; Bergeles, C.; Nelson, B.J. Characterizing the swimming properties of artificial bacterial flagella. *NanoLett.* **2009**, *9*, 3663–3667.

12. Fahrni, F.; Prins, M.W.J.; van Ijzendoorn, L.J. Magnetization and actuation of polymeric microstructures with magnetic nanoparticles for application in microfluidics. *J. Magn. Magn. Mater.* **2009**, *321*, 1843–1850.

13. Weissleder, R.; Pittet, M.J. Review article imaging in the era of molecular oncology. *Nature* **2008**,*452*, 580–589.

14. Na, H.B.; Song, I.C.; Hyeon, T. Inorganic nanoparticles for MRI contrast agents. *Adv. Mater.* **2009**,*21*, 2133–2148.

15. Shen, L.; Laibinis, P.E.; Hatton, T.A. Bilayer surfactant stabilized magnetic fluids: Synthesis and interactions at interfaces. *Langmuir* **1999**, *15*, 447–453.

16. Kohler, N.; Sun, C.; Fichtenholtz, A.; Gunn, J.; Fang, C.; Zhang, M. Methotrexate-immobilized poly(ethylene glycol) magnetic nanoparticles for MR imaging and drug delivery. *Small* **2006**, *2*, 785–792.

17. Harisinghani, M.G.; Barentsz, J.; Hahn, P.F.; Deserno, W.M.; Tabatabaei, S.; van de Kaa, C.H.; de la Rosette, J.; Weissleder, R. Noninvasive detection of clinically occult lymph-node metastases in prostate cancer. *N. Engl. J. Med.* **2003**, *348*, 2491–2499.

18. Nasongkla, N.; Bey, E.; Ren, J.; Ai, H.; Khemtong, C.; Guthi, J.S.; Chin, S.F.; Sherry, A.D.; Boothman, D.A.; Gao, J. Multifunctional polymeric micelles as cancer-targeted, MRI-ultrasensitive drug delivery systems. *NanoLett.* **2006**, *6*, 2427–2430.

19. Raynal, I.; Prigent, P.; Peyramaure, S.; Najid, A.; Rebuzzi, C.; Corot, C. Macrophage endocytosis of superparamagnetic iron oxide nanoparticles: Mechanisms and comparison of ferumoxides and ferumoxtran-10. *Invest. Radiol.* **2004**, *39*, 56–63.

20. Sun, C.; Veiseh, O.; Gunn, J.; Fang, C.; Hansen, S.; Lee, D.; Sze, R.; Ellenbogen, R.G.; Olson, J.; Zhang, M. *In vivo* MRI detection of

gliomas by chlorotoxin-conjugated superparamagneticnanoprobes. *Small* **2008**, *4*, 372–379.

21. Brazel, C.S. Magnetothermally-responsive nanomaterials: Combining magnetic nanostructures and thermally-sensitive polymers for triggered drug release. *Pharm. Res.* **2009**, *26*, 644–656.

22. Mouli, S.K.; Zhao, L.C.; Omary, R.A.; Thaxton, C.S. Lymphotropicnanoparticle enhanced MRI for the staging of genitourinary tumors. *Nat. Rev. Urol.* **2010**, *7*, 84–93.

23. Will, O.; Purkayastha, S.; Chan, C.; Athanasiou, T.; Darzi, A.W.; Gedroyc, W.; Tekkis, P.P. Diagnostic precision of nanoparticle-enhanced MRI for lymph-node metastases: A meta-analysis.*Lancet Oncology* **2006**, *7*, 52–60.

24. Neuwelt, E.A.; Hamilton, B.E.; Varallyay, C.G.; Rooney, W.R.; Edelman, R.D.; Jacobs, P.M.; Watnick, S.G. Ultrasmallsuperparamagnetic iron oxides (USPIOs): A future alternative magnetic resonance (MR) contrast agent for patients at risk for nephrogenic systemic fibrosis (NSF)? *Kidney Int.* **2009**, *75*, 465–474.

25. Vellinga, M.M.; Engberink, O.; Raoul, D.; Seewann, A.; Pouwels, P.J.W.; Wattjes, M.P.; van der Pol, S.; Pering, C.; Polman, C.H.; de Vries, H.E. Pluriformity of inflammation in multiple sclerosis shown by ultra-small iron oxide particle enhancement. *Brain* **2008**, *131*, 800–807.

26. Corot, C.; Robert, P.; Idée, J.M.; Port, M. Recent advances in iron oxide nanocrystal technology for medical imaging. *Adv. Drug Delivery Rev.* **2006**, *58*, 1471–1504.

27. Lee, H.Y.; Lee, S.H.; Xu, C.; Xie, J.; Lee, J.H.; Wu, B.; Leen Koh, A.; Wang, X.; Sinclair, R.; Wang, S.X. Synthesis and characterization of PVP-coated large core iron oxide nanoparticles as an MRI contrast agent. *Nanotechnology* **2008**, *19*, 165101:1–165101:6.

28. Tang, T.Y.; Howarth, S.P.S.; Miller, S.R.; Graves, M.J.; Patterson, A.J.; U-King-Im, J.M.; Li, Z.Y.; Walsh, S.R.; Brown, A.P.; Kirkpatrick, P.J. The ATHEROMA (atorvastatin therapy: Effects on reduction of macrophage activity) study. Evaluation using ultrasmall superparamagnetic iron oxide-enhanced magnetic resonance imaging in carotid disease. *J. Am. Coll. Cardiol.* **2009**, *53*, 2039–2050.

29. Warren, S.C.; Messina, L.C.; Slaughter, L.S.; Kamperman, M.; Zhou, Q.; Gruner, S.M.; DiSalvo, F.J.; Wiesner, U. Ordered mesoporous

materials from metal nanoparticle—block copolymer self-assembly. *Science* **2008**, *320*, 1748–1752.

30. Biswas, S.; Belfield, K.D.; Das, R.K.; Ghosh, S.; Hebard, A.F. Block copolymer-mediated formation of superparamagnetic nanocomposites. *Chem. Mater.* **2009**, *21*, 5644–5653.

31. Belfield, K.D.; Zhang, L. Norbornene-functionalized diblock copolymers via ring-opening metathesis polymerization for magnetic nanoparticle stabilization. *Chem. Mater.* **2006**, *18*, 5929–5936.

32. Anderson, B.F.; Baker, H.M.; Norris, G.E.; Rice, D.W.; Baker, E.N. Structure of human lactoferrin: Crystallographic structure analysis and refinement at 2.8 A resolution. *J. Mol. Biol.* **1989**, *209*, 711–734.

33. James, N.G.; Berger, C.L.; Byrne, S.L.; Smith, V.C.; MacGillivray, R.T.A.; Mason, A.B. Excited-state lifetime studies of the three tryptophan residues in the N-lobe of human serum transferrin.*Biochemistry* **2007**, *46*, 10603–10611.

34. Bielawski, C.W.; Grubbs, R.H. Living ring-opening metathesis polymerization. *Prog. Polym. Sci.***2007**, *32*, 1–29.

35. Buchmeiser, M.R. *Handbook of Ring-Opening Polymerization*, 1st; Dubois, P., Coulemibier, O., Raquez, J.M., Eds.; Wiley-VCH: Berlin, Germany, 2009.

36. Hilf, S.; Kilbinger, A.F.M. Functional end groups for polymers prepared using ring-opening metathesis polymerization. *Nat. Chem.* **2009**, *1*, 537–546.

37. Merrett, K.; Liu, W.; Mitra, D.; Camm, K.D.; McLaughlin, C.R.; Liu, Y.; Watsky, M.A.; Li, F.; Griffith, M.; Fogg, D.E. Synthetic neoglycopolymer-recombinant human collagen hybrids as biomimetic crosslinking agents in corneal tissue engineering. *Biomaterials* **2009**, *30*, 5403–5408.

38. Kabanov, A.V.; Batrakova, E.V.; Alakhov, V.Y. Pluronic block copolymers as novel polymer therapeutics for drug and gene delivery. *J. Control. Release* **2002**, *82*, 189–212.

39. Moghimi, S.M. Mechanisms regulating body distribution of nanospheres conditioned with pluronic and tetronic block co-polymers. *Adv. Drug Delivery Rev.* **1995**, *16*, 183–193.

40. Wadhwa, S.; Paliwal, R.; Paliwal, S.R.; Vyas, S.P. Nanocarriers in ocular drug delivery: An update review. *Curr. Pharm. Des.* **2009**, *15*, 2724–2750.

41. Lim, J.K.; Majetich, S.A.; Tilton, R.D. Stabilization of superparamagnetic iron oxide core-gold shell nanoparticles in high ionic strength media. *Langmuir* **2009**, *25*, 13384–13393.

42. Qin, J.; Asempah, I.; Laurent, S.; Fornara, A.; Muller, R.N.; Muhammed, M. Injectablesuperparamagneticferrogels for controlled release of hydrophobic drugs. *Adv. Mater.***2009**, *21*, 1354–1357.

43. Gonzales, M.; Krishnan, K.M. Phase transfer of highly monodisperse iron oxide nanocrystals with Pluronic F127 for biomedical applications. *J. Magn. Magn. Mater.* **2007**, *311*, 59–62.

44. Lin, J.J.; Chen, J.S.; Huang, S.J.; Ko, J.H.; Wang, Y.M.; Chen, T.L.; Wang, L.F. Folic acid-Pluronic F127 magnetic nanoparticle clusters for combined targeting, diagnosis, and therapy applications.*Biomaterials* **2009**, *30*, 5114–5124.

45. Morales, M.A.; Jain, T.K.; Labhasetwar, V.; Leslie-Pelecky, D.L. Magnetic studies of iron oxide nanoparticles coated with oleic acid and Pluronic block copolymer. *J. Appl. Phys.* **2005**, *97*, 10Q905:1–10Q905:3.

46. Biswas, S.; Belfield, K.D. Self–assembled block copolymer/superparamagneticnanoparticlenanocomposites. *Pol ym. Prepr. Am. Chem. Soc. Div. Polym. Chem.* **2008**, *49*, 1063–1064.

47. Bohorquez, M.; Koch, C.; Trygstad, T.; Pandit, N. A study of the temperature-dependent micellization of pluronic F127. *J. Colloid Interface Sci.* **1999**, *216*, 34–40.

48. Sakai, T.; Alexandridis, P. Mechanism of gold metal ion reduction, nanoparticle growth and size control in aqueous amphiphilic block copolymer solutions at ambient conditions. *J. Phys. Chem. B***2005**, *109*, 7766–7777.

49. *MDI Jade, version 7*; MDI: Livermore, CA, USA, 2009; PDF No. 00-039-1346.

50. Xia, H.; Foo, P.; Yi, J. Water-dispersible spherically hollow clusters of magnetic nanoparticles.*Chem. Mater.* **2009**, *21*, 2442–2451.

51. Li, D.; Teoh, W.Y.; Selomulya, C.; Woodward, R.C.; Munroe, P.; Amal, R. Insight into microstructural and magnetic properties of flame-made γ-Fe_2O_3nanoparticles. *J. Mater. Chem.***2007**, *17*, 4876–4884.

52. Jain, T.K.; Reddy, M.K.; Morales, M.A.; Leslie-Pelecky, D.L.; Labhasetwar, V. Biodistribution, clearance, and biocompatibility of iron oxide magnetic nanoparticles in rats. *Mol. Pharm.* **2008**, *5*, 316–327.

53. Zhao, X.; Kim, J.; Cezar, C.A.; Huebsch, N.; Lee, K.; Bouhadir, K.; Mooney, D.J. Active scaffolds for on-demand drug and cell delivery. *Proc. Natl. Acad. Sci.* **2011**, *108*, 67–72.

7

CHAPTER

Polyfluorene-Polytriarylamine Block Copolymer as an Additive for Electroluminescent Devices Based on Polymer Blends

Mehdi Jahanfar, Ying Tan, Kousuke Tsuchiya, Takeshi Shimomura, Kenji Ogino*

Graduate School of Bio-Applications and Systems Engineering, Tokyo University of Agriculture and Technology, Tokyo, Japan

ABSTRACT

Electroluminescent characteristics were investigated for the blue emitting devices fabricated with the blend systems consisting of hole transporting polytriarylamine (PTAA), electron transporting polyfluorene (PF), and a block copolymer with both segments (PF-b-PTAA) as an active layer in order to elucidate the relationship between the chemical nature and morphology of the active layer, and EL performance. The addition of PF-b-PTAA to PF homopolymer afforded the hole injecting and/or electron blocking ability to increase the efficiency. The addition to PF/PTAA blend keeping the chemical composition constant also improved the performance by controlling the morphology and/or the domain size in phase-separated films.

Keywords: Polyfluorene; Polytriarylamine; Block Copolymer; Polymer Blend; Polymer-Light Emitting Diode

1. INTRODUCTION

Polyfluorenes (PFs) have been attracting much attention for a blue polymeric light emitting material because of their superior properties such as highly efficient luminescence, excellent thermal stability, and good solubility in common organic solvents [1,2]. PFs often have a high ionization potential resulting in an injection barrier for holes and low electroluminescent (EL) efficiency [3]. To improve the injection and conduction of holes in PFs, triarylamine (TAA) derivatives are commonly used, which are known to be good hole conductors [4-7]. Multilayered devices with fluorene-TAA copolymers as interlayers between the conductive polymer, PEDOT:PSS and the emitting layer showed much higher performance than those without the interlayer [8]. Recently we reported a different approach utilizing block copolymers consisting of PF and polytriarylamine (PTAA) segments with lower ionization potential in order to improve the EL efficiency due to the effective hole injection and/or electron blocking [9,10]. Hydrophilic trioxyethylene (TEO) group was attached to the side chain of PTAA. Therefore, it is considered that PTAA segment segregate into the interface by hydrogen bonding interaction with PEDOT:PSS layer.

Block copolymers assemble into microor nano-phase separated structures with various domain shapes such as lamella, cylinder, or sphere. Exploiting nanostructures of block copolymers with appropriate designs can improve performance of applications due to allocation of functionality to each domain [11]. Besides the above mentioned reports, several block copolymers for EL applications have been reported in order to improve injection of hole and electron into the emitting layer as well as recombination of charges [12-15].

We also investigated the correlation between device performance and morphology in polymer layer by evaluation of the devices based on the polymer blend (hole and electron transporting polymers), to which various contents of block copolymer are added to control phaseseparated structure, and found that the current efficiency significantly increased with the addition of the block copolymer accompanied with the decrease of surface roughness [16]. In this paper, EL devices were fabricated based on the blend systems consisting of PTAA, PF, and a block copolymer with both segments (PF-b-PTAA), and the relationship between chemical nature and morphology of the active layer, and EL performance was elucidated.

2. EXPERIMENTAL

2.1. Materials

PF, PTAA homopolymers and PF-b-PTAA block copolymers were synthesized as reported [9,10]. Chemical structures of these polymers are shown in **Figure 1**. Characteristics of utilized polymers are listed on **Table 1**.

2.2. Characterization

The performance of the EL devices based on blend systems with the device structure of ITO/PEDOT:PSS (30 nm)/polymer (30 nm)/BCP (50 nm)/LiF (0.5 nm)/Al (100 nm) was evaluated. As a hole blocking layer, 2,9-dimethyl-4,7-diphenyl-1,10-phenanthroline(bathocuproine, BCP) was used between the emission layer and cathode. EL devices were fabricated as described in a previous report [9]. Atomic force microscopic (AFM) measurements were performed on JEOL JSPM 4200 system in trapping mode (phase and topographic modes) with an MPP-11100-10 silicon probe (resonant frequency: 300 kHz, force constant: 40 N/m). All thin film of polymers were spin-cast onto glass slide with a MIKASA 1H-D7 spin coater from THF solutions at 1500 rpm for 30 s.

3. RESULTS AND DISCUSSION

3.1. PF1/PF-PTAA1 Blend System

A series of EL devices based on PF1/PF-PTAA1 blend system were characterized to investigate the effect of the block copolymer. **Figure 2** showed current efficiency vs. current density characteristics. The EL characteristics were summarized in **Table 2**.

With the increase of the content of the block copolymer, the luminance (at 30 mA/m^2) and the maximum current efficiency increased, and were almost equal to those of the block only device (100 wt%). As discussed in a previous report [10], it can be speculated that PTAA segment with hydrophilic moiety segregates at the interface between PEDOT:PSS and the active layer. Therefore, it is considered that the bipolar block copolymer effectively enhances the hole injection and electron blocking ability. It is noteworthy that blend system (77 wt%) showed higher current efficiency in the region of high current density than block only device. In the blend film, the size of PF domain is considered to be larger compared with the block copolymer film [17].

The large domain seemed to enhance the recombinetion probability and/or improve the carrier balance especially in the high current density region. As shown in **Table 3**, the device with PF2/PTAA showed the worst performance due to the severe phase separation (see **Figure 4**). It is revealed that the utilization of the block copolymer is much favorable in order to enhance the hole injection ability in blend systems.

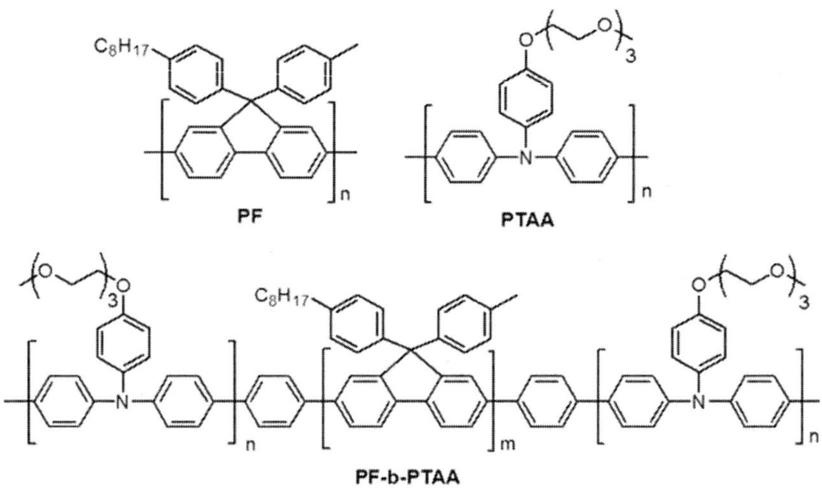

Figure 1. Chemical structure of PF, PTAA, and PF-bPTAA.

Table 1. Characteristics of utilized polymers.

Polymer	Molar ratio of PF/PTAA[a]	M_n[b]	PDI[b]	T_g/°C[c]
PF1	-	13000	2.4	118
PF-b-PTAA1	55/45	23000	2.7	118, 174
PF2		26000	3.5	119
PF-b-PTAA2	60/40	43000	3.3	119, 173
PTAA		9700	2.8	175

[a] Determined by ^{1}H-NMR, [b] determined by GPC, [c] determined by DSC.

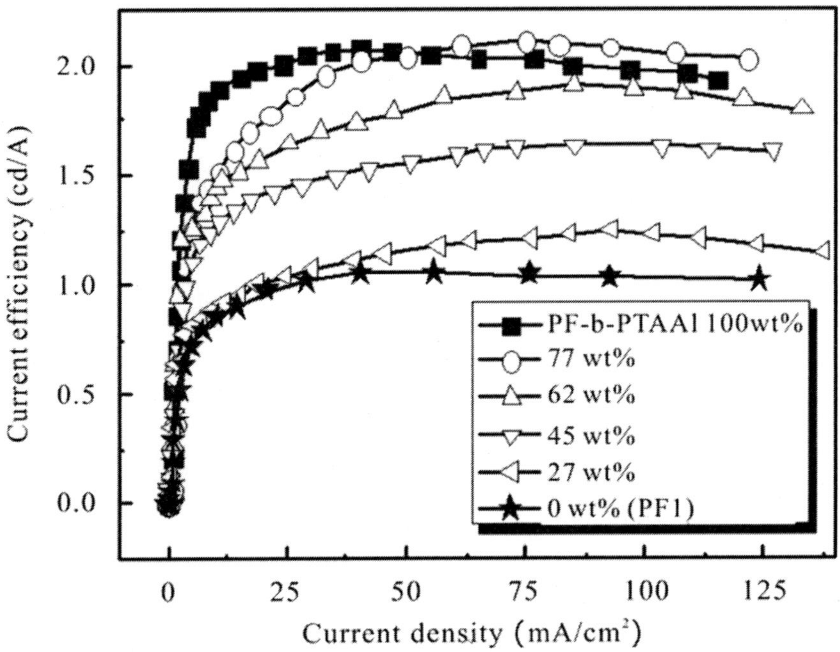

Figure 2. Current efficiency vs. current density characteristics for the devices based on PF1/PF-b-PTAA1 blend systems.

3.2. PF2/PTAA/PF-PTAA2 Blend System

In order to investigate the effect of the morphology or domain size of phase separated systems on EL performance, we examined the ternary blend systems consisting of both homopolymers and the block copolymer keeping the chemical composition constant (fluorene:triarylamine = 60:40). The addition of corresponding block copolymer is one of the effective strategies for more stable, and finer scale morphologies [18,19]. **Figure 3** showed current efficiency vs. current density characteristics for the

Table 2. EL characteristics for the devices based on PF1/PF-b-PTAA1 blend systems.

wt% of PF-b-PTAA1 (%)	Molar ratio of (PF/PTAA)/PF	Luminance at at 30 mA/cm^2 (cd/m^2)	Maximum current efficiency (cd/A)
100	(55/45)/0	626	2.10 (14 V)
77	(55/45)/28	648	2.12 (15 V)
62	(55/45)/55	547	1.91 (15 V)
45	(55/45)/110	443	1.63 (15.5V)
27	(55/45)/220	398	1.25 (16 V)
0	0/1	297	1.07 (13 V)

Table 3. EL characteristics for the devices based on PF2/PTAA/PF-b-PTAA2 blend systems.

wt% of PF-b-PTAA2 (%)	Molar ratio of (PF/PTAA)/PF2/PTAA	Luminance at at 30 mA/cm^2 (cd/m^2)	Maximum current efficiency (cd/A)
100	(60/40)/0/0	724	2.15 (15 V)
80	(60/40)/15/10	648	2.28 (16.5 V)
60	(60/40)/40/27	521	2.07 (16 V)
40	(60/40)/90/60	478	1.78 (17 V)
20	(60/40)/240/160	407	1.51 (17 V)
0	0/60/40	235	0.88 (14 V)

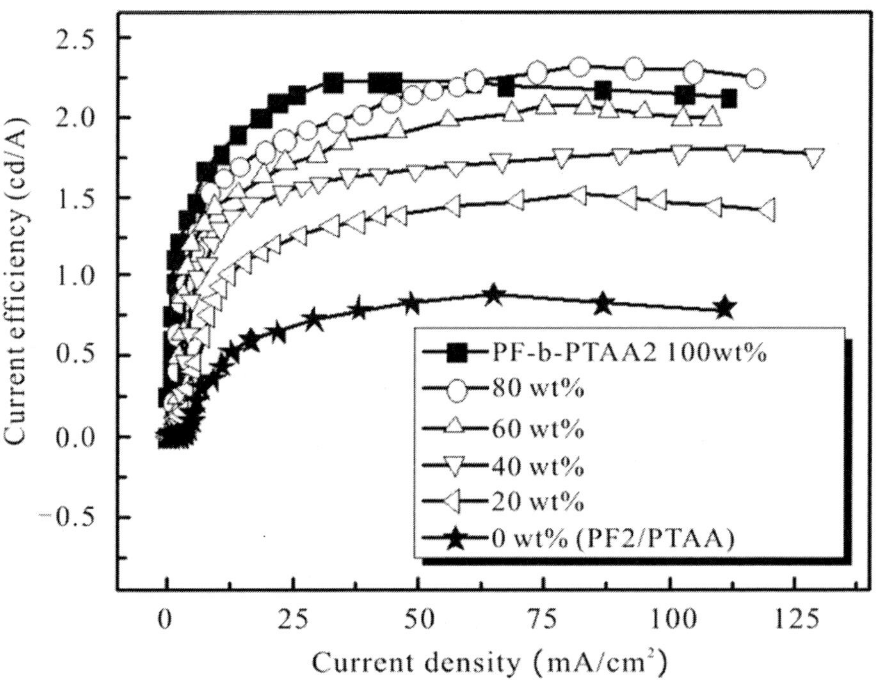

Figure 3. Current efficiency vs. current density characteristics for the devices based on PF2/PTAA/PF-b-PTAA2 blend systems.

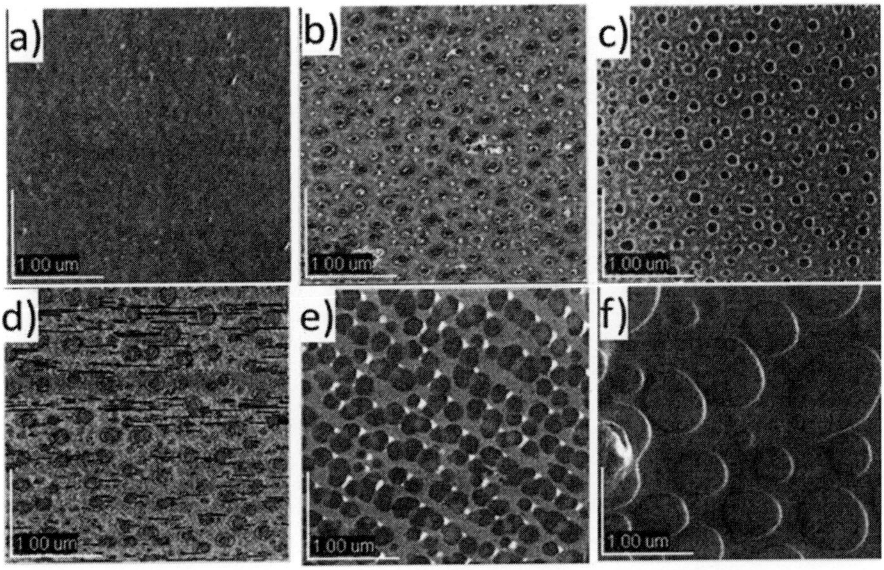

Figure 4. AFM images for PF2/PTAA/PF-b-PTAA2 blend films on PEDOT/PSS (tapping mode). PF-b-PTAA2 content is: a) 100 wt%; b) 80 wt%; c) 60 wt%; d) 40 wt%; e) 20 wt%; and f) 0 wt%. All thin films (ca. 30 nm thick) were annealed at 120°C for 1 h.

PF2/PTAA/PF-b-PTAA2 blend systems. The EL characteristics were summarized in **Table 3**. Although it is found that the addition of PTAA component to the active layer improved the performance as shown in Figure 2 and **Table 2**, the simple blend of both homopolymers afforded the worst performance we examined as above mentioned.

This result strongly suggests that the morphology in the active layer also plays an important role as well as its chemical characters. In fact, the maximum current efficiency and luminescence at 30 mA/cm^2 increased with the increase of the content of PF-b-PTAA2. The similar effect of a block

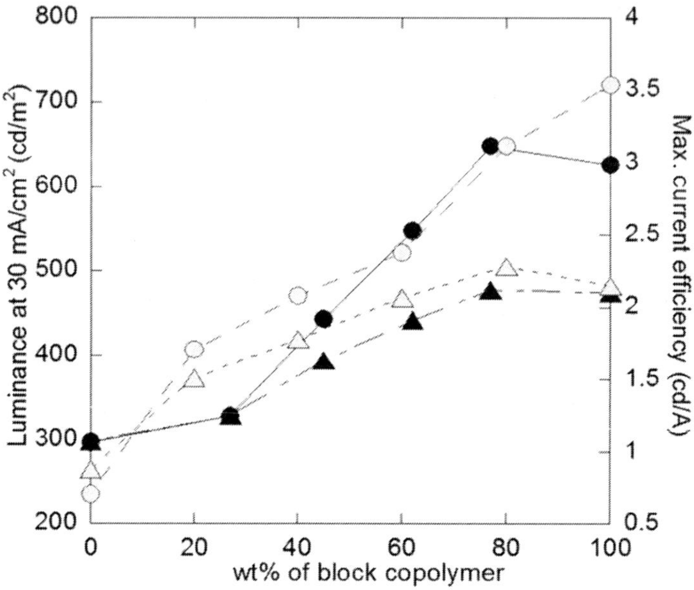

Figure 5. Relationship between the content of block copolymer and EL performance. Filled and open symbols are for the devices based on PF1/PF-b-PTAA1, and PF2/PTAA/ PF-b-PTAA2 blend systems, respectively. Circle and traingle represent luminescent at 30 mA/cm^2 and maximum current efficiency, respectively.

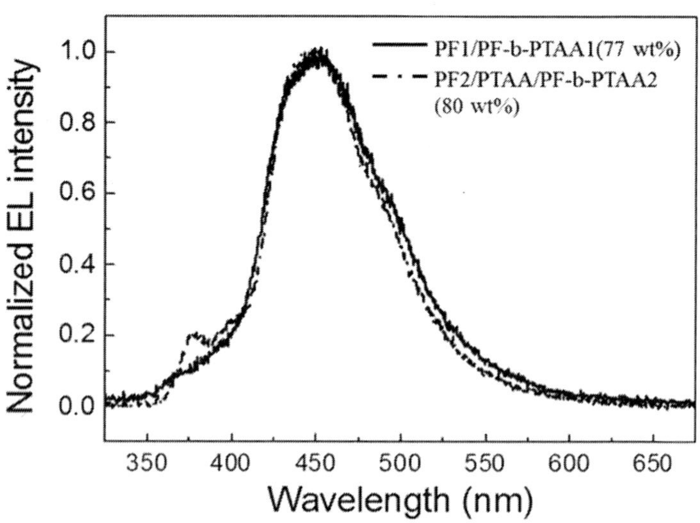

Figure 6. Electroluminescence spectra for devices based on both blend systems at 14 V.

copolymer as an additive was observed for electro-phosphorescent devices based on the blend systems consisting of two types of vinyl homopolymers (hole and electron transporting), and bipolar block copolymer with both segments [16]. **Figure 4** represents the AFM images of blend films with the thickness of ca. 30 nm on PEDOT/PSS in a tapping mode. In the simple blend (**Figure 4**(f)), ca. 1 μm of domains were observed, and the domain size decreased as the content of PF-bPTAA2 increased. These results indicate that the finer domain size was achieved by the addition of the block copolymer as expected, and was favorable for better EL performance. Higher current efficiency was observed at the higher current density region for the blend system (80 wt% of block copolymer) than block only device. As discussed above, this is probably due to the appropriate domain size for the effective recombination and/or carrier balance in the blend system.

The relationship between the block copolymer content and EL performance for both blend systems are shown in **Figure 5**. For both cases, the maximum current efficien cies increased with the content of the block copolymer, and reached almost constant values at 80 wt%. The addition of block copolymers altered mainly the chemical nature of the active layer in the first blend system, and changed the morphological nature in the second system.

As shown in **Figure 5**, it is found that EL performance is governed by the both factors which can be controlled by the utilization of block copolymers as the additive to PF or PF/PTAA blend systems.**Figure 6** shows EL spectra for typical two types of devices (PF1/PF-bPTAA1 77 wt%, and PF2/PTAA/PF-b-PTAA 80 wt%). These profiles are almost the same as that observed for the block copolymer based device [6]. Other devices also exhibited a similar blue EL profiles without unnecessary green-emitting bands.

4. CONCLUSION

We demonstrated the method how to improve the EL performance of blue-emitting devices based on polymer

blends. PF-b-PTAA block copolymer afforded hole injecting and/or electron blocking ability to PF, and it worked as a compatibilizer for PF/PTAA blend to control the morphology of blend films. In order to optimize morphology and EL performance, further studies are underway including the investigation of the effect of the molecular weights on both homopolymers and block copolymer, and the chemical composition of the latter.

REFERENCES

1. Y. Ohmori, M. Uchida, K. Muro and K. Yoshino, "Blue Electroluminescent Diodes Utilizing Poly (Alkylfluorene)," Japanese Journal of Applied Physics, Vol. 30, No. 11B, 1991, pp. 1941-1943.

2. M. Leclerk, "Polyfluorenes: Twenty Years of Progress," Journal of Polymer Science Part A: Polymer Chemistry, Vol. 39, No. 17, 2001, pp. 2867-2873.

3. S. K. Lee, T. Ahn and N. S. Cho, "Synthesis of New Polyfluorene Copolymers with a Comonomer Containing Triphenylamine Units and Their Applications in WhiteLight-Emitting Diodes," Journal of Polymer Science Part A: Polymer Chemistry, Vol. 45, No. 7, 2007, 45, pp. 1199-1209. [Paper reference:1]

4. M. Redecker, D. D. Bradley, M. Inbasekaran, W. W. Wu and E. P. Woo, "High Mobility Hole Transport FluoreneTriarylamine Copolymers," Advanced Materials, Vol. 11, No. 3, 1999, pp. 241-246.

5. C. Ego, A. Grimsdale, F. Uckert, G. Yu, G. Srdanov and K. Müllen, "Triphenylamine-Substituted Polyfluorene— A Stable Blue-Emitter with Improved Charge Injection for Light-Emitting Diodes," Advanced Materials, Vol. 14, No. 11, 2002, pp. 809-811.

6. T. Miteva, A. Meisel, W. Knoll, H. G. Nothofer, U. Scherf, D. C. Müller, K. Meerholz, A. Yasuda and D. Neher, "Improving the Performance of Polyfluorene-Based Organic Light-Emitting Diodes via End-capping," Advanced Materials, Vol. 13, No. 8, 2001, pp. 565-570.

7. H. T. Nicolai, A. Hof, J. L. M. Oosthoek and P. W. M. Blom, "Charge Transport and Recombination in Polyspirobifluorene Blue Light-Emitting Diodes," Advanced Functional Materials, Vol. 21, No. 8, 2011, pp. 1505- 1510.

8. L. Duan, B. D. Chin, N. C. Yang, M.-H. Kim, H. D. Kim, S. T. Lee and H. K. Chung, "Multilayer Blue Polymer Light-Emitting Devices with Spin-Coated Interlayers," Synthetic Metals, Vol. 157, No. 8-9, 2007, pp. 343-346.]

9. Y. Tan, Z. Gu, K. Tsuchiya and K. Ogino, "Synthesis and Luminescent Properties of Block Copolymers Based on Polyfluorene and Polytriphenylamine," Polymer, Vol. 53, No. 7, 2012, pp. 1444-1452.

10. Y. Tan, K. Tsuchiya and K. Ogino, "Synthesis of Polyfluorene Block Copolymers and Effect of Side Chain Group on Electroluminescent Device Performance," Chemistry Letters, Vol. 41, No. 3, 2012, pp. 257-259.

11. C. Park, J. Yoon and E. L. Thomas, "Enabling Nano-technology with Self Assembled Block Copolymer Patterns," Polymer, Vol. 44, No. 22, 2003, pp. 6725-6760.

12. L. Deng, P. T. Furuta, S. Garon, J. Li, D. Kavulak, M. E. Thompson and J. M. J. Fréchet, "Living Radical Polymerization of Bipolar Transport Materials for Highly Efficient Light Emitting Diodes," Chemistry of Materials, Vol. 18, No. 2, 2006, pp. 386-395.

13. B. Ma, B. J. Kim, L. Deng, D. A. Poulsen, M. E. Thompson and J. M. J. Fréchet, "Bipolar Copolymers as Host for Electroluminescent Devices: Effects of Molecular Structure on Film Morphology and Device Performance," Macromolecules, Vol. 40, No. 23, 2007, pp. 8156-8161.

14. K. Tsuchiya, K. Sakaguchi, H. Kasuga, A. Kawakami, H. Taka, H. Kita and K. Ogino, "Synthesis of Charge Transporting Block Copolymers Containing 2,7-Dimethoxycarbazole Units for Light Emitting Device," Polymer, Vol. 50, No. 3, 2010, pp. 616-622.

15. K. Tsuchiya, H. Kasuga, A. Kawakami, H. Taka, H. Kita and K. Ogino, "Synthesis of Bipolar Charge Transporting Block Copolymers and Characterization for Organic Light-Emitting Diode," Journal of Polymer Science Part A: Polymer Chemistry, Vol. 48, No. 7, 2010, pp. 1461-1468.

16. K. Tsuchiya, K. Sakaguchi, A. Kawakami, H. Taka, H. Kita, T. Shimomura and K. Ogino, "Charge Transporting Block Copolymer for Morphological Control in Light Emitting Device Based on Polymer Blends," Synthetic Metals, Vol. 160, No. 15-16, 2010, pp. 1679-1682.

17. D. J. Kinning, E. L. Thomas and L. J. Fetters, "Morphological Studies of Micelle Formation in Block Copolymer/Homopolymer Blends," Journal of Chemical Physics, Vol. 90, No. 10, 1989, pp. 5806-5825.

18. C. Koning, M. V. Duin, C. Pagnoulle and R. Jerome, "Strategies for Compatibilization of Polymer Blends," Progress in Polymer Science, Vol. 23, No. 4, 1998, pp. 707-757.

19. J. A. Gallowaya, H. K. Jeonb, J. R. Bell and C. W. Macosko, "Block Copolymer Compatibilization of Cocontinuous Polymer Blends," Polymer, Vol. 46, No. 1, 2005, pp. 183-181.

8

CHAPTER

Surface Roughening of Polystyrene and Poly(methyl methacrylate) in Ar/O2 Plasma Etching

Yuk-Hong Ting [1], Chi-Chun Liu [2], Sang-Min Park [2], Hongquan Jiang [3], Paul F. Nealey [2] and Amy E. Wendt [1,*]

[1] *Department of Electrical and Computer Engineering, University of Wisconsin-Madison, Madison, WI 53706, USA*

[2] *Department of Chemical and Biological Engineering, University of Wisconsin-Madison, Madison, WI 53706, USA; E-Mails:*

[3] *Department of Materials Science and Engineering, University of Wisconsin-Madison, Madison, WI 53706, USA*

ABSTRACT

Selectively plasma-etched polystyrene-block-poly(methyl methacrylate) (PS-b-PMMA) diblock copolymer masks present a promising alternative for subsequent nanoscale patterning of underlying films. Because mask roughness can be detrimental to pattern transfer, this study examines roughness formation, with a focus on the role of cross-linking, during plasma etching of PS and PMMA. Variables include ion bombardment energy, polymer molecular weight and etch gas mixture. Roughness data support a proposed model in which surface roughness is attributed to polymer aggregation associated with cross-linking induced by energetic ion bombardment. In this model, RMS roughness peaks when cross-linking rates are comparable to chain scissioning rates, and drop to negligible levels for either very low or very high rates of cross-linking. Aggregation is minimal for very low rates of cross-linking, while very high rates produce a continuous cross-linked surface layer with low roughness. Molecular weight shows a negligible effect on roughness, while the introduction of H and F atoms suppresses roughness, apparently by terminating dangling bonds. For PS etched in Ar/O_2 plasmas, roughness decreases with increasing ion energy are tentatively attributed to the formation of a continuous cross-linked layer, while roughness increases with ion energy for PMMA are attributed to increases in cross-linking from negligible to moderate levels.

Keywords: diblock copolymer; plasma etching; surface roughness; cross-linking; polymer aggregation

1. INTRODUCTION

Nanoscale structures have many applications, including integrated circuit manufacturing and biological applications. For example, biomimetic nanostructures with surface topography are used to study how living cells or micro-organisms respond to their environments [1,2,3,4,5,6]. In order to find the lower limit of biomimetic cellular response to surface topography, self-assembly regulated domain nanostructures of diblock copolymer have been exploited as a way to create templates for nanoscale patterning beyond the limit of conventional optical lithography [7,8,9,10,11].

Diblock copolymer structures contain two chemically distinct polymer blocks. In liquid form, microphase separation between the two is observed, while in annealed solid thin films, the two blocks spontaneously self-assemble into nanoscale domains (~5 nm to 50 nm) that exhibit ordered morphology at equilibrium. The chemical distinction between the two polymer domains allows selective removal of one component of the structure by either wet etching or plasma etching. However, capillary forces associated with wet etching have been found to cause remaining structures to distort and collapse [10,12]. One of the most commonly used diblock copolymer is polystyrene-block-poly(methyl methacrylate) (PS-b-PMMA). We have previously developed a plasma etch process with high PMMA/PS etch selectivity and low surface roughness for selective removal of PMMA to produce high quality PS templates for patterning underlying materials.

In our previous study, a survey of etch gas mixtures [12] showed that an Ar/O_2 plasma in combination with a high substrate bias voltage at −110 V (~125 eV ions) produced the best combination of etch selectivity and PS mask surface and line edge smoothness. Two particular results underlie this roughness study. First, PS and PMMA exhibit different surface roughness dependence on ion energy in O_2 and Ar/O_2plasma etching. Surface roughness of PS, as measured with atomic force microscopy (AFM) after plasma etching, decreases with increasing ion bombardment energy from ~25 eV to ~125 eV, while the opposite is observed for PMMA. Second, the surface roughness of PS after Ar/O_2 etching with high ion bombardment energy is lower than after O_2 plasma etching, suggesting that the higher ion to radical flux ratio expected with the Ar admixture plays a role in the reduced surface roughness.

The aim of this study is to identify mechanisms of surface roughening in plasma etching of polymers. Ar/O_2 plasma etching of PS and PMMA is chosen as a case study, and the focus is on the role of cross-linking induced aggregation.

Surface cross-linking has been proposed as a factor in degradation during plasma etching of polymer materials [13,14,15,16]. During plasma exposure, ion bombardment causes breakage of the polymer backbone, side chain scissioning and depolymerization, collectively enhancing etch rate and mobility [17,18] in the surface layer. However, removal of H atoms and scissioning of chemical bonds in polymer chains also produces dangling bonds that may recombine to cross-link neighboring polymer chains. The resulting degree of cross-linking, associated with increased etch resistance, depends on the chemical structure of the polymer, the plasma gas mixture and ion bombardment energy. For example, one of the observed differences between PS and PMMA is the degree of cross-linking following plasma exposure. In the case of PS, surface cross-linking is prevalent, while in PMMA, chain scissioning and depolymerization dominate [13,14,16]. While chemical structures such as the aromatic ring in PS and the methyl ester group in PMMA are commonly believed to play important roles in etching behaviors such as etch rate, cross-linking and chain scission, most studies reporting a correlation between chemical structures and etch behaviors of polymers are based on indirect observations, including change in molecular weight [16], etch rate [13,19], interface adhesion [14,16] or diffusion coefficient [20]. Details of the mechanisms involved have yet to be elucidated.

In addition, while O_2 plasmas are found to enhance cross-linking during PS etching through formation of C–O–C bonds, H and F atoms were shown to suppress cross-linking during plasma etching [19]. H and F atoms form a single bond with C atoms, terminating dangling bonds and preventing cross-linking between C atoms in neighboring chains.

Cross-linking in plasma etching is also affected by ion bombardment energy. Tead *et al.* [20] studied the diffusion properties of O_2 plasma treated PS and suggested that cross-linking could cause reductions in mobility (ability of polymer to diffuse) and permeability at the surface layer of PS etched in O_2 plasmas. They demonstrated that higher ion energy could lead to greater cross-linking, resulting in reduced mobility and permeability.

Studies of the role of cross-linking in surface roughening of polymer etching are few. A correlation between cross-linking and aggregation leading to nano-scale inhomogeneity in mass and density has been proposed by Sumiya *et al.* to explain surface roughening in polymer films [21,22,23]. Polymer aggregation is commonly understood as the state when polymer chains are massed into dense tangled clusters. Studies of photoresist development in conventional optical lithography also suggest a correlation between cross-linking and aggregation in polymer materials. Yoshimura *et al.* showed that cross-linking in photoresist materials could be exploited to create nanoscale microgel (aggregate like) structures [24], similar to the model proposed by Sumiya *et al.* [21]. It has also been reported that molecular weight (MW) affects the size of intrinsic polymer aggregates [26,27].

A recent study of Ar etching of 193 nm photoresist by Chung *et al.* [17] showed that enhanced cross-linking could reduce surface roughening while higher levels of chain scissioning and mobility in the polymer could cause the opposite [17,18]. But rather than cross-linking induced aggregation as proposed in previous studies, a buckling theory [17,28] is proposed by Bruce *et al.* to explain the mechanism behind this observation. In the proposed model, a stressed modified surface layer could buckle and form wrinkles and thus surface roughness during plasma etching. Cross-linking could increase the Young's modulus of the modified surface layer and thus suppress buckling.

Different roughening mechanisms in plasma etching can be distinguished by characteristics such as form and size of roughness features during initial roughness formation and evolution. For example, during initial roughness formation, roughness features created by ion damage or deposition of sputtered materials are random in size and shape [29,30]. In contrast, roughness features induced by aggregation in polymer materials are more consistent in size and shape [25,31,32,33]. Roughness amplification mechanisms may also contribute to feature characteristics during roughness evolution. For example, in the "shadowing effect," reactive radicals generated in the plasma have a lower flux in the "valleys" of surface topography than at the "peaks," which block access to the valleys. Shadowing leads to either an increase or a decrease in the vertical dimension of existing roughness features depending on whether the radicals involved contribute to deposition or etching, respectively [34,35,36]. In contrast, "flux re-emission" involves deflection or re-emission of energetic ions or radicals from feature sidewalls, increasing their relative flux at feature bottoms. Simulation results show smaller roughness features in valleys may disappear over time due to flux re-emission, leading to a decrease in concentration of features on the substrate surface as it evolves [30,34,37,38]. In "micro-masking", small regions of greater etch resistance form on film surface, typically caused by redeposition of material sputtered from chamber walls. However, inhomogeneities induced in the modified film surface during plasma exposure may have a similar effect [39].

The approach taken here to investigate the roles of cross-linking and aggregation includes examination of the role of aggregation in surface roughening by comparing initial roughness formation and evolution, using characteristics such as form and size of roughness features of Ar/O_2 etched PS and PMMA. In addition, the opposite dependence on ion bombardment energy of PS and PMMA surface roughness in Ar/O_2 plasma etching [21] may offer insight into the role of cross-linking. A unique capability to independently select ion bombardment energy is utilized here in an examination of the role of cross-linking in surface roughening. Higher ion bombardment energy is known to enhance cross-linking in O_2 plasma etching of PS [20], but has minimal effect on cross-linking in PMMA etching, in which chain scissioning dominates. Surface roughness of PS and PMMA films are characterized in this study with atomic force microscopy (AFM). Etch times are varied to observe initial roughness formation as well as further evolution resulting from plasma exposure, with the following variables: ion bombardment energy, monomer molecular weights and etch times. Studies of the effect of the energy of bombarding ions make use of a novel system in which the energy of bombarding ions is selected while maintaining a narrow energy spread compared to typical etch systems [40].

Another set of experiments with Ar/H_2 and Ar/F_2 etching of PS and PMMA are designed to suppress the effects of cross-linking through the termination of dangling bonds by F or H atoms [19]. Post-etch roughness of PS samples with different MWs are also compared to investigate the contribution of intrinsic aggregation in surface roughening. A model of surface roughening during plasma etching that involves cross-linking-induced aggregation is presented. In this model, differences between Ar/O_2plasma treated PS and PMMA are attributed to modified layers with varying degrees of cross-linking on the surface of the etched polymer films.

Most reported studies of surface roughening compare surface roughness induced in plasma etching with different plasma conditions and with etch time or ion fluence held fixed [41,42,43], resulting in considerable variation in corresponding etch depth due to etch rate differences. Etch rate [41,44] and roughening rate [37,44] might also change with time before a steady state is reached during etching. Here, samples are etched to the same etch depth (~150 nm) in most of the tests (except those examining roughness initiation) for studying the effect of gas mixtures and ion bombardment energy. This way, results are applicable to process optimization of etching processes with constant etch depth. In addition, the size of the roughness features on etched PS and PMMA samples is in a range up to ~20 nm. Etch depths greater than the size of the roughness are required to observe the mechanisms that contribute to the surface evolution after initial roughness is formed. As different gas mixtures, like O_2 and pure Ar, may produce differences in etch rate up to a factor of 100 [12], a constant etch depth ensures comparison in the same etching/roughening regime.

2. EXPERIMENT SETUP

All etching experiments were carried out in a helicon plasma tool [40] equipped with a diode laser thin film interferometer for *in situ* etch depth measurement [45]. One advantage of the helicon plasma etch tool used in this study compared to more conventional reactive ion etch (RIE) tools is that the plasma source power and self-bias power are decoupled [46], so that ion bombardment energy at the substrate can be varied independent of ion flux, by changing the self-bias voltage. A substrate bias system equipped with a combination of a programmable voltage wave form generator, a broad-band power amplifier, and a coupling capacitor is used to produce a specially tailored bias voltage wave form, described in more detail in Reference [40]. The tailored wave form was introduced as a method to produce a narrow distribution of ion energies at the substrate, and has shown to be effective in improving etch selectivity of SiO_2 over Si [47] and organosilicate glass (OSG) over Si_3N_4 and SiC [48].

The ion bombardment energy (E_i) is determined by the difference between plasma potential V_p and the DC self bias voltage on the wafer (V_p-V_{bias}). The DC self-bias voltage (V_{bias}) is measured at the power cable between the blocking capacitor and the substrate holder. V_p in our process is estimated at ~15V while the self-bias voltage at the substrate without bias power output is −10V. Therefore the minimum ion bombardment energy is (V_p-V_{bias}) e = ~25 eV. The stage is described as "floating" when no bias power is applied. A floating rather than grounded electrode is chosen for the lowest ion energy level, as it is not possible to ground the surface of the insulating polymer films. The ion bombardment energy ranges from ~25 eV (V_{bias} = −10 V) to ~125 eV (V_{bias} = −110 V) in this study.

All samples were etched at a pressure of 10 mTorr. The substrate electrode temperature was controlled at 5 °C with 7 sccm of He flow to the backside of the wafer to ensure good thermal contact with the cooled electrode. For Ar/O_2, Ar/F_2 and Ar/H_2 gas mixtures, the flow rate ratios between Ar and O_2 and F_2 and H_2, respectively, is 10:1 with total flow of 55 sccm. The helicon source power was set at 300 W, producing a sufficiently low etch rate to enable good control of the etch depth [12].

The PS homopolymer, with M_n of 60 kg/mole and PDI of 1.03, was purchased from Polymer Source. The PMMA solution is 4 wt% in anisole, bought from MicroChem Corp., with M_n of 950 kg/mole. PS test wafer are made by spin coating a 5 wt% PS solution in toluene at 2,000 RPM for 1 minute, followed by a 1 minute, 160 C bake, to produce a ~300 nm thick film. PMMA test wafers were made in the same way except the solution was used as purchased. Blanket PS and PMMA film samples of size 1 ° 1 cm^2, cut from the 4"D wafers, are used for etch surface roughness measurements. All samples are mounted on a 4-inch bare silicon wafer with electrically and thermally conductive graphite adhesive before loading onto the wafer stage for plasma etching. The surface roughness of the etched samples was measured using AFM with a Veeco MultiMode Scanning Probe Microscope system in tapping mode. Etched silicon tips with force constant of 3.5 N/m have been used in the AFM measurement.

3. RESULTS

To characterize roughness initiation during plasma etching, PS blanket film samples etched with bias voltage of −10 V and etch time from 0 second to 266 seconds were measured with AFM. Figure 1 andFigure 2 show the AFM images and RMS roughness measurement results, respectively, for etched PS samples. Although surface roughness measurements show that RMS roughness increases almost linearly with time, roughness formation is divided into two stages base on AFM images. During the initial roughness formation (etch time from 0 seconds to 66 seconds), cone shaped roughness features appear and increase gradually in number on the surface [Figure 1 (a–e)]. The roughness features show good uniformity in lateral size (~20 nm) and form, thus resembling the polymer aggregates observed in photoresist materials during development [25,31,32,33]. Initial roughness formation is very different from the initial roughness induced by common ion bombardment or redeposition of sputtered materials from the chamber wall that normally leads to craters or clusters of random size and shape. No metal impurities were detected by XPS measurements on plasma etched polymer samples, excluding the possibility of redeposition of chamber wall material. These results leave open the possibility that polymer aggregation may be involved in the initial roughness formation during PS etching.

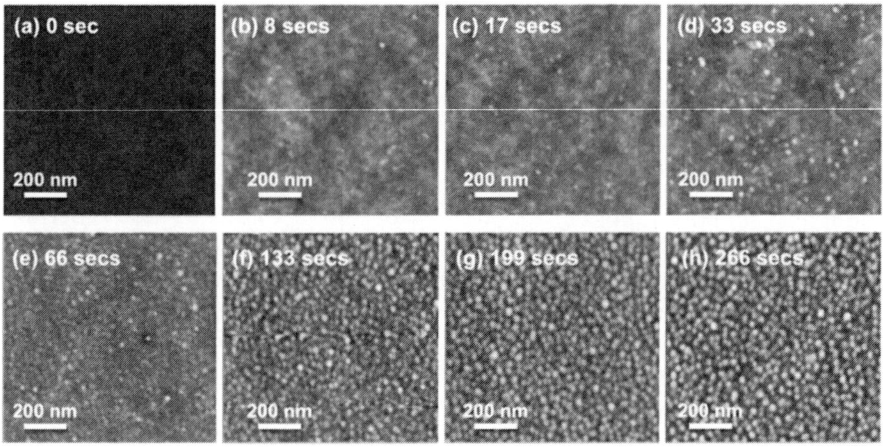

Figure 1. AFM images of Ar/O_2 etched PS with bias voltage of −10 V and different etch times from (a) 0 seconds to (h) 266 seconds, corresponding to approximate etch depths (**a**) 0 nm, (**b**) 9 nm, (**c**) 18 nm, (**d**) 35 nm, (**e**) 70 nm, (**f**) 142 nm, (**g**) 212 nm and (**h**) 284 nm. Roughness features (white spots) of uniform size (~20 nm) and form initially increase in number over time during plasma etching, and subsequently increase in amplitude without further increase in number.

There are two possible causes of aggregation. It may be induced intrinsically throughout the film during polymer sample preparation [31,32,33] or induced at the surface due to plasma exposure [21]. If polymer aggregation is intrinsic, the inhomogeneity could induce etch rate non-uniformity immediately upon plasma exposure. Surface roughness features during initial roughness formation are expected to appear on the surface and saturate in number in short time after etching begins [25]. But Figure 1 shows that roughness features appear to start to saturate in number after etch times greater than 66 seconds, corresponding to etch depths of more than 70 nm, much larger than the size scale (~20 nm) of the roughness features. This result leads to the conclusion that roughness observed following Ar/O_2 plasma etching of PS does *not* result from intrinsic inhomogeneity of the spun on film.

For etch times of 133 seconds and longer [Figure 1 (f–h)], the number of roughness features on the surface has saturated. Increases in RMS (root mean square) roughness (Figure 2) during this period are consistent with magnification of the surface roughness due to micro-masking. The consistency of the size and number of roughness features also supports the hypothesis that micro-masking plays a dominant role, and argues against flux re-emission [39].

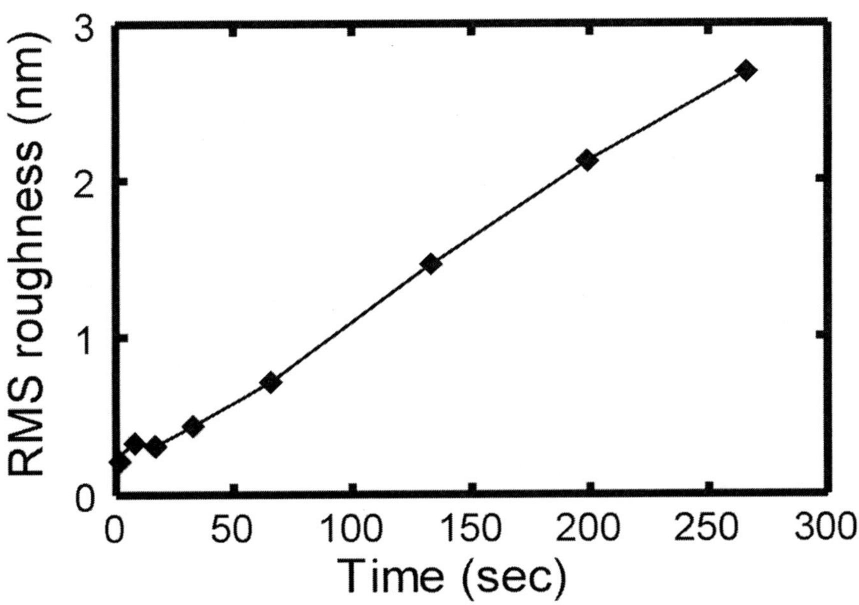

Figure 2. RMS roughness of Ar/O_2 etched PS with bias voltage of −10 V and etch times from (a) 0 seconds to (h) 266 seconds, corresponding to approximate etch depths (a) 0 nm, (b) 9 nm, (c) 18 nm, (d) 35 nm, (e) 70 nm, (f) 142 nm, (g) 212 nm and (h) 284 nm.

Etching experiments with polymers of different molecular weight (MW) were carried out as a further test for the contribution of intrinsic aggregation to roughness formation during etching. It has been reported that MW affects the size of intrinsic polymer aggregates [26,27]. Therefore, the sizes of surface roughness features on blanket PS films of different MWs etched with Ar/O_2 plasmas under identical conditions and etch duration are compared. Figure 3 shows that etched PS samples of

MWs of 60K g/mole and 1.88M g/mole exhibit surface roughness features of very similar height and lateral dimension, providing further evidence that the roughness observed is not due to intrinsic inhomogeneity.

The role of cross-linking is studied by observing the dependence of surface roughnesses of PS and PMMA on ion energy. Following both Ar/O_2 and O_2 plasma exposure, the surface roughnesses of PS and PMMA films show opposite dependence on ion energy [12]. PS roughness decreases with increasing ion bombardment energy, while that of PMMA increases. A major difference between PS and PMMA is the level of cross-linking during plasma exposure, with the degree of cross-linking in PS is higher than in PMMA during plasma etching. Cross-linking in PS plasma etching may be further enhanced by increasing ion energy, leading to observed differences in surface roughness dependence on ion bombardment energy.

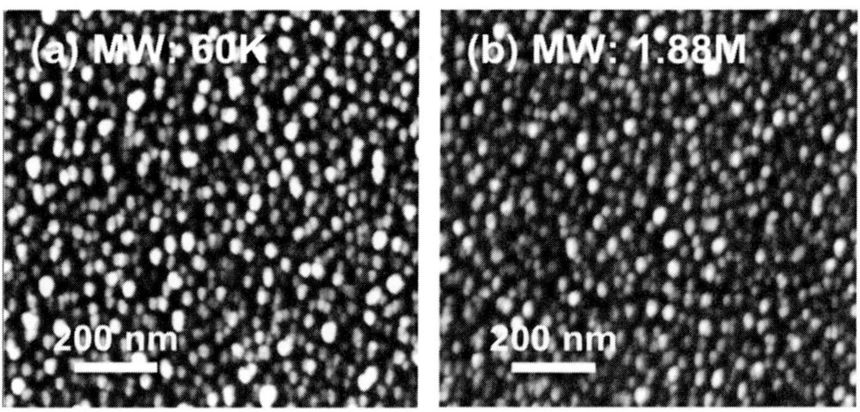

Figure 3. AFM images of Ar/O_2 etched PS with bias voltage of −10 V and molecular weights: (**a**) 60K g/mole (RMS: 4.92 nm), and (**b**) 1.88M g/mole (RMS: 4.19 nm) show similar surface morphology.

AFM images of Ar/O_2 etched PS and PMMA samples for three bias voltages: −10 V, −60 V and −110 V (ion energy: ~25 eV, −75 eV and 125 eV) are shown in Figure 4 and Figure 5, respectively. For PS, suppression of surface roughening is observed with increasing bias voltage. A proposed explanation is that for sufficiently high levels of cross-linking (higher bias voltages), a continuous etch resistant surface layer forms that suppresses roughness formation. In contrast, lower ion bombardment energies may produce a lower degree of cross-linking, such that cross-linking induced aggregation leads to heterogeneous etch resistance and thus greater surface roughness. For PMMA, surface roughening vanishes when the bias voltage is reduced to −10 V, perhaps because cross-linking, which has been shown to be lower in plasma etched PMMA than PS, is reduced to negligible levels under these conditions [17,18]. The formation of a continuous cross-linked etch resistant layer in PS during Ar/O_2etching with high bombardment energy is consistent with etch rate measurements. When bias voltage is increased from −10 V to −110 V (ion energy from 25 eV to 125 ev), the PMMA etch rate increase by 345% while PS etch rate increase by only by 221% as shown in Figure 6. This difference is mostly attributed to the enhanced cross-linking in PS at higher ion bombardment energy. While the dependence of surface roughness on ion energy is opposite for PS and PMMA etched in Ar/O_2, the surface roughness features are similar in size and form [12] suggesting that the same roughening mechanism is involved.

Figure 4. AFM images of Ar/O_2 etched PS film with bias voltages of (**a**) −10 V, (**b**) −60 V and (**c**) −110 V.

Figure 5. AFM images of Ar/O_2 etched PMMA film with bias voltages of (**a**) −10 V, (**b**) −60 V and (**c**) −110 V.

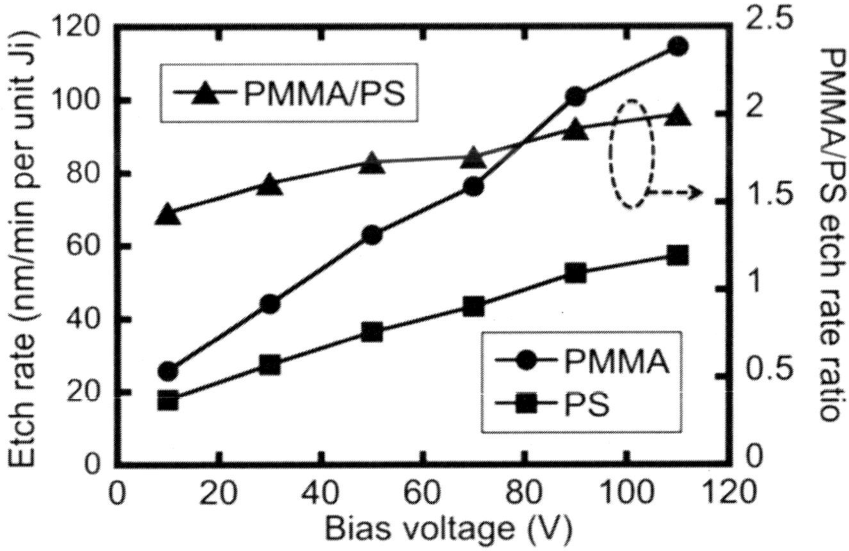

Figure 6. PMMA etch rates in Ar/O_2 show a higher increase (345%) than those for PS (221%), for bias voltage magnitudes of 10 V to 110 V.

The role of cross-linking is further explored through the use of alternate etch gas mixtures that are expected to suppress cross-linking [19]. Surface roughness of Ar/H_2 and Ar/F_2 etched PS and PMMA samples with four substrate bias voltages: −30 V to −150 V (ion energy: 45 eV to 165 eV) are shown inFigure 7. Low RMS surface roughness levels of ~0.5 nm are observed in all samples tested, comparable to the unetched sample (~0.3 nm). Surface roughness features observed in Ar/O_2 etched PS and PMMA samples are not observed or significantly reduced. The suppression of surface roughness is thus attributed to the reduction in cross-linking associated with these etch gas mixtures.

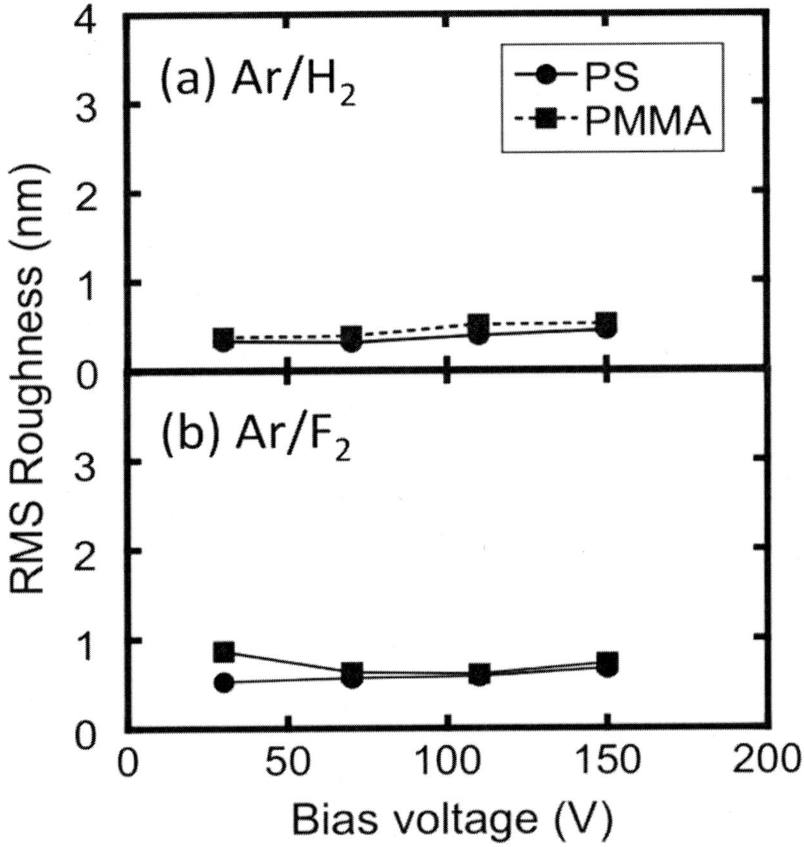

Figure 7. RMS roughness of (**a**) Ar/H_2 plasma and (**b**) Ar/F_2 plasma etched PS films (circles) and PMMA films (squares) etched with different substrate bias voltages.

4. DISCUSSION

The data presented have been interpreted in the context of a model in which aggregation and cross-linking dominate surface roughening during polymer etching have been examined. The data may also be examined in the context of surface buckling, an alternative theory that predicts surface roughness with uniform size and shape [17,28], in which the highly-stressed carbon-rich modified surface layer buckles, producing wrinkles. The predicted surface roughness magnitude is affected by the wavelength of the wrinkles, the lateral dimension of the surface roughness features. Results of initial roughness formation and evolution are consistent with our speculation that aggregation is likely to be the origin of the surface roughening during plasma etching. The theory was previously supported by a correlation between the thickness of the modified layer and the lateral dimension of the roughness [28].

Three observations reported herein are, however, not consistent with buckling theory predictions. First, initial roughness development in Ar/O_2 etching of PS shows that features are constant in size but increase in number during roughness formation (Figure 1), while buckling theory predicts that roughness feature size should evolve during this transient period along the thickness of the modified layer. Second, ion bombardment energy during Ar/O_2 etching of PS and PMMA affects RMS roughness but not lateral feature dimensions, while buckling theory predicts that RMS roughness is correlated with lateral wrinkle dimension. Third, similar roughness features were observed on both blanket films and on the top surface of freestanding 25 nm lamellar PS structures following Ar/O_2 etching [12], as shown in Figure 8. A highly stressed surface layer on a freestanding structure is expected to exhibit different wrinkling behavior compared to a blanket film, as the added surface topography enhances stress release.

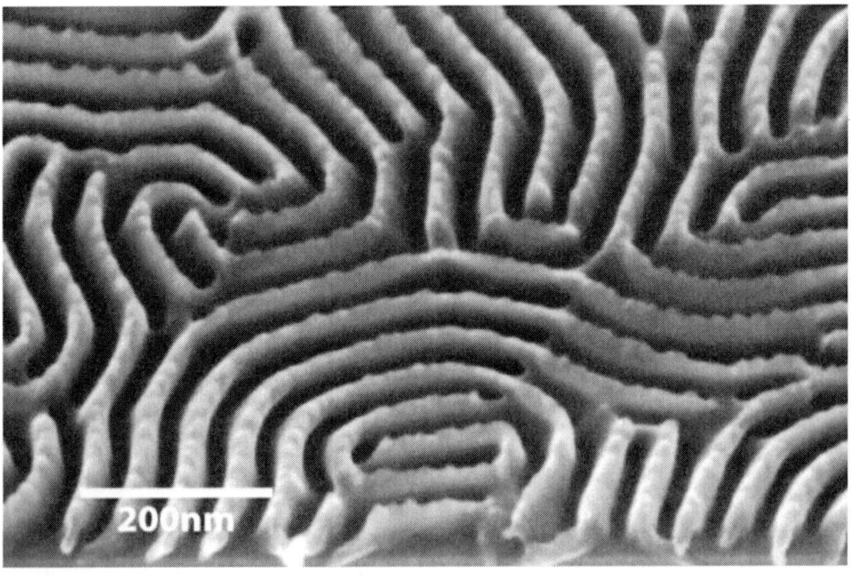

Figure 8. Roughness features observed on the top surface of the freestanding lamellar PS structure after PMMA removal in block copolymer lithography have comparable dimensions to those on full film PS samples.

Conclusions from the data presented are summed up in a proposed cross-linking induced aggregation model to explain surface roughening in polymer etching. As mentioned in Section I, during plasma etching, cross-linking and chain scissioning happen simultaneously at the film surface Breakage of polymer backbone bonds leads to a reduction in molecular weight and increased mobility [17,18], while cross-linking between adjacent polymer chains causes the opposite. In this picture, the balance between cross-linking and chain scission are a crucial factor in aggregation. If the rates of chain scissioning are comparable those of cross-linking, the mobility in this modified layer will be relatively high [18]. The "stickiness" [49] and recombination of dangling bonds in local cross-linking could trigger movement that leads to tangling or bundling of these polymer chains into aggregates on the surface [21]. When, on the other hand, the rate of cross-linking dominates so that a continuous cross-linked layer forms, mobility as well as aggregation will be suppressed. If cross-linking between polymer chains is suppressed, for example, by passivation of dangling bonds with F or H atoms, aggregation may also be suppressed.

In this model, surface roughness following plasma etching is categorized in three regimes associated with the degree of cross-linking (Figure 9). In Regime I, the rate of cross-linking is relatively low comparing to chain scissioning, so that aggregation and thus surface roughness are minimal. In Regime II, the rate of cross-linking is comparable to the rate of chain scissioning, inducing aggregation that leads to surface roughening. In Regime III, the rate of cross-linking is high, leading to an extensive cross-linked layer, suppressing mobility and aggregation, thus resulting in low surface roughness.

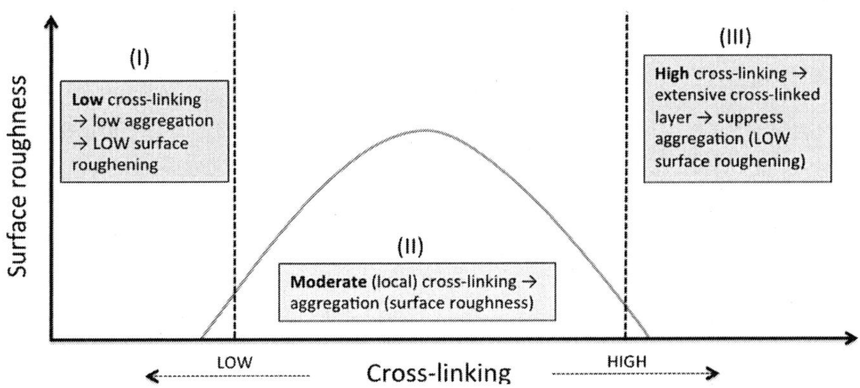

Figure 9. Cross-linking induced aggregation model of formation of surface roughness during plasma etching of polymers.

According to this model, surface roughness could be reduced by shifting the process away from Regime II to either Regime I or III. For example, in Ar/H_2 and Ar/F_2 etching of PS and PMMA, H and F atoms reduce the rate of cross-linking and suppress cross-linking induced aggregation, while a high rate of chain scissioning induced by ion bombardment is maintained (Regime I). Indeed, surface roughness vanishes in the samples etched in these gas mixtures.

In the case of Ar/O_2 plasma etching of PS, increasing ion bombardment energy enhances cross-linking in PS and shifts the process from Regime II toward Regime III. When the degree of cross-linking reaches a threshold where an extensive cross-linked layer is formed, aggregation is suppressed and etched films show lower surface roughness, as observed for ion bombardment energies of 125 eV and above in this study (Figure 4).

For Ar/O_2 etching of PMMA, in contrast, depolymerization is known to dominate over cross-linking under typical etching conditions. While cross-linking may be negligible at the lowest ion energies (Regime I), the balance between cross-linking and chain scissioning rates may shift sufficiently with increasing ion energy to produce low levels of aggregation and thus surface roughness (Regime II). Thus a possible explanation of the opposite roughening trends between PMMA and PS in Ar/O_2 etching is that their respective etching regimes are different for a given ion energy.

Although a connection between surface roughness, cross-linking and aggregation is suggested by data presented here, improved methods are needed to quantify cross-linking and aggregation. Some groups have detected changes in cross-linking through the measurement of molecular weight (MW) by gel permeation chromatography (GPC) [16]. However, attempts with this technique on the plasma etched polymers described herein failed because only the surface layer with thickness less than ~10 nm is modified by plasma etching. Therefore, the MW signal is dominated by the unmodified bulk of the polymer sample, masking any changes in MW in the surface layer. As more refined quantitative methods become available, a quantitative description of the relationships between surface roughness, cross-linking and aggregation will be possible.

5. CONCLUSIONS

Several conclusions about surface roughening of PS and PMMA during Ar/O_2 plasma etching arise from the data presented. First, it appears that intrinsic inhomogeneity of the polymer films is not responsible for roughness, as monomer molecular weight shows a negligible effect on roughness following etching. Data are also inconsistent with the predictions of buckling theory, in which roughness arises as wrinkle produced when a surface layer modified during plasma exposure buckles due to stress. The data are most consistent with a model based on surface cross-linking induced during plasma exposure. For PS, roughness decreases with increasing ion energy are tentatively attributed to the formation of a continuous cross-linked layer, while roughness increases with ion energy for PMMA are attributed to increases in cross-linking from negligible to moderate levels. Furthermore, the introduction of H and F atoms suppresses roughness, apparently by terminating dangling bonds.

ACKNOWLEDGEMENTS

The authors acknowledge the support of the National Science Foundation sponsored UW MRSEC (Materials Research Science and Engineering Center).

REFERENCES

1. Dalby, M.J. Topographically induced direct cell mechanotransduction. *Med. Eng. Phys.* **2005**, *27*, 730–742.
2. Dalby, M.J.; McCloy, D.; Robertson, M.; Agheli, H.; Sutherland, D.; Affrossman, S.; Oreffo, R.O.C. Osteoprogenitor response to semi-ordered and random nanotopographies. *Biomaterials* **2006**, *27*, 2980–2987.
3. Berry, C.C.; Dalby, M.J.; Oreffo, R.O.C.; McCloy, D.; Affrosman, S. The interaction of human bone marrow cells with nanotopographical features in three dimensional constructs. *J. Biomed. Mater. Res. A* **2006**, *79A*, 431–439.
4. Karuri, N.W.; Liliensiek, S.; Teixeira, A.I.; Abrams, G.; Campbell, S.; Nealey, P.F.; Murphy, C.J. Biological length scale topography enhances cell-substratum adhesion of human corneal epithelial cells. *J. Cell. Sci.* **2004**, *117*, 3153–3164.
5. Teixeira, A.I.; Abrams, G.A.; Murphy, C.J.; Nealey, P.F. Cell behavior on lithographically defined nanostructured substrates. *J. Vac. Sci. Technol. B* **2003**, *21*, 683–687.

6. Teixeira, A.I.; Nealey, P.F.; Murphy, C.J. Responses of human keratocytes to micro- and nanostructured substrates. *J. Biomed. Mater. Res. A* **2004**, *71A*, 369–376.

7. Black, C.T.; Guarini, K.W.; Milkove, K.R.; Baker, S.M.; Russell, T.P.; Tuominen, M.T. Integration of self-assembled diblock copolymers for semiconductor capacitor fabrication. *Appl. Phys. Lett.* **2001**,*79*, 409–411.

8. Harrison, C.; Park, M.; Chaikin, P.M.; Register, R.A.; Adamson, D.H. Lithography with a mask of block copolymer microstructures. *J. Vac. Sci. Technol. B* **1998**, *16*, 544–552.

9. Stoykovich, M.P.; Nealey, P.F. Block copolymers and conventional lithography. *Mater. Today* **2006**,*9*, 20–29.

10. Liu, C.C.; Nealey, P.F.; Ting, Y.H.; Wendt, A.E. Pattern transfer using poly(styrene-block-methyl methacrylate) copolymer films and reactive ion etching. *J. Vac. Sci. Technol. B* **2007**, *25*, 1963–1968.

11. Thurn-Albrecht, T.; Steiner, R.; DeRouchey, J.; Stafford, C.M.; Huang, E.; Bal, M.; Tuominen, M.; Hawker, C.J.; Russell, T. Nanoscopic templates from oriented block copolymer films. *Adv. Mater.***2000**, *12*, 787–791.

12. Ting, Y.H.; Park, S.M.; Liu, C.C.; Liu, X.S.; Himpsel, F.J.; Nealey, P.F.; Wendt, A.E. Plasma etch removal of poly(methyl methacrylate) in block copolymer lithography. *J. Vac. Sci. Technol. B* **2008**,*26*, 1684–1689.

13. Zekonyte, J.; Zaporojtchenko, V.; Faupel, F. Investigation of the drastic change in the sputter rate of polymers at low ion fluence. *Nucl. Instrum. Meth. Phys. Res. Sect. B* **2005**, *236*, 241–248.

14. Zaporojtchenko, V.; Zekonyte, J.; Faupel, F. Effects of ion beam treatment on atomic and macroscopic adhesion of copper to different polymer materials. *Nucl. Instrum. Meth. Phys. Res. Sect. B* **2007**, *265*, 139–145.

15. Vegh, J.J.; Nest, D.; Graves, D.B.; Bruce, R.; Engelmann, S.; Kwon, T.; Phaneuf, R.J.; Oehrlein, G.S.; Long, B.K.; Willson, C.G. Near-surface modification of polystyrene by Ar+: Molecular dynamics simulations and experimental validation. *Appl. Phys. Lett.* **2007**, *91*, 233113.

16. Friedrich, J.; Unger, W.; Lippitz, A.; Koprinarov, I.; Ghode, A.; Geng, S.; Kuhn, G. Plasma-based introduction of monosort

functional groups of different type and density onto polymer surfaces. Part 1: Behaviour of polymers exposed to oxygen plasma. *Compos. Interf.* **2003**, *10*, 139–171.

17. Chung, T.; Nest, D.; Graves, D.B.; Weilnboeck, F.; Bruce, R.L.; Oehrlein, G.S.; Wang, D.; Li, M.; Hudson, E.A. Electron, ion and vacuum ultraviolet photon effects in 193 nm photoresist surface roughening. *J. Phys. D Appl. Phys.* **2010**, *43*, 272001.

18. Nest, D.; Chung, T.; Vegh, J.J.; Graves, D.B.; Bruce, R.L.; Lin, T.; Phaneuf, R.J.; Oehrlein, G.S.; Long, B.K.; Willson, C.G. Role of polymer structure and ceiling temperature in polymer roughening and degradation during plasma processing: A beam system study of P4MS and P alpha MS. *J. Phys. D Appl. Phys.* **2010**, *43*, 085204.

19. Vegh, J.J.; Nest, D.; Graves, D.B.; Bruce, R.; Engelmann, S.; Kwon, T.; Phaneuf, R.J.; Oehrlein, G.S.; Long, B.K.; Willson, C.G. Molecular dynamics simulations of near-surface modification of polystyrene: Bombardment with Ar+ and Ar+. *J. Appl. Phys.* **2008**, *104*, 034308.

20. Tead, S.F.; Vanderlinde, W.E.; Marra, G.; Ruoff, A.L.; Kramer, E.J.; Egitto, F.D. Polymer diffusion as a probe of damage in ion or plasma-etching. *J. Appl. Phys.* **1990**, *68*, 2972–2982.

21. Sumiya, M.; Bruce, R.; Engelmann, S.; Weilnboeck, F.; Oehrlein, G.S. Study of 193 nm photoresist degradation during short time fluorocarbon plasma exposure. I. Studies of modified layer formation. *J. Vac. Sci. Technol. B* **2008**, *26*, 1637–1646.

22. Sumiya, M.; Bruce, R.; Engelmann, S.; Weilnboeck, F.; Oehrlein, G.S. Study of 193 nm photoresist degradation during short time fluorocarbon plasma exposures. II. Plasma parameter trends for photoresist degradation. *J. Vac. Sci. Technol. B* **2008**, *26*, 1647–1653.

23. Sumiya, M.; Bruce, R.; Engelmann, S.; Weilnboeck, F.; Oehrlein, G.S. Study of 193 nm photoresist degradation during short time fluorocarbon plasma exposure III. Effect of fluorocarbon film and initial surface condition on photoresist degradation. *J. Vac. Sci. Technol. B* **2008**, *26*, 1978–1986.

24. Yoshimura, T.; Shiraishi, H.; Yamamoto, J.; Terasawa, T.; Okazaki, S. Fabrication of polymer nanodots with single electron patterning technology. *Appl. Phys. Lett.* **1996**, *68*, 1799–1801.

25. Yamaguchi, T.; Namatsu, H.; Nagase, M.; Yamazaki, K.; Kurihara, K. A new approach to reducing line-edge roughness by using a

cross-linked positive-tone resist. *Jpn. J. Appl. Phys. Part 1 Reg. Pap. Short Notes Rev. Pap.* **1999**, *38*, 7114–7118.

26. Patsis, G.P.; Gogolides, E.; Van Werden, K. Effects of photoresist polymer molecular weight and acid-diffusion on line-edge roughness. *Jpn. J. Appl. Phys. 1* **2005**, *44*, 6341–6348.

27. Yamaguchi, T.; Yamazaki, K.; Namatsu, H. Influence of molecular weight of resist polymers on surface roughness and line-edge roughness. *J. Vac. Sci. Technol. B* **2004**, *22*, 2604–2610.

28. Bruce, R.L.; Weilnboeck, F.; Lin, T.; Phaneuf, R.J.; Oehrlein, G.S.; Long, B.K.; Willson, C.G.; Vegh, J.J.; Nest, D.; Graves, D.B. Relationship between nanoscale roughness and ion-damaged layer in argon plasma exposed polystyrene films. *J. Appl. Phys.* **2010**, *107*, 084310.

29. Chan, W.L.; Chason, E. Making waves: Kinetic processes controlling surface evolution during low energy ion sputtering. *J. Appl. Phys.* **2007**, *101*, 12.

30. Zhao, Y.P.; Drotar, J.T.; Wang, G.C.; Lu, T.M. Roughening in plasma etch fronts of Si(100). *Phys. Rev. Lett.* **1999**, *82*, 4882–4885.

31. Yamaguchi, T.; Namatsu, H.; Nagase, M.; Yamazaki, K.; Kurihara, K. Nanometer-scale linewidth fluctuations caused by polymer aggregates in resist films. *Appl. Phys. Lett.* **1997**, *71*, 2388–2390.

32. Yamazaki, K.; Yamaguchi, T.; Namatsu, H. Edge-enhancement writing for electron beam nanolithography. *Jpn. J. Appl. Phys. 1 Reg. Pap. Short Notes Rev. Pap.* **2003**, *42*, 3833–3837.

33. Ishikawa, A.; Tanji, T.; Kawai, A. Cohesion property of polymer aggregate depending on hardening treatment. *J. Photopolym. Sci. Technol.* **2004**, *17*, 99–102.

34. Drotar, J.T.; Zhao, Y.P.; Lu, T.M.; Wang, G.C. Surface roughening in shadowing growth and etching in 2+1 dimensions. *Phys. Rev. B* **2000**, *62*, 2118–2125.

35. Hunt, A.W.; Orme, C.; Williams, D.R.M.; Orr, B.G.; Sander, L.M. Instabilities in MBE growth. *Europhys. Lett.* **1994**, *27*, 611–616.

36. Kokkoris, G.; Constantoudis, V.; Angelikopoulos, P.; Boulousis, G.; Gogolides, E. Dual nanoscale roughness on plasma-etched Si surfaces: Role of etch inhibitors. *Phys. Rev. B* **2007**, *76*, 193405.

37. Drotar, J.T.; Zhao, Y.P.; Lu, T.M.; Wang, G.C. Mechanisms for plasma and reactive ion etch-front roughening. *Phys. Rev. B* **2000**, *61*, 3012–3021.

38. Petri, R.; Brault, P.; Vatel, O.; Henry, D.; Andre, E.; Dumas, P.; Salvan, F. Silicon roughness induced by plasma-etching. *J. Appl. Phys.* **1994**, *75*, 7498–7506.

39. Gogolides, E.; Boukouras, C.; Kokkoris, G.; Brani, O.; Tserepi, A.; Constantoudis, V. Si etching in high-density SF6 plasmas for microfabrication: Surfaceroughness formation. *Microelectron. Eng.***2004**, *73–74*, 312–318.

40. Wang, S.B.; Wendt, A.E. Control of ion energy distribution at substrates during plasma processing. *J. Appl. Phys.* **2000**, *88*, 643–646.

41. Ling, L.; Hua, X.; Li, X.; Oehrlein, G.S.; Hudson, E.A.; Lazzeri, P.; Anderle, M. Investigation of surface modifications of 193 and 248 nm photoresist materials during low-pressure plasma etching. *J. Vac. Sci. Technol. B* **2004**, *22*, 2594–2603.

42. Orf, B.J.; Walton, S.G.; Leonhardt, D.; Oehrlein, G.S. Study of photoresist etching and roughness formation in electron-beam generated plasmas. *J. Vac. Sci. Technol. B* **2007**, *25*, 779–784.

43. Pargon, E.; Nest, D.; Graves, D.B. Ar+ bombardment of 193 nm photoresist: Morphological effects.*J. Vac. Sci. Technol. B* **2007**, *25*, 1236–1243.

44. Engelmann, S.; Bruce, R.L.; Kwon, T.; Phaneuf, R.; Oehrlein, G.S.; Bae, Y.C.; Andes, C.; Graves, D.; Nest, D.; Hudson, E.A.; Lazzeri, P.; Lacob, E.; Anderle, M. Plasma-surface interactions of model polymers for advanced photoresists using C4F8. *J. Vac. Sci. Technol. B* **2007**, *25*, 1353–1364.

45. Hayes, T.R.; Heimann, P.A.; Donnelly, V.M.; Strege, K.E. Maskless laser interferometric monitoring of InP/InGaAsP heterostructure reactive ion etching. *Appl. Phys. Lett.* **1990**, *57*, 2817–2819.

46. Lieberman, M.A.; Lichtenberg, A.J. *Principles of Plasma Discharges and Processing*; John Wiley & Sons: New York, NY, USA, 1994; p. 20.

47. Wang, S.B.; Wendt, A.E. Ion bombardment energy and SiO2/Si fluorocarbon plasma etch selectivity. *J. Vac. Sci. Technol. A* **2001**, *19*, 2425–2432.

48. Silapunt, R.; Wendt, A.E.; Kirmse, K.; Losey, L.P. Ion bombardment energy control for selective fluorocarbon plasma etching of organosilicate glass. *J. Vac. Sci. Technol. B* **2004**, *22*, 826–831.

49. Metcalfe, A.M.; Stoll, S.; Burd, A. The effect of inhomogeneous stickiness on polymer aggregation.*J. Colloid Interface Sci.* **2006**, *298*, 629–638.

9

CHAPTER

The Mechanical Properties of Epoxy Composites Filled with Rubbery Copolymer Grafted SiO2

Jianing Gao [1,*], Junting Li [2], Brian C. Benicewicz [2], Su Zhao [3], Henrik Hillborg [3] and Linda S. Schadler [1]

[1]*Department of Materials Science & Engineering, Rensselaer Polytechnic Institute, 110 8th Street, Troy, NY 12180, USA*

[2]*Department of Chemistry and Biochemistry, University of South Carolina, Columbia, SC 29208, USA*

[3]*ABB Corporate Research, Forskargränd 7, 72226 Västerås, Sweden*

ABSTRACT

This study demonstrated a method for toughening a highly crosslinked anhydride cured DGEBA epoxy using rubbery block copolymer grafted SiO_2 nanoparticles. The particles were synthesized by a sequential reversible addition-fragmentation chain transfer (RAFT) polymerization. The inner rubbery block poly(n-hexyl methacrylate) (PHMA) had a glass transition temperature below room temperature. The outer block poly(glycidyl methacrylate) (PGMA) was matrix compatible. A rubbery interlayer thickness of 100% and 200% of the particle core radius was achieved by grafting a 20 kg/mol and a 40 kg/mol PHMA at a graft density of 0.7 chains/nm^2 from the SiO_2 surface. The 20 kg/mol rubbery interlayer transferred load more efficiently to the SiO_2cores than the 40 kg/mol rubbery interlayer and maintained the epoxy modulus up to a loading of 10 vol% of the rubbery interlayer. Both systems enabled cavitation or plastic dilatation. Improvement of the strain-to-break and the tensile toughness was found in both systems. We hypothesize that plastic void growth in the matrix is the primary mechanism causing the improvement of the ductility.

Keywords: toughened epoxy; rubbery interlayer; nanocomposites; RAFT polymerization

1. INTRODUCTION

Epoxies are widely used in industry. However, many epoxies are brittle and have poor fracture toughness, weak impact strength and low resistance to fatigue crack propagation [1]. Epoxies can be toughened by adding rubbery fillers, such as liquid rubbers [2,3,4] and preformed core/shell rubber particles [5,6,7,8]. Rubber cavitation happens under local hydrostatic stress. The cavitation releases the local hydrostatic stresses and causes plastic void growth and shear banding in the matrix. However, the addition of rubbers decreases the modulus, strength and glass transition temperature of epoxies [9]. Rigid particles such as ceramic nanoparticles [10,11], clay [12], carbon nanotubes [13] and graphene [14], have also been found to toughen epoxies by crack deflection, crack pinning, particle debonding, and plastic void growth, *etc.*, while increasing the modulus and strength and maintaining the glass transition temperature. However, they do not improve the matrix ductility and large volumetric loading of rigid fillers is required to improve the material toughness.

In order to combine the advantages of both rubbery and rigid fillers, Kinloch *et al.* [15] incorporated both SiO_2 nanoparticles and CTBN liquid rubber in an epoxy and found that the fracture energy increased synergistically with the use of both fillers. The modulus of the liquid rubber/epoxy composites was improved by adding SiO_2 nanoparticles up to 15.4 wt%. However, the modulus was still lower than the unmodified epoxy. The same modulus drop in the hybrid toughened epoxy was also observed by Liang and Pearson [16]. Lee and Yee [17] encapsulated micron sized glass beads with rubber at a ratio of rubber coating thickness to glass beads radius of less than 0.06. The tensile modulus of all coated glass bead filled epoxy composites was higher than the unmodified epoxy. However, the aggregation of particles during the encapsulating process, the poor bond between the rubber coating and the epoxy matrix, and the "thin" rubbery coating led to only a moderate toughening efficiency. This suggests that a strong matrix-rubber bond and good dispersion of the fillers are important. Recently, Zhao *et al.* [11] treated 50 nm Al_2O_3 nanoparticles with amino terminated tri-functional silane and found that 10 wt% of the silane treated Al_2O_3 significantly increased the ductility of the epoxy matrix with moderate improvement of the Young's modulus. The paper suggests that the tri-functional silane generated a soft silane network around the nanoparticle cores, which played an important role in improving the matrix ductility and tensile toughness due to plastic void growth around the particles. However, no cavitation was found in the soft silane region and there was no increase in high strain rate fracture toughness. This suggests that the silane soft layer is beneficial in improving tensile toughness and ductility, but a thicker interlayer is needed to generate cavitation.

In this paper, we will explore the use of diblock copolymer modified SiO_2 nanoparticles in an epoxy. The grafted copolymer consists of a thick rubbery inner block (poly(n-hexylmethacrylate), PHMA) and a matrix compatible outer block (poly(glycidylmethacrylate), PGMA) with well-controlled molecular weight and graft density. This design aims at taking an advantage of a well bonded soft layer around the particle that can transfer tensile load to the silica core, maintaining the modulus and strength of the composites. Above a certain load, the rubbery layer will begin to fail, cavitating and toughening the epoxy. The thickness of the rubbery interlayer needs to be controlled so it can be thin enough to effectively transfer load and thick enough to cause cavitation. The PGMA outer block was chosen to be matrix compatible, allowing the nanoparticles to be well dispersed. A generalized scheme of the rubbery copolymer grafted SiO_2 nanoparticles is shown in Figure 1(a).

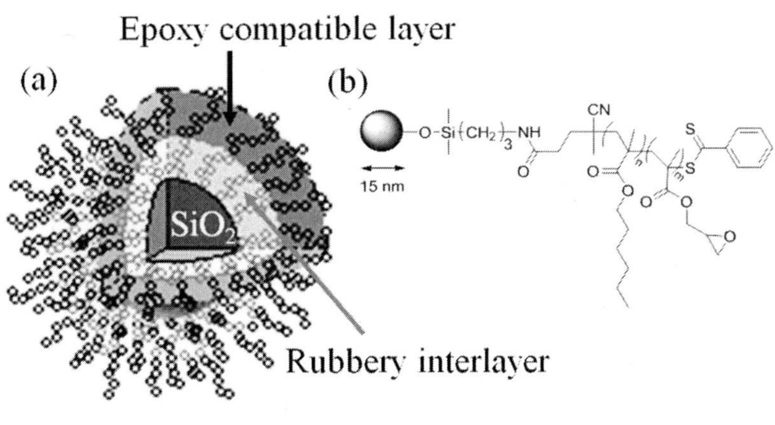

Figure 1. A schematic of (**a**) a polymer grafted nanoparticle; and (**b**) the chemistry of the rubbery copolymer grafted nanoparticle.

2. EXPERIMENTAL SECTION

2.1. Materials

Huntsman Araldite® epoxy was used as the thermosetting matrix polymer. The system includes (i) Araldite F- bisphenol A liquid epoxy resin; (ii) HY905-modified dicarboxylic anhydride hardener; and (iii) DY062-amine catalyst. The SiO_2 nanoparticles were ORGANOSILICASOL™ purchased from Nissan Chemical Co. Aminopropyldimethylethoxysilane (95%) was purchased from Gelest and used as received. Tetrahydrofuran (THF) (99.9%, Acros) was dried over CaH_2 overnight and distilled before use. 4-Cyanopentanoic acid dithiobenzoate (CPDB) was purchased from Strem Chemical Inc. and used as received. Both HMA and GMA were passed through a neutral alumina column to remove inhibitors before the polymerization. Unless otherwise specified, all chemicals were purchased from Fisher Scientific and used as received.

2.2. Composites Preparation and Characterization

The process of modifying the SiO_2 nanoparticles with chain transfer agents has been reported in the previous literature [17] and is briefly discussed here. SiO_2 nanoparticles (30 wt% in methyl isobutyl ketone) were diluted in dried THF, grafted with amino-functionalized silane coupling agents, and reacted with the chain transfer agents (activated CPDB) sequentially. UV-vis absorption spectra were acquired for graft density determination, and the nanoparticles were then dried and weighed.

The CTA anchored nanoparticles, monomer (HMA or GMA), AIBN and THF were added to a dried Schlenk tube. The mixture was degassed by three freeze-pump-thaw cycles, backfilled with nitrogen, and then placed in an oil bath at 60 °C for various intervals. The polymerization was quenched in ice water. A sample of grafted chains was cleaved from the surface using hydrofluoric acid (HF). Typically, about 40 mg of nanoparticles was mixed with 4 mL of THF and 0.5 mL of HF, and the solution was allowed to stir at room temperature overnight, and then dried for GPC tests. If the polymers contained GMA residue, around 40 mg of nanoparticles were dissolved in 4 mL of methylene chloride, and then 4 mL of water, 0.5 mL of HF and one drop of Aliquot® 336 were added. After stirring overnight, the organic layer was taken out and evaporated for GPC tests. Molecular weights and molecular weight distributions (PDI) were determined by gel permeation chromatography conducted on a Varian 390-LC system equipped with refractive index, viscometry and light scattering detectors, and 3× PLgel 10 μm mixed-B LS columns (300 × 7.5 mm). THF was used as eluent and measurements were made at 30 °C and a flow rate of 1.0 mL/min. The GPC system was calibrated with poly (methyl methacrylate) standards obtained from Polymer Laboratories. The remainder of the particles were precipitated in methanol to remove unreacted monomers, and then either re-dispersed in THF to polymerize another block, or re-dispersed in methylene chloride.

The methylene chloride solution was then mixed with epoxy resin. The solvent in the mixture was evaporated until the mass of the mixture remained unchanged, indicating that all the solvent was gone. The hardener and the catalyst were then added to the particle-resin mixture and thoroughly mixed using a Hauschild Dual Asymmetric Centrifugal mixer. The mixture was cured at 80 °C for 10 hours and 135 °C for 10 h in a silicone mold. Nanocomposites with 0.1 vol%, 0.3 vol%, 0.6 vol% and 1 vol% SiO_2cores were prepared. The dispersion was characterized using a DI-NanoScope III atomic force microscope (AFM) from Veeco. The curing process of the epoxy matrix was studied by an isoconversional method on a Modulated Differential Scanning Calorimeter (MDSC) model Q100 from TA Instruments. About 10 mg uncured sample wax cured in closed aluminum pans in the MDSC at heating rates of 2.5, 5, 7.5, 10 and 12.5 °C/min. Activation energies were calculated according to the Kissinger equation. The glass transition temperatures of the nanocomposites were measured by DSC at a heating rate of 10 °C/min and dynamic mechanical analysis on a DMTA V from Rheometric Scientific from −150 to 200 °C at 2 °C/min.

2.3. Tensile Tests

Tensile tests were carried out using an Instron 4201 at room temperature and constant cross-head speed of 0.1 mm/min. Dog bone shaped specimens were made. Young's modulus, ultimate tensile stress and strain-to-break of epoxy and epoxy nanocomposites were measured. At least 5 composite samples were tested for each loading according to ASTM standard D638-08.

2.4. Fracture Analysis

The fracture surface of the neat epoxy and nanocomposites was analyzed by a JEOL 6330F field emission scanning electron microscope (SEM). The sub surface failure was characterized by a JOEL CM12 transmitted electron microscope (TEM) on the vertical plane beneath the fracture surface.

3. RESULTS AND DISCUSSION

3.1. Synthesis

Block-copolymer grafted SiO_2 nanoparticles were synthesized by sequential reversible addition-fragmentation chain transfer (RAFT) polymerization from the surface of the nanoparticles [18]. RAFT provides quantitative control over the molecular weight and polydispersity (PDI) of each block. The graft density can be adjusted from approximately 0.01 chains/nm^2 to 0.7 chains/nm^2 [19]. To ensure that the epoxy compatible layer is compact enough to reduce the epoxy monomer penetration and shield the rubbery interlayer, nanoparticles with high graft density of 0.7 chains/nm^2 were prepared.

The polymerization of each monomer on SiO_2 nanoparticles was studied separately prior to block copolymer (PHMA-*b*-PGMA) synthesis. The results of the RAFT polymerization of HMA (hexylmethacrylate) and GMA (glycidylmethacrylate) are shown in Figure 2. This figure shows that the PHMA molecular weight increased almost linearly with the reaction time. However, the PDI of the PHMA did not vary significantly during the reaction process. The polymerization of GMA on the SiO_2nanoparticles exhibited a similar linear relationship between molecular weight and reaction time with slightly higher PDI.

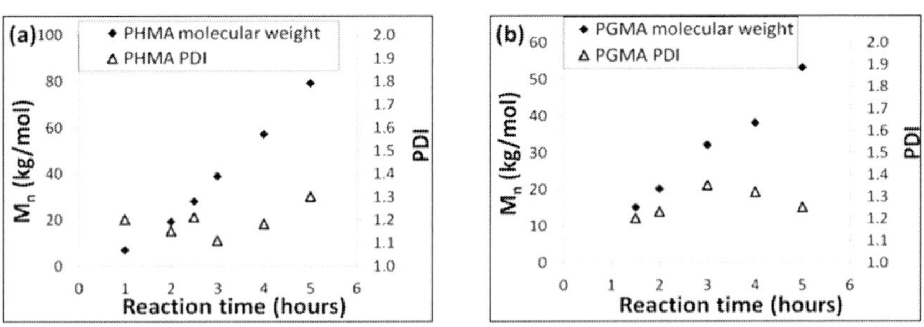

Figure 2. Plots showing the dependence of molecular weight and PDI on reaction time for the RAFT polymerization of HMA (**a**) and GMA (**b**) (5 M in THF) at 60 °C with AIBN as initiator (1.5×10^{-5} M) mediated with RAFT agent anchored silica nanoparticles (1.5×10^{-4}M; 0.61 chains/nm^2). The gel permeation chromatography (GPC) system was calibrated with poly(methyl methacrylate) standards.

The rubbery block was synthesized at two molecular weights, 20 kg/mol and 40 kg/mol, to study the effect of the rubbery interlayer thickness on the mechanical properties of the copolymer grafted SiO_2nanocomposites. To ensure good dispersion of the rubbery block grafted SiO_2 nanoparticles, a 20 kg/mol PGMA block was grown from the 20 kg/mol PHMA block. Similarly, a 40 kg/mol PGMA block was grown from the 40 kg/mol PHMA block. The nanocomposites are therefore denoted as 20k20k for the 20 kg/mol PHMA and 20 kg/mol PGMA grafted SiO_2 filled epoxy and 40k40k for the 40 kg/mol PHMA and 40 kg/mol PGMA grafted SiO_2 filled epoxy.

Good dispersion was achieved in both the 20k20k and the 40k40k systems as shown in the AFM images in Figure 3. To show that the PHMA is acting as a rubbery interlayer, AFM images were taken comparing the PHMA-*b*-PGMA systems to silica nanoparticles with only PGMA attached. Nanocomposites with a loading of 0.3 vol% SiO_2 cores were used for AFM observation to ensure that the grafted particles were isolated. A sharp phase change between the SiO_2 cores and the epoxy matrix was observed in the solely PGMA grafted SiO_2 epoxy nanocomposites (see Figure 3(a,b)). When the PHMA is present, a softer interphase can be seen in both the 20k20k and the 40k40k systems (see Figure 3(c–f)). This softer interphase appears darker than the epoxy matrix in the phase images (see Figure 3(d,f)).

The ratio of the rubbery interlayer thickness to the SiO_2 radius of the 20k20k and the 40k40k systems was analyzed based on at least 10 particles from each AFM image using an image processing program ImageJ [20]. The measured values were compared to the ratio calculated using the rubbery block molecular weight, graft density and bulk PHMA density of 1.0007 g/cm^3. The comparison is summarized in Figure 4.

Both the calculated and the measured results showed that the rubbery interlayer thickness increases as expected with increasing molecular weight. The AFM images were taken using a tip with a radius of 7 nm. Because the tip radius was on the order of the radius of the particle, the apparent particle size in the AFM image is greater than the real particle size. This contributes to the discrepancy between the calculation and the measured interlayer thickness to particle radius ratio. Due to this limitation of the AFM imaging, the calculated rubbery interlayer to particle core ratio was used when calculating the volume percent of the rubbery interlayer in the nanocomposites. The volumetric loading of the SiO_2cores in the current paper, together with the volumetric ratio of the PHMA rubbery interlayer to the SiO_2 cores and the volumetric fraction of the SiO_2 cores in the polymer-SiO_2 nanoparticles are summarized in Table 1.

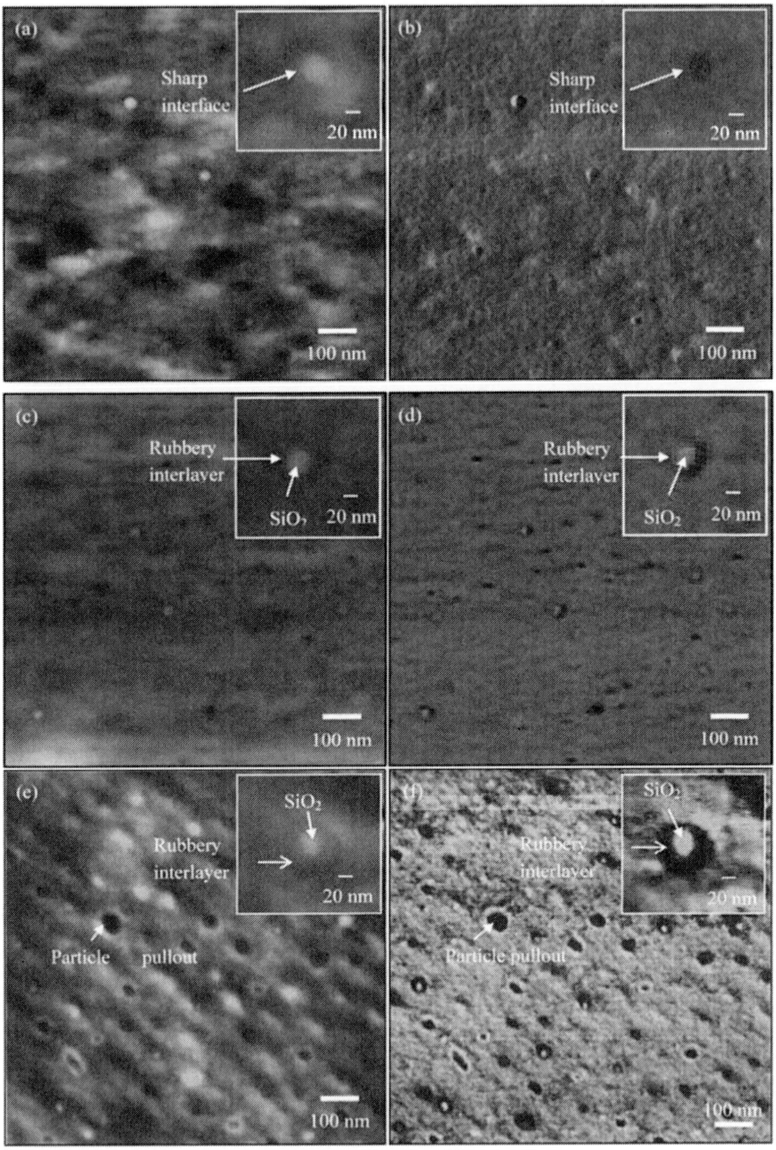

Figure 3. (**a**) AFM height image of 0.3 vol% PGMA-SiO_2/epoxy; (**b**) AFM phase image of 0.3 vol% PGMA-SiO_2/epoxy; (**c**) AFM height image of 0.3 vol% 20k20k-SiO_2/epoxy; (**d**) AFM phase image of 0.3 vol% 20k20k-SiO_2/epoxy; (**e**) AFM height image of 0.3 vol% 40k40k-SiO_2/epoxy; (**f**) AFM phase image of 0.3 vol%40k40k-SiO_2/epoxy.

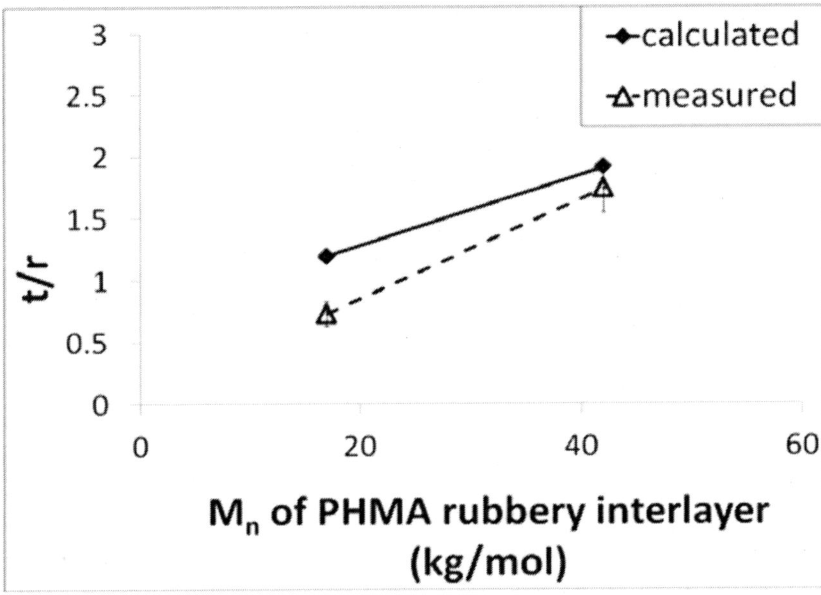

Figure 4. A comparison between the measured and calculated ratio of the rubbery interlayer thickness (t) and the radius (r) of the particle cores for particles with a 20 kg/mol and 40 kg/mol PHMA rubbery interlayer.

Table 1. Volumetric loading of the copolymer grafted SiO_2/epoxy nanocomposites prepared in the paper. The volume ratio of the PHMA rubbery interlayer to that of the SiO_2, and the volumetric fraction of the SiO_2 core in the polymer grafted SiO_2 nanoparticles for both 20k20k and 40k40k are also calculated and summarized.

Sample	t_{PHMA}/r_{SiO_2}	V_{PHMA}/V_{SiO_2}	$V_{SiO_2}/V_{polymer-SiO_2}$	Vol% of SiO_2 Cores
20k20k	1	13.8	3.50%	0.1%
				0.3%
				0.6%
				1.0%
40k40k	2	34.2	1.44%	0.1%
				0.2%
				0.3%

3.2. Curing and Glass Transition Temperature

The PGMA outer block of the copolymer grafted SiO_2 nanoparticles introduces additional epoxy groups to the composite, which can influence the curing process of the matrix by decreasing the stoichiometric anhydride hardener to epoxy ratio. This can impact the glass transition temperature and the mechanical properties of the nanocomposites. The curing can be further complicated when one considers that the PGMA layer might scavenge epoxy and anhydride monomers (because the PGMA layer is epoxy compatible), probably at different efficiency. The difference in the curing process between the nanocomposites and the neat epoxy alone can impact the glass transition temperature and mechanical properties. However, we can still study the influence of the rubbery interlayer on the glass transition temperature and the mechanical properties of the nanocomposites by separating the effect of the interlayer from the influence of the anhydride/epoxy ratio. The impact of the anhydride to epoxy ratio on the glass transition temperature and the mechanical properties of the neat epoxy was studied. The curing of neat epoxy with an anhydride to epoxy ratio from 0.8 to 1.2 was analyzed using modulated differential scanning calorimetry. The influence of the ratio on the curing mechanism (evaluated by curing activation energy) and the degree of curing completion (indicated by glass transition temperature) was quantified. The activation energy for curing was calculated from the Kissinger equation [21,22,23],

$$\frac{d\left(\ln\frac{q}{T_{q,p}^2}\right)}{d\left(\frac{1}{T_{q,p}}\right)} = -\frac{E}{R} \tag{1}$$

where $T_{q,p}$ is the temperature at the heat flow peak generated by a heating rate of q. E is the activation energy of the curing reaction and R is the gas constant. The activation energy was measured from the slope of $1/T_{q,p}^2$ *vs.* $q/T_{q,p}^2$. Table 2 shows the average activation energy to cure the neat epoxy with an anhydride to epoxy ratio (denoted as R in Table 2) of 0.8, 1 and 1.2. The activation energy of the three systems was similar (within 0.2%). However, at a ratio of 1.2 there is a 7% decrease in the T_g from the 1:1 ratio. Therefore, changing the anhydride to epoxy ratio by 20% from the stoichiometric ratio does not change the curing mechanism, but does change the extent of curing completion.

Table 2. The glass transition temperature of the epoxy systems.

Material	vol% of SiO_2 Cores	R *	Activation Energy (J/g)	T_g (°C)
Neat epoxy		0.8	74.3	110
		1	75.8	112
		1.2	76.7	105
20k20k				
	0.1%	0.99		109
	0.3%	0.98		110
	0.6%	0.96		105
	1%	0.93		100
40k40k				
	0.1%	0.98		110
	0.2%	0.97		110
	0.3%	0.95		109

* R is the number ratio of anhydride reacting groups and epoxy groups from the epoxy resin and the PGMA epoxy compatible layer.

The impact of the anhydride to epoxy ratio on the mechanical properties of unfilled epoxy was also determined. Figure 5 shows the mechanical properties (Young's modulus and strain-to-break) of the neat epoxy with anhydride to epoxy ratio from 0.4 to 1.2. It is observed that the modulus and strain-to-break of the neat epoxy remain unchanged in the window of 0.8 to 1, suggesting that the mechanical properties of the epoxy matrix are insensitive to the degree of curing completion when the anhydride to epoxy ratio is between 0.8 and 1 in the current study. Similar results have been found for other anhydride cured epoxies [24].

Because the glass transition temperature and the mechanical properties of the neat epoxy are not significantly altered for the anhydride to epoxy ratios from 0.8 to 1; the addition of the PGMA layer only alters the potential anhydride/epoxy ratio from 0.9 to 1 (also calculated in Table 2); and the amount of epoxy and anhydride monomer that penetrates the brush is unknown, the anhydride to epoxy ratio was not altered for the nanocomposites. Based on the study above, the impact of the grafted polymer on the glass transition temperature could be extracted from the difference between the relationship of the anhydride to epoxy ratio and the glass transition temperature of the nanocomposites (Figure 6(a)). It is interesting to note that the glass transition temperature of the composites is always lower than the neat epoxy, which is due to the existence of the grafted polymer. As shown in Figure 6(b), the 20k20k system exhibited a larger decrease in T_g at the same volumetric loading of the PHMA rubbery layer as the 40k40k composites. This suggests that the drop in T_g of the 20k20k composites is not due to the rubbery volume percent alone. The reason for this interesting phenomenon is unclear.

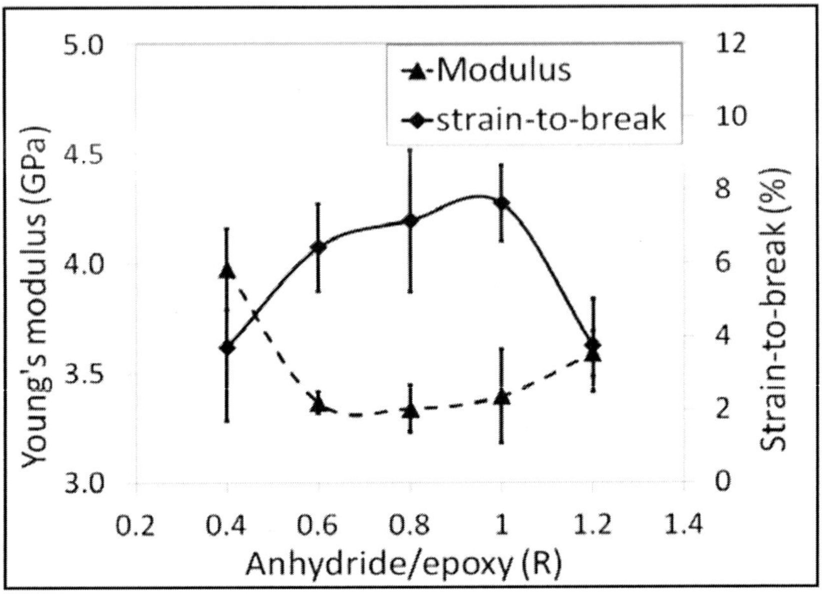

Figure 5. Modulus and strain-to-break of the neat epoxy with different anhydride/resin ratios.

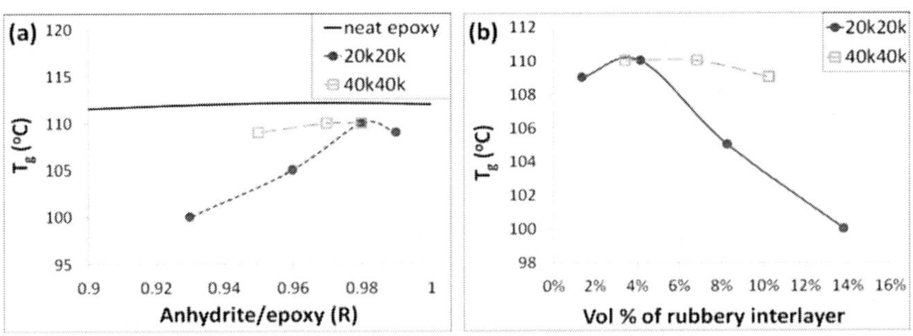

Figure 6. Plots of glass transition temperature *vs.* (**a**) anhydride/epoxy ratio R of different epoxy systems; (**b**) the rubbery interlayer volume percent in the 20k20k and the 40k40k systems.

3.3. Mechanical Properties of HMA-b-GMA-SiO_2/Epoxy Nanocomposites

To understand whether the rubbery interlayer thickness in the current study will alter the mechanical properties of the rubbery interlayer, temperature sweep dynamic mechanical analysis was performed on the neat epoxy, 0.3 vol% and 1 vol% 20k20k-SiO_2 and 0.3 vol% 40k40k-SiO_2/epoxy nanocomposites. The temperature sweep curves for all four systems are shown in Figure 7. There are two peaks in the loss modulus curve: the glass transition and β-transition for the unmodified epoxy. An additional peak was observed in the loss modulus curves of the 1 vol% 20k20k-SiO_2/epoxy around 0 °C. The β-transition broadens at around 0 °C for the 0.3 vol% 40k40k-SiO_2/epoxy. Neither a peak nor a broadening at 0 °C is obvious in the 0.3 vol% 20k20k-SiO_2. The peak around 0 °C is due to the existence of the rubbery interlayer, and matches the glass transition temperature of the PHMA, which has been reported to be around 0 °C [25,26]. A plot of the loss modulus increase at 0 °C *vs.* the volume percent of the rubbery interlayer is shown in Figure 8, showing that the increase in the loss modulus is almost linearly proportional to the volume percent of the rubbery interlayer. This data suggests that the dynamic mechanical properties of the PHMA interlayer are insensitive to the molecular weights that were studied in the paper.

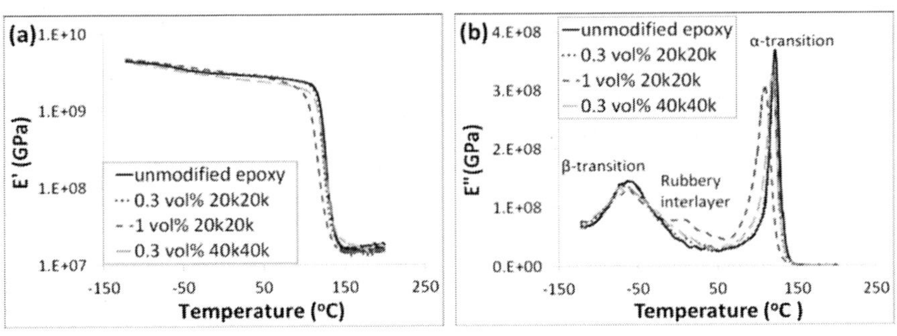

Figure 7. A temperature sweep of the neat epoxy and PHMA-b-PGMA grafted SiO_2 epoxy nanocomposites. (**a**) Storage modulus *vs.*temperature; (**b**) Loss modulus *vs.* temperature.

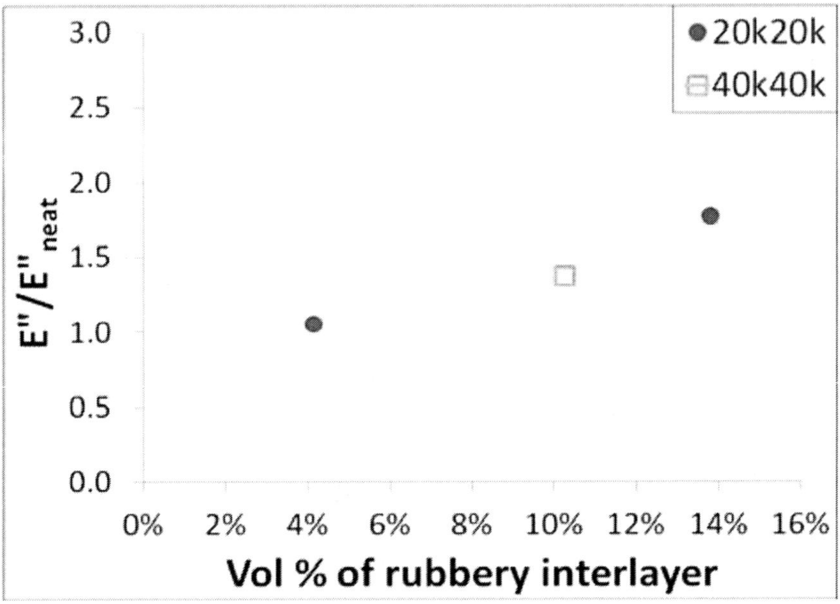

Figure 8. The rubbery interlayer volume fraction *vs.* the ratio of loss modulus of the nanocomposites and that of the neat epoxy at 0 °C.

Important information about the matrix molecular weight between crosslinks M_c and toughenability can be drawn from the DMA data. It is believed that lightly crosslinked epoxies ($M_c > 500$ g/mol) can be toughened more efficiently by rubbers than highly crosslinked epoxies ($M_c < 500$ g/mol). The toughening effect levels off as the molecular weight between crosslinks increases ($M_c > 3000$ g/mol) [27]. The M_c of the epoxy matrix can be calculated from the following equation based on the theory of rubber elasticity [28]:

$$M_c = \frac{\rho RT}{E_r} \tag{2}$$

where ρ is the density of an epoxy at temperature T, R is the gas constant and E_r is the storage modulus of the epoxy at 50 °C above its glass transition temperature. A comparison of the toughenability of the studied DGEBA/anhydride epoxy to that of a highly crosslinked and a lightly crosslinked epoxy from the literature is shown in Table 3. The fracture energy of the lightly crosslinked epoxy matrix with $M_c = 679$ g/mol improved by a factor of 3 with the addition of 5 wt% rubber, while at the same filler concentration the fracture energy increased by a factor of about 2.5 for the highly crosslinked epoxy matrix with M_c = 260 g/mol. The studied epoxy with M_c = 312 g/mol is a highly crosslinked epoxy with small toughenability.

Table 3. A comparison of the toughenability of the epoxy used in this study to some literature systems.

System	M_c (g/mol)	G_{Ic} (J/m^2)	G_{Ic} at 5 wt% Rubber (J/m^2)
DGEBA/anhydride (the system used in the current work)	312	281	
DGEBA+DDS [27]	260	150	400
DGEBA+DDS [27]	679	4500	14000

The tensile properties, including Young's modulus, tensile strength and strain-to-break of the neat epoxy, 20k20k-SiO_2, 40k40k-SiO_2 and untreated SiO_2 filled epoxy nanocomposites with different loadings are summarized in Table 4. The strain-to-break of the rubbery grafted SiO_2/epoxy nanocomposites increased with increasing loading. A *t*-test shows that the strain-to-break of the rubbery grafted SiO_2/epoxy nanocomposites is significantly larger than the neat epoxy at a confidence level of 95% (except at 0.1% SiO_2 core), while the strain-to-break of the bare SiO_2/epoxy nanocomposites showed no statistical difference from the neat epoxy. This suggests that the increase in the copolymer grafted SiO_2/epoxy nanocomposites was not due to the existence of the SiO_2 cores but the existence of the rubbery layer. Representative stress strain curves of all three composites studied are shown in Figure 9.Figure 10 shows the relationship between the rubbery interlayer volume percent and the mechanical properties of the nanocomposites. Both the 40k40k and 20k20k systems improve the strain-to-break and the tensile toughness with a similar trend (Figure 10(a)), with the 40k40k showing a better improvement below 8 vol% of the rubbery interlayer. However, an obvious decrease in the modulus and tensile strength in the 40k40k systems occurs at a lower rubbery interlayer volume fraction than in the 20k20k system (Figure 10(b)). Therefore, the 20k20k system can achieve similar improvement in ductility and tensile toughness of the matrix but maintain better modulus and tensile strength than the 40k40k system. It is thus possible that an improvement in the strain-to-break, while maintaining modulus and tensile strength, can be achieved by reducing the interlayer thickness and increasing the particle loading.

Table 4. The tensile properties of the nanocomposites and the neat epoxy. Error estimates are the measured standard deviations.

Sample	$f_{\text{silica cores}}$	ε_b (%)	(*t*-test)	UTS (MPa)	E (GPa)	Area under σ-ε Curve W (MPa or 10^{-6} J/m^3)
Neat		7.36 ± 0.41		83.73 ± 1.15	3.30 ± 0.07	5.3 ± 1.6
Bare- SiO_2	0.3	6.99 ± 0.34	-	83.54 ± 0.34	3.40 ± 0.05	5.3 ± 0.4
	1.6	7.19 ± 0.96	9%	83.46 ± 1.10	3.43 ± 0.10	5.7 ± 1.0
	3	7.24 ± 1.74	19%	84.24 ± 0.94	3.67 ± 0.09	6.0 ± 0.9
20k20k-SiO_2	0.1	7.18 ± 0.35	74%	82.35 ± 1.15	3.17 ± 0.09	5.6 ± 1.2
	0.3	9.49 ± 1.02	98%	79.60 ± 1.37	3.18 ± 0.08	6.1 ± 1.4
	0.6	9.07 ± 0.30	93%	75.72 ± 1.00	3.26 ± 0.13	5.8 ± 1.0
	1	11.52 ± 0.45	99%	66.67 ± 1.33	2.81 ± 0.05	6.4 ± 1.3
40k40k-SiO_2	0.1	7.68 ± 0.54	93%	78.62 ± 0.70	3.32 ± 0.08	5.6 ± 0.8
	0.2	12.50 ± 0.76	98%	77.93 ± 0.22	2.94 ± 0.05	6.7 ± 0.3
	0.3	6.91 ± 1.59	97%	76.63 ± 2.05	3.02 ± 0.19	5.4 ± 1.9

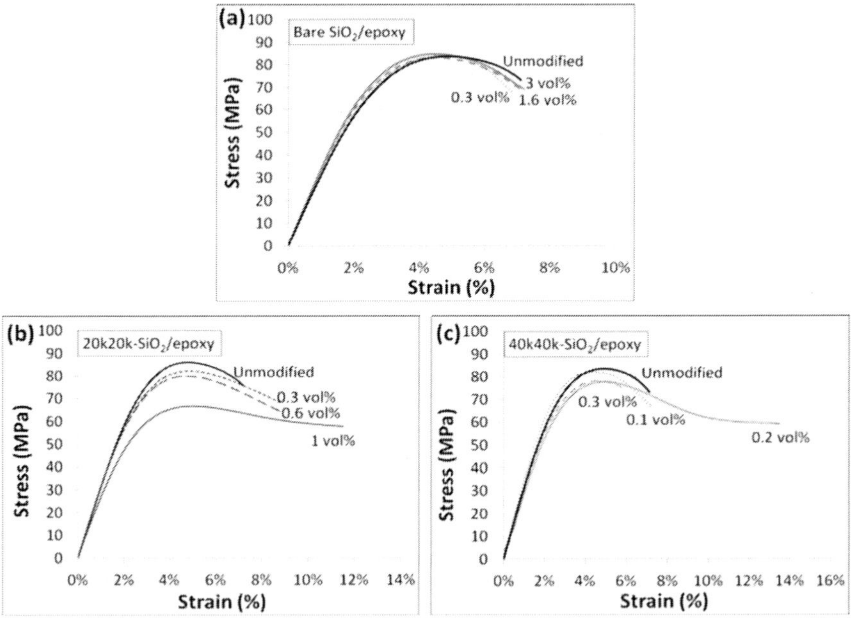

Figure 9. Representative stress strain curves for (**a**) bare SiO_2/epoxy nanocomposites; (**b**) 20k20k-SiO_2/epoxy nanocomposites; and (**c**) 40k40k-SiO_2/epoxy nanocomposites at different volumetric loadings of SiO_2 cores.

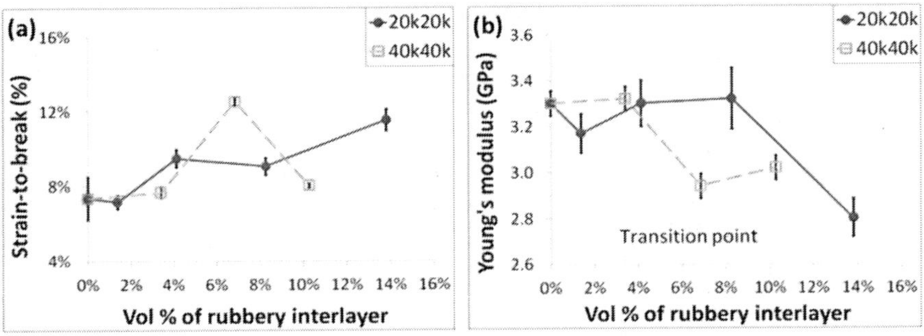

Figure 10. A plot of (**a**) strain-to-break; and (**b**) modulus *vs.* vol% of PHMA rubbery interlayer for the 20k20k and 40k40k systems.

A comparison of the normalized modulus and strain-to-break of the new rubbery copolymer grafted SiO_2/epoxy nanocomposites to some conventional liquid rubber filled epoxy composites at similar testing conditions and with a similar matrix is shown in Figure 11. The grafted SiO_2/epoxy nanocomposites exhibit a slightly larger extent of strain-to-break increase at low loadings of the rubbery phase than the conventional rubbers (see Figure 11(a)), but maintain the Young's modulus better than some conventional rubber fillers, especially for the 20k20k system.

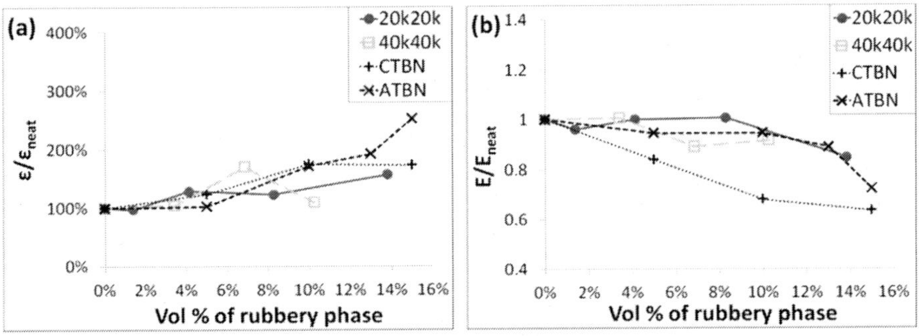

Figure 11. (**a**) Comparison of the improvement in the strain-to-break*vs.* the volumetric loading of the rubbery phase between the rubbery copolymer grafted SiO_2/epoxy nanocomposites and conventional rubber (CTBN and ATBN) filled epoxy composites; (**b**) Comparison of the Young's modulus *vs.* the volumetric loading of the rubbery phase between the rubbery copolymer grafted SiO_2/epoxy nanocomposites and conventional rubber (CTBN and ATBN) filled epoxy composites [29].

3.4. Fracture Mechanisms

A brittle fracture surface usually consists of a mirror zone, mist zone and hackle zone [11]. The mirror zone is the slow crack propagation region next to the crack initiation site. It is usually smooth because the crack is propagating within a plane. As the crack starts to grow faster, a narrow mist zone is developed, where the crack is still propagating in the original crack growth plane. Right after the mist zone, cracks began to bifurcate into 3D planes, leading to a rough hackle zone [30]. The fracture surfaces of the neat epoxy, 20k20k-SiO_2/epoxy at 0.3 and 1 vol% SiO_2 cores and 40k40k-SiO_2/epoxy at 0.1 and 0.2 vol% SiO_2 cores after tensile failure are shown in Figure 12–Figure 14. The three zones are observed in all the materials. The mirror zones are marked in white circles on the overview SEM images in Figure 12.

Cavitation or plastic void growth of the rubber layer can be seen in the magnified mirror zones of the nanocomposites (see Figure 13(b–e)), which promotes plastic deformation in the matrix, hinders crack growth, and generates a rougher mirror zone surface. This leads to the improvement in ductility and toughness in the rubbery copolymer grafted SiO_2 nanoparticle composites as shown in Table 4 andFigure 9 and Figure 10. Note that the 20 kg/mol PHMA and 40 kg/mol PHMA grafted rubbery interlayer exhibit cavitation or void growth (Figure 13(b–i)), which might result in similar extent of the improvement in ductility of the epoxy matrix at the same volume percent of the rubbery interlayer.

The features in the hackle zone for the neat epoxy and low concentration copolymer grafted SiO_2/epoxy nanocomposites (marked as region "C" in Figure 12 and shown in Figure 14) have random 3D features. This is because crack branching produces severe surface roughening; however, it doesn't effectively retard crack growth [31]. Unlike the neat epoxy and the nanocomposites at low loadings, the 1 vol% 20k20k and 0.2 vol% 40k40k show features of parabolic patterned markings (see Figure 14(c,e)). The formation of the parabolic markings is believed to be a result of interaction between the crack front and the cavitation, which has also been observed in this region (see Figure 14(f,g)) [32]. This interaction might also contribute to the improvement of the ductility and tensile toughness of the epoxy nanocomposites at those loadings, as they achieved the most improvement compared to other loadings, which do not show the parabolic markings in the hackle zone (see Table 3).

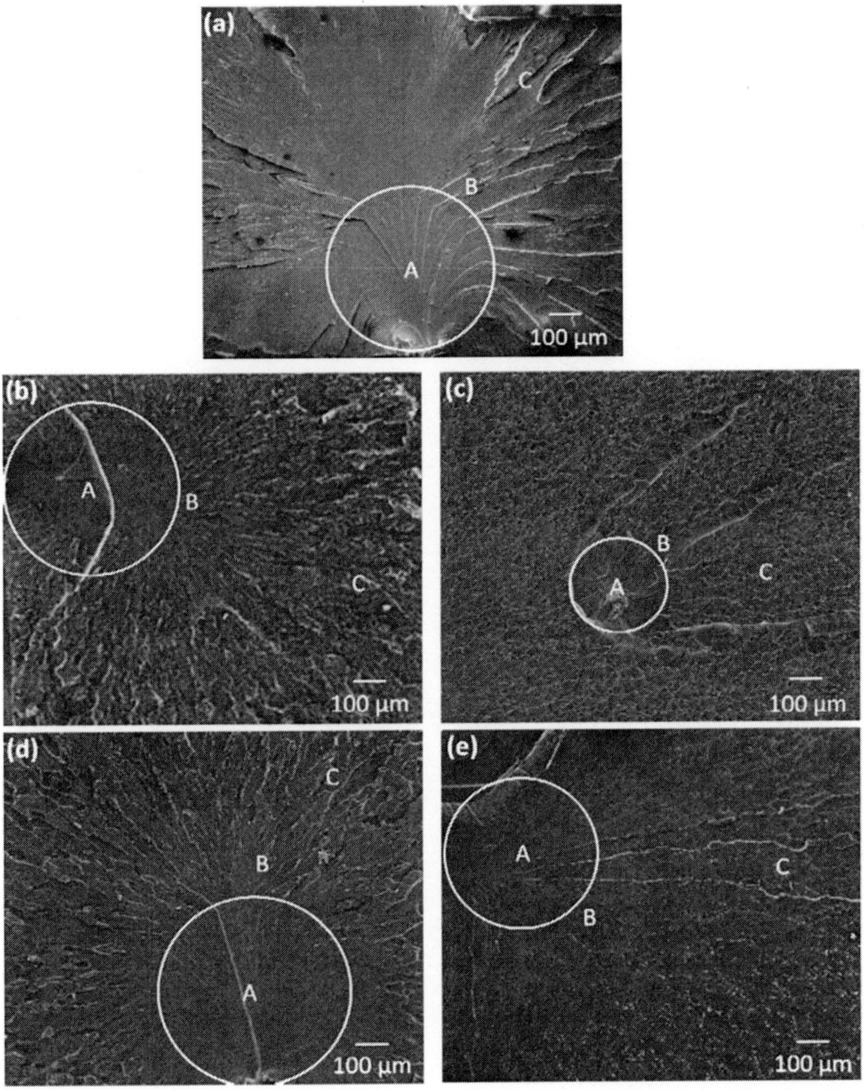

Figure 12. Scanning Electron Micrographs showing the tensile fracture surfaces of (**a**) the neat epoxy; (**b**) 0.3 vol% 20k20k-SiO_2nanocomposites; (**c**) 1 vol% 20k20k-SiO_2 nanocomposites; (**d**) 0.1 vol% 40k40k-SiO_2 nanocomposites; and (**e**) 0.2 vol% of 40k40k-SiO_2nanocomposites. "A": mirror zone, "B": mist zone and "C": hackle zone.

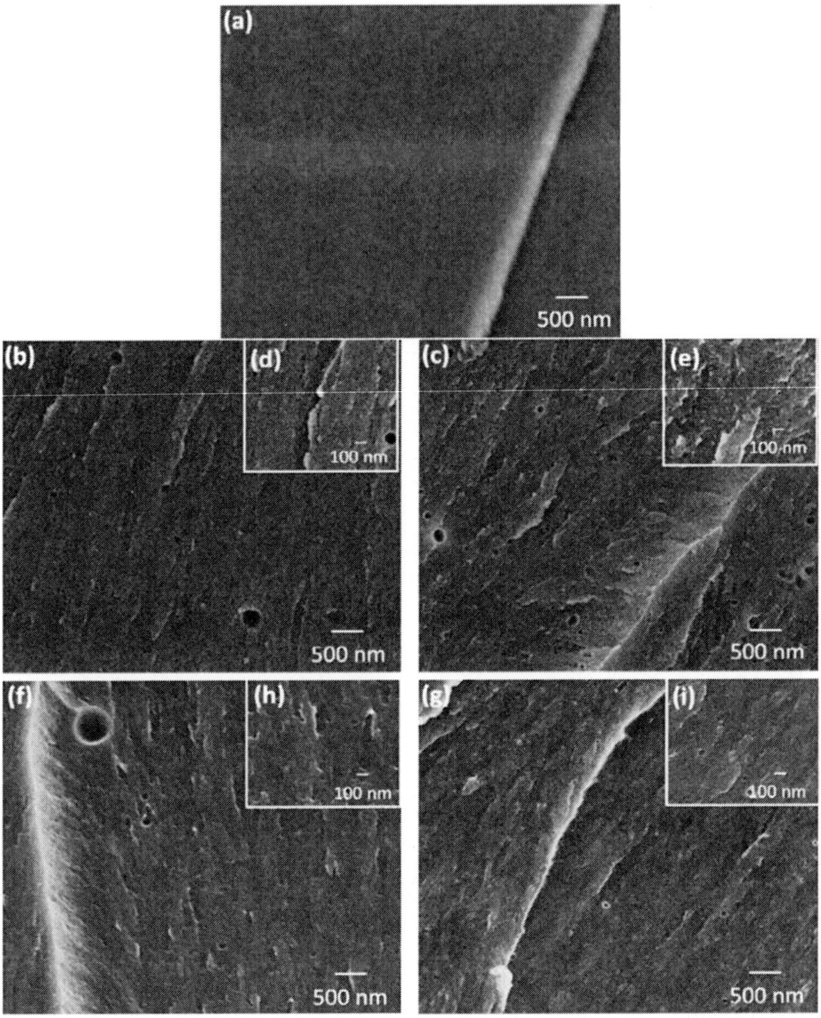

Figure 13. Scanning Electron Micrographs showing the 'mirror' zone on tensile fracture surfaces of (**a**) neat epoxy; (**b**), (**d**) 0.3 vol% of 20k20k-SiO_2 nanocomposites; (**c**), (**e**) 1 vol% of 20k20k-SiO_2nanocomposites; (**f**), (**h**) 0.1 vol% 40k40k-SiO_2 nanocomposites; and (**g**), (**i**) 0.2 vol% of 40k40k-SiO_2 nanocomposites.

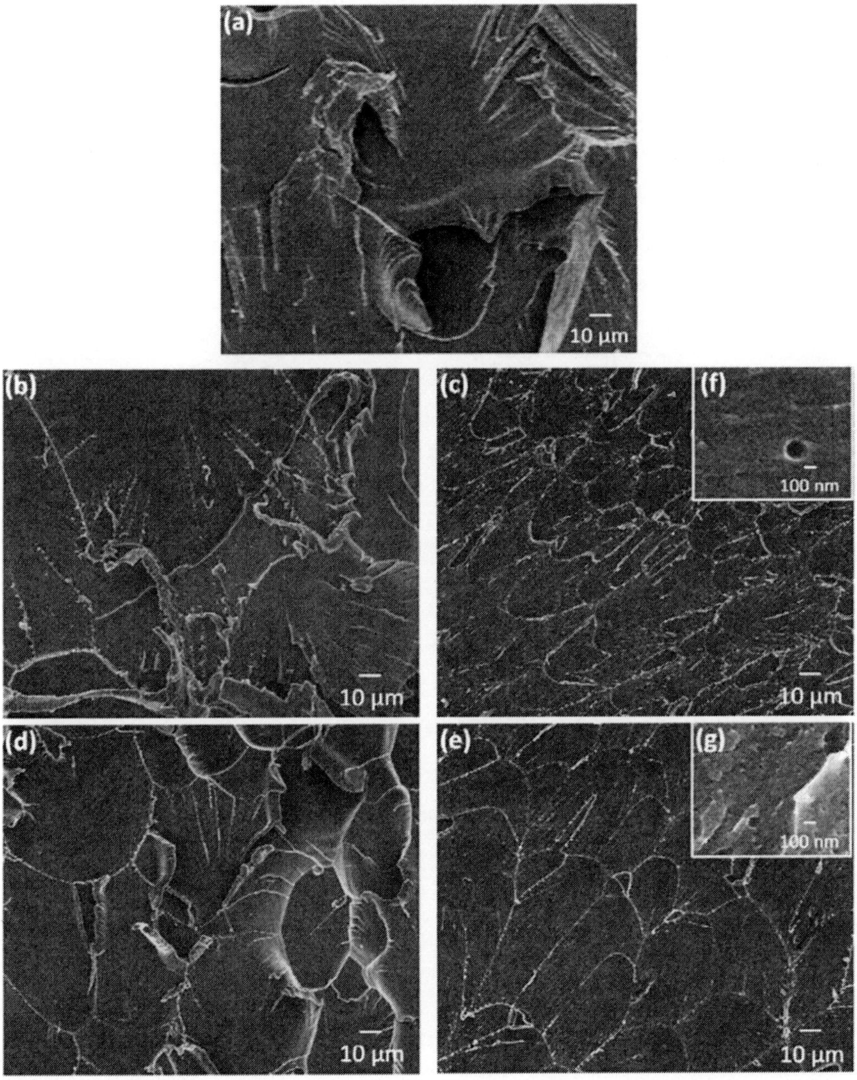

Figure 14. SEM images of the hackle zones of (**a**) the neat epoxy and the nanocomposites with (**b**) 0.3 vol% of 20k20k-SiO_2; (**c**) 1 vol% of 20k20k-SiO_2; (**d**) 0.1 vol% of 40k40k-SiO_2; (**e**) 0.2 vol% of 40k40k-SiO_2; (**f**) higher magnification for 1 vol% 20k20k; and (**g**) higher magnification for 0.2 vol% 40k40k.

To understand the leading mechanisms for the improvement of the mechanical properties of the epoxy matrix by the rubbery grafted SiO_2 nanoparticles, transmission electron microscopy (TEM) micrographs of thin slices of material perpendicular to the fracture surface were taken. The TEM micrographs of the fracture damage zone of the 1 vol% 20k20k nanocomposite and the 0.3 vol% 40k40k nanocomposites are shown in Figure 15. There are voids around the SiO_2 nanoparticles and it is caused by the rubbery interlayer cavitation and deformation. Similar images taken of non-deformed samples did not show the contrast exhibited here.

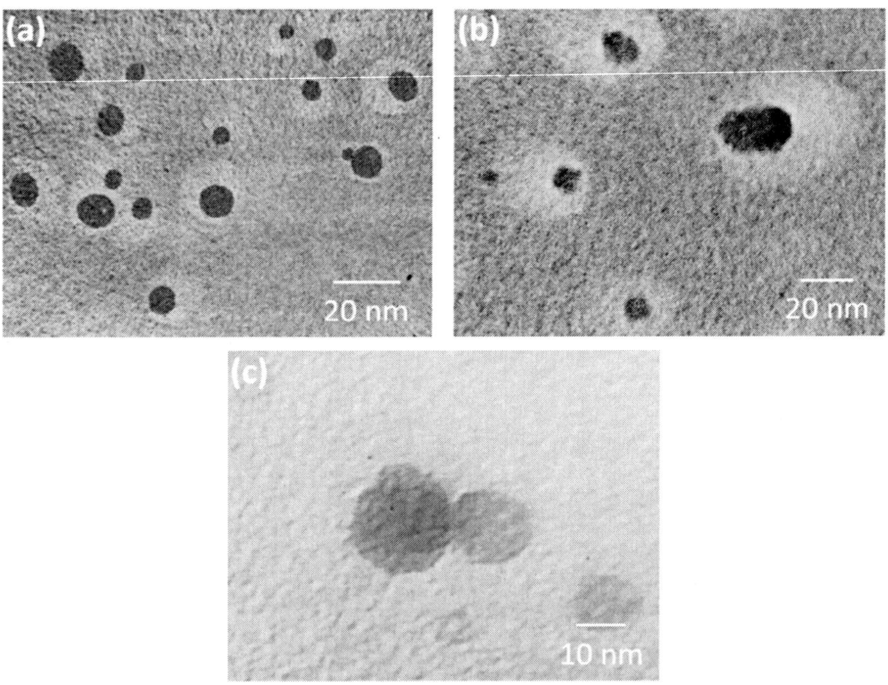

Figure 15. Transmission electron micrographs near the fracture surface of (**a**) 1 vol% 20k20k nanocomposites; (**b**) 0.3 vol% 40k40k nanocomposites; and (**c**) 1 vol% 20k20k as prepared.

The cavitation causes permanent failure in the rubbery interlayer, which results in energy release. The energy released by cavitation is negligible and does not contribute to the toughening directly but makes it easy for the matrix to deform plastically, as proven by Bagheri and Pearson *et al.* who found that the toughening efficiency is similar for rubbers with different cavitation resistance and pre-formed voids, which did not involve any cavitation energy [33]. Rubber cavitation, however, is usually followed by plastic void growth extending to the matrix and local shear yielding [34], both of which are believed to be primary toughening mechanisms in rubber filled epoxy. The irreversible plastic void-growth in the matrix dissipates energy and contributes to the enhanced fracture energy, and the local shear yielding can promote plastic deformation in the matrix, which also improves the fracture energy [35].

There are well-established models from Huang and Kinloch predicting the contribution of plastic void growth and shear banding to the fracture energy in a process zone at the tip of a pre-crack under plane strain uniaxial tension [38]. In this study, we have tested the tensile properties and thus the model cannot be used directly. Because energy density is an intensive property, which is independent of volume of the process zone, the dissipated strain-energy density instead of the fracture energy of the tensile fracture is theoretically calculated to compare to the measured tensile strain-energy density. The dissipated strain-energy density (U_v) for a void in the epoxy to undergo plastic void growth was quantitatively studied by Huang and Kinloch [38] and is expressed as:

$$U_v = p(f_v - f_r) \tag{11}$$

where p is the local hydrostatic stress, and f_v and f_r are the volumetric loading of the voids and rubbery interlayer. Since f_v and f_r of the current studied system can be measured and calculated, the only unknown parameter is the local hydrostatic stress, which is different from that in the process zone at a crack tip as given by Huang and Kinloch [38]. Once necking has occurred in a specimen under a uniaxial plane strain tensile test, there is a triaxial stress state in the onset region of the neck [36,37,38]. When the neck profile is concave is there a hydrostatic stress state. The hydrostatic stress at the concave profile region in the uniaxial tension can be calculated from the following equations [40,41,42]:

$$p = \sigma_{yt}\left(\frac{1}{F_T} - 1\right) \tag{12}$$

$$F_T = \frac{1}{\left(1 + \frac{2R}{a}\right) \cdot ln\left(1 + \frac{a}{2R}\right)} \tag{13}$$

$$\frac{a}{R} = 3.2\left[1 - e^{\frac{-(\varepsilon_b - \varepsilon_n)}{3}}\right] \tag{14}$$

where σ_{yt} is the yield stress of the material; F_T is the local triaxiality factor; a/R is the ratio of half of the cross-section edge a to the profile radius R of the neck as shown in Figure 16. There is an empirical relationship [42] between a/R and ε_b and ε_n which are the strain at break and the strain at the onset of necking, respectively. ε_b and ε_n were determined experimentally.

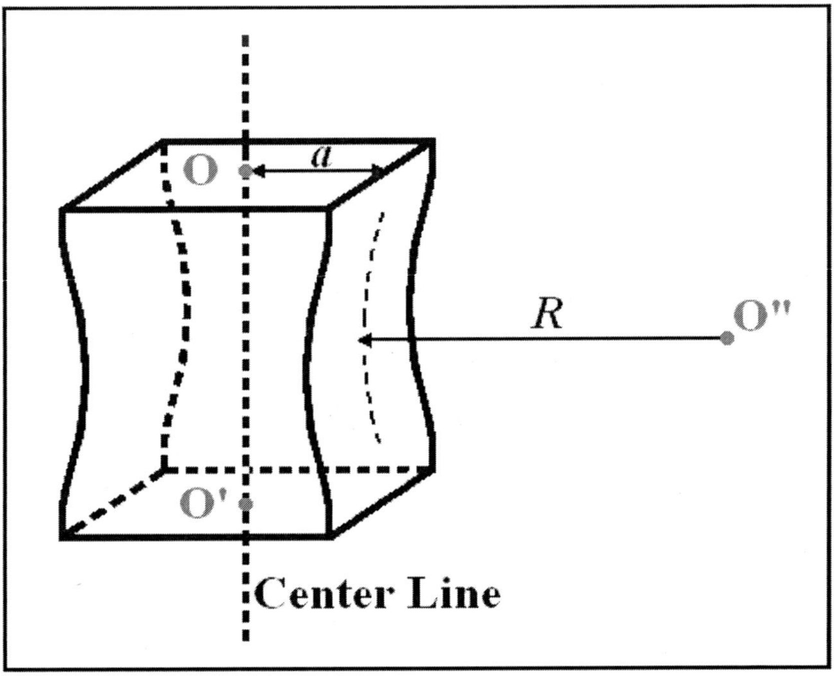

Figure 16. Neck region of a tensile bar.

The strain energy density of a tensile test can be calculated as the area underneath the stress strain curve (Equation (15)), which is also known as the tensile toughness:

$$W = \int_0^{\varepsilon_b} \sigma \cdot \mathrm{d}\varepsilon \tag{15}$$

where ε_b is the strain-to-break; σ and ε are stress and strain of a tensile curve; and W is the strain energy density (or tensile toughness). The difference in the tensile toughness of the composites and the epoxy matrix is equal to the strain energy density contribution from all operating toughening mechanisms. Since the plastic void growth mechanism is obviously happening as shown in Figure 15, the following equation holds

$$W_c - W_u = \Delta W = U_v + \sum U_{1,2...n} \quad (16)$$

where the subscript c and u mean the composite and unmodified epoxy; and U_v is the strain energy density contributed by plastic void growth. The sum is the sum of the strain energy contribution by other operating mechanisms.

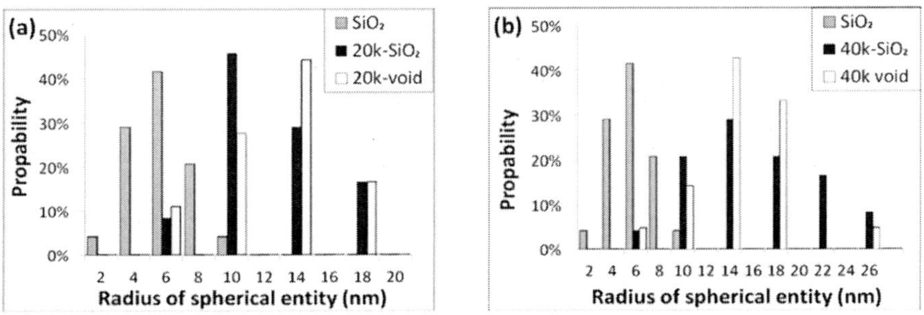

Figure 17. (**a**) Distribution of SiO_2 core size, 20 kg/mol PHMA-$SiO_2$and void surrounded SiO_2 in 20k20k systems; (**b**) Distribution of SiO_2 core size, 40 kg/mol PHMA-SiO_2 and void surrounding SiO_2 in 40k40k systems.

Characterization of the deformation due to plastic void growth and the size distribution of SiO_2nanoparticles were measured from 5 TEM micrographs and are summarized in Figure 17. The distribution of rubbery layer thickness is given in Figure 4. The distribution of the void sizes of both systems is overlapping; however the weighted mean of the void size in both systems shows that the voids are slightly bigger than the rubbery interlayer size as summarized in Table 6. The volume difference between the void and the rubbery layer was used to calculate the plastic void growth contribution to the tensile toughness.

By substituting the volume increase that occurs after rubbery interlayer cavitation in Equations (11) and (16), the fracture mechanism involving plastic void growth is compared with the experimental data of both systems in Figure 18. There is a good match between the theoretical model and experimental data for the 40k40k system of all three tested loadings and the 20k20k system above 4 vol% of the rubbery interlayer. The plastic void growth was not yet measured for the 20k20k system at low loadings, so no comparison is made. From this analysis, the plastic void growth mechanism is believed to be the leading mechanism for improving the tensile toughness in the studied copolymer grafted SiO_2/epoxy nanocomposites.

Table 6. Weighted mean of rubbery interlayer size and void size in both 20k20k and 40k40k systems.

	vol% of SiO_2 Cores	Weight Mean of Rubbery Interlayer Thickness (nm)	vol% of Rubbery Interlayer	Weight Mean of Void Size from the Surface of the SiO_2 to the Edge of the Void	vol% of Void
20k20k	0.1%	6.33	1.03%	6.83	Not measured yet
	0.3%		3.08%		Not measured yet
	0.6%		6.16%		6.98%
	1.0%		10.26%		11.63%
40k40k	0.1%	10.16	2.46%	11.47	3.07%
	0.2%		4.93%		6.13%
	0.3%		7.39%		9.20%

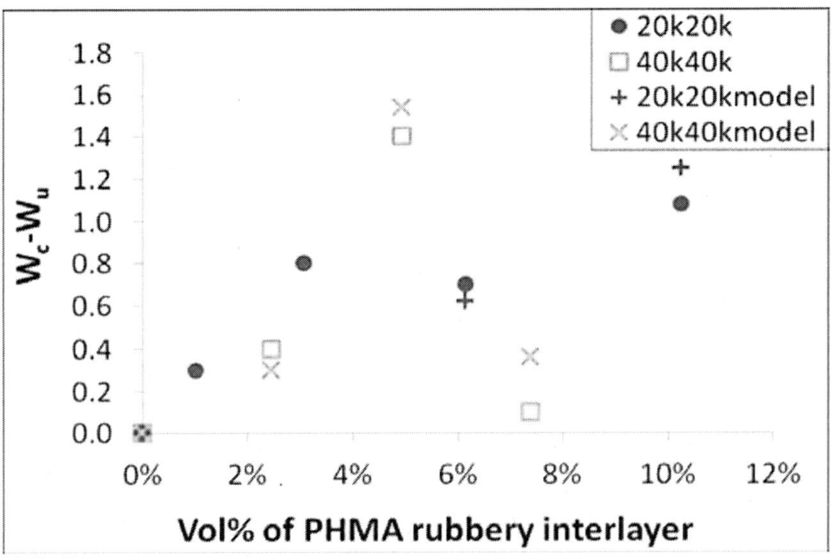

Figure 18. Plastic void growth mechanism in rubbery interlayer grafted SiO_2/epoxy.

Localized shear yielding [39,40,41,42] is another possible phenomenon following rubbery interlayer cavitation. Localized shear yielding will decrease the composites yield strength. The tensile results in Table 4 show that the yield strength decreases as the rubbery interlayer volume percent increases. In addition, the voids around the SiO_2 nanoparticles in both 20k20k and 40k40k systems shown in Figure 15 are elliptical although not oriented. However, the shear bands were not visible in a TEM, and thus the contribution from shear banding is unclear.

Other fracture mechanisms such as crack bridging and pinning [43] and crack deflection [44,45] might also be operating. Crack bridging is unlikely due to the small dimensions of the rubbery layer. Crack pinning is a possible mechanism, since the crack might propagate at slower velocity when it meets the plastic void growth region around the SiO_2 nanoparticles. However there is no clear crack pinning feature on the fracture surface of the nanocomposites in Figure 12–Figure 14, indicating the region of plastic deformation around the SiO_2 nanoparticles might not be big enough to launch this mechanism. Wetzel *et al.* [46] found that crack deflection increases with increasing content of rigid nanoparticles. However, in the current study, the volume percent of the SiO_2 cores is less than 1 vol%, which is unlikely to cause significant crack deflection. At such low particle volume percent, Wetzel *et al.* observed a large discrepancy between the crack deflection model and experimental data for 13 nm Al_2O_3nanoparticle filled epoxy. Therefore, the energy release by crack deflection is probably negligible in the current study.

4. SUMMARY

Copolymer grafted SiO_2 nanoparticles with a rubbery PHMA inner layer and a matrix compatible PGMA outer layer were prepared by RAFT polymerization and were found to disperse well in an epoxy matrix. The rubbery layer thickness can be tailored by varying the molecular weight of the rubbery block. An improvement of the strain-to-break was found to be proportional to the volume percent of the rubbery interlayer. Both the 20 kg/mol and the 40 kg/mol PHMA rubbery interlayer are thick enough to cavitate and cause plastic void growth. The plastic void growth model fit the data reasonably well indicating that the mechanism is operating. Shear banding might also be occurring, but was not visible. Other common toughening mechanisms such as crack bridging, crack pinning and crack deflection are not likely to operate in the studied systems. Though the modulus and tensile strength of the composites decreased with the addition of the rubbery polymer grafted SiO_2, they are somewhat better than for rubber filled epoxy composites. Thinner rubbery layers lead to equivalent improvements of strain to failure, but less of a reduction in Young's modulus of the composites.

ACKNOWLEDGMENTS

This work was supported by ABB Corporate Research and Swedish Research Council (IFA 2007-5095). It is also supported by the Nanoscale Science and Engineering Initiative of the National Science Foundation under NSF Award Number DMR-0117792.

REFERENCES

1. Lee, H.; Neville, K. *Handbook of Epoxy Resins*, 1st ed; McGraw-Hill: New York, NY, USA, 1967.
2. Kunz-Douglass, S.; Beaumont, P.W.R.; Ashby, M.F. A model for the toughness of epoxy-rubber particulate composites. *J. Mater. Sci.* **1980**, *15*, 1109–1123.
3. Kunz, S.C.; Sayre, J.A.; Assink, R.A. Morphology and toughness characterization of epoxy resins modified with amine and carboxyl terminated rubbers. *Polymer* **1982**, *23*, 1897–1906.
4. Hwang, J.F.; Manson, J.A.; Hertzberg, R.W.; Miller, G.A.; Sperling, J.H. Structure-property relationships in rubber-toughened epoxies. *Polym. Eng. Sci.* **1989**, *29*, 1466–1476.
5. Qian, J.Y.; Pearson, R.A.; Dimonie, V.L.; Ei-Aasser, M.S. Synthesis and application of core-shell particles as toughening agents for epoxies. *J. Appl. Polym. Sci.* **1995**, *88*, 439–448.
6. Lovell, P.A.; McDonald, J.; Saunders, D.E.J.; Young, R.J. Studies of rubber-toughened poly(methyl methacrylate): 1. Preparation and thermal properties of blends of poly (methyl methacrylate) with multiple-layer toughening particles. *Polymer* **1993**, *34*, 61–69.
7. Day, R.J.; Lovell, P.A.; Wazzan, A.A. Toughened carbon/epoxy composites made by using core/shell particles. *Compos. Sci. Technol.* **2001**, *61*, 41–56.

8. Johnsen, B.B.; Kinloch, A.J. Toughening mechanisms of nanoparticle-modified epoxy polymers.*Polymer* **2007**, *48*, 530–541.

9. Balakrishnan, S.; Start, P.R.; Raghavan, D.; Hudson, S.D. The influence of clay and elastomer concentration on the morphology and fracture energy of preformed acrylic rubber dispersed clay filled epoxy nanocomposites. *Polymer* **2005**, *46*, 11255–11262.

10. Wetzel, B.; Haupert, F.; Zhang, M.Q. Epoxy nanocomposites with high mechanical and tribological performance. *Compos. Sci. Technol.* **2003**, *63*, 2055–2067.

11. Zhao, S.; Schadler, L.S.; Hillborg, H.; Auletta, T. Improvements and mechanisms of fracture and fatigue properties of well-dispersed alumina/epoxy nanocomposites. *Compos. Sci. Technol.* **2008**,*68*, 2976–2982.

12. Lan, T.; Pinnavaia, T.J. Clay-reinforced epoxy nanocomposites. *Chem. Mater.* **1994**, *6*, 2216–2219.

13. Allaoui, A.; Bai, S.; Cheng, H.M.; Bai, J.B. Mechanical and electrical properties of a MWNT/epoxy composite. *Compos. Sci. Technol.* **2002**, *62*, 1993–1998.

14. Rafiee, M.A.; Rafiee, J.; Zhou, W.; Song, H.H.; Yu, Z.Z.; Koratkar, N. Enhanced mechanical properties of nanocomposites at low graphene content. *ACS Nano* **2009**, *3*, 3884–3890.

15. Kinloch, A.J.; Mohammed, R.D.; Taylor, A.C.; Eger, C.; Spenger, S.; Egan, D. The effect of silica nano particles and rubber particles on the toughness of multiphase thermosetting epoxy polymers. *J. Mater. Sci.* **2005**, *40*, 5083–5086.

16. Liang, Y.L.; Pearson, R.A. The toughening mechanism in hybrid epoxy-silica-rubber nanocomposites (HESRN). *Polymer* **2010**, *51*, 4880–4890.

17. Lee, J.; Yee, A.F. Effect of rubber interlayers on the fracture of glass bead/epoxy composites. *J. Mater. Sci.* **2001**, *36*, 7–20.

18. Li, C.Z.; Han, J.W.; Ryu, C.Y.; Benicewicz, B.C. A versatile method to prepare RAFT agent anchored substrates and the preparation of PMMA grafted nanoparticles. *Macromolecules* **2006**, *39*, 3175–3183.

19. Li, C.Z.; Benicewicz, B.C. Synthesis of well-defined polymer brushes grafted onto silica nanoparticles via surface reversible addition-fragmentation chain transfer polymerization.*Macromolecules* **2005**, *38*, 5929–5936.

20. Rasband, W.S. *ImageJ*; US National Institutes of Health: Bethesda, MD, USA, 1997. Available online: http://imagej.nih.gov/ij/ (accessed on 19 October 2011).

21. Kissinger, H.E. Reaction kinetics in differential thermal analysis. *Anal. Chem.* **1957**, *29*, 1702–1706.

22. Ton-That, M.T.; Ngo, T.D.; Ding, P.; Fang, G.; Cole, K.C.; Hoa, S.V. Epoxy nanocomposites: Analysis and kinetics of cure. *Polym. Eng. Sci.* **2004**, *44*, 1132–1141.

23. Montserrat, S.; Flaque, C.; Calafell, M.; Andreu, G.; Malek, J. Influence of the accelerator concentration on the curing reaction of an epoxy-anhydride system. *Thermochim. Acta* **1995**, *213*, 269–270.

24. Guerrero, P.; De la Caba, K.; Vales, A.; Corcuera, M.A.; Mondragon, I. Influence of cure schedule and stoichiometry on the dynamic mechanical behavior of tetrafunctional epoxy resins cured with anhydrides. *Polymer* **1996**, *37*, 2195–2200.

25. Beiner, M.; Kahle, S.; Hempel, E.; Schroter, K.; Donth, E. Crossover region of dynamic glass transition in poly (n-hexyl methacrylate) by heat capacity spectroscopy. *Macromolecules* **1998**, *31*, 8973–8980.

26. Child, W.C., Jr; Ferry, J.D. Dynamic mechanical properties of poly-*n*-hexyl methacrylate. *J. Colloid Sci.* **1957**, *12*, 389–399.

27. Thompson, Z.J.; Hillmyer, M.A.; Liu, J.; Sue, H.J.; Dettloff, M.; Bates, F.S. Block copolymer toughened epoxy: Role of cross-link density. *Macromolecules* **2009**, *42*, 2333–2335.

28. Van der Sanden, M.C.M.; Meijer, H.E.H. Deformation and toughness of polymeric systems: 3. Influence of crosslink density. *Polymer* **1993**, *34*, 5063–5072.

29. Chikhi, N.; Fellahi, S.; Bakar, M. Modification of epoxy resin using reactive liquid (ATBN) rubber.*Eur. Polym. J* **2002**, *38*, 251–254.

30. Hull, D. *Fractography–Observing, Measuring and Interpreting Fracture Surface Topography*, 1st ed; Cambridge University Press: Cambridge, UK, 1999; pp. 121–150.

31. Lawn, B. *Fracture of Brittle Solids*, 2nd ed; Cambridge University Press: Cambridge, UK, 1993; p. 95.

32. Fineberg, J.; Marder, H.M. Instability in dynamic fracture. *Phys. Rep.* **1999**, *313*, 1–108.

33. Bagheri, R.; Pearson, R.A. Role of particle cavitation in rubber-toughened epoxies: 1. Microvoid toughening. *Polymer* **1996**, *17*, 4529–4538.

34. Huang, Y.; Kinloch, A.J. The role of plastic void growth in the fracture of rubber-toughened epoxy polymers. *J. Mater. Sci. Lett.* **1992**, *11*, 484–487.

35. Ma, J. Study of epoxy toughened by *in situ* formed rubber nanoparticles. *J. Appl. Polym. Sci.* **2008**,*110*, 304–312.

36. Nicolaou, P.D.; Semiatin, S.L. The effect of stress triaxiality on tensile behavior of cavitating specimens. *J. Mater. Sci.* **2001**, *36*, 5155–5159.

37. Lombard, C.M.; Goetz, R.L.; Semiatin, S.L. Numerical analysis of the hot tension test. *Metall. Trans. A* **1993**, *24A*, 2039–2047.

38. Ragab, A.R. Prediction of ductile fracture in axisymmetric tension by void coalescence. *Int. J. Fract.* **2000**, *105*, 391–409.

39. Kinloch, A.J. Relationship between the microstructure and fracture behavior of rubber-toughened thermosetting polymers. In *Rubber-Toughened Plastics*; Riew, C.K., Ed.; American Chemical Society: Washington, DC, USA, 1989; Volume 222, pp. 67–91.

40. Pearson, R.A.; Yee, A.F. Toughening mechanisms in elastomer-modified epoxies. *J. Mater. Sci.***1986**, *21*, 2462–2474.

41. Bagheri, R.; Pearson, R.A. Role of particle cavitation in rubber-toughened epoxies. *Polymer* **1996**,*17*, 4529–4538.

42. Yee, A.F.; Li, D.M.; Li, X.W. The importance of constraint relief caused by rubber cavitation in the toughening of epoxy. *J. Mater. Sci.* **1993**, *28*, 6392–6398.

43. Kunz, S. A model for the toughness of epoxy-rubber particulate composites. *J. Mater. Sci.* **1980**, *15*, 1109–1123.

44. Faber, K.T.; Evans, A.G. Crack deflection process—I. Theory. *Acta Metall.* **1983**, *31*, 565–576.

45. Faber, K.T.; Evans, A.G. Crack deflection process—II. Experiment. *Acta Metall.* **1983**, *31*, 577–584.

46. Wetzel, B.; Rosso, P.; Haupert, F.; Friedrich, K. Epoxy nanocomposites—Fracture and toughening mechanisms. *Eng. Fract. Mech.* **2006**, *73*, 2375–2398.

10
CHAPTER

Synthesis of Polyfluorene-Polytriarylamine Block Copolymer with Emitting Part at Junction Point for Light Emitting Applications

Mehdi Jahanfar, Kenta Suwa, Kousuke Tsuchiya, Kenji Ogino*

Graduate School of Bio-Applications and Systems Engineering, Tokyo University of Agriculture and Technology, Tokyo, Japan

ABSTRACT

A block copolymer consisting of polyfluorene (PF) and polytriarylamine (PTAA) functionalized with green emitting phenoxazine moiety at the junction point of two blocks was designed and prepared for electroluminescent application. PF homopolymer was synthesized by Suzuki coupling polymerization, and was reacted with brominated phenoxazine. In the presence of the resulting PF functionalized with phenoxazine, C-N coupling polymerization of 4-(4'-bromophenyl)-4''-butyldiphenylamine was carried out to afford a triblock copolymer, PTAA-phenoxazine-PF-phenoxazinePTAA (PF-Ph-PTAA). Two types of random copolymers were also synthesized with fluorene and phenoxazine (PF2) by Suzuki coupling polymerization for comparison. All the polymers were soluble in common organic solvents and readily formed thin films by a solution processing. Prepared polymers exhibited similar UV absorption and PL emission in chloroform solutions. In a film state, the existence of phenoxazine unit drastically changed PL spectra. Although the content of phenoxazine unit in PF-Ph-PTAA was relatively high (13 mol%), it showed similar PL spectrum to that of PF2(phenoxazine content, 0.2 mol%) indicating that phenoxazine unit is isolated in single polymer chain nevertheless the high content. EL device based on PF-Ph-PTAA showed green-emission, suggesting that emission sites predominantly located in the vicinity of phenoxazine moiety because of its shallow HOMO level.

Keywords: Polyfluorene; Polytriarylamine; Block Copolymer; Junction Point; Polymer Light-Emitting Diode

1. INTRODUCTION

Polymer light-emitting diodes (PLEDs) have attracted much scientific and technological research interest since their first discovery in 1990 [1]. Utilization of electroluminescent polymeric materials shows several advantages over organic small molecules for use in PLEDs: better processability, and high flexibility [2]. Furthermore, inexpensive wet-processes such as spin-coating and ink-jet can be applied for the fabrication of PLED devices, which is essential in order to apply PLEDs to display and lighting technologies.

Unfortunately, it is generally recognized that polymeric devices show lower performance (efficiency, life time) compared with devices fabricated with a vacuum process based on low-molecular weight materials. Low efficiency in polymeric devices is partially due to difficulty in fabricating the devices with a layered-structure. To overcome the drawbacks, a breakthrough is necessary from the point of the molecular design and the control the morphology in the active layer. We have showed the advantage of block copolymers consisting of hole and electron transporting blocks as the host materials in phosphorescent devices [3-5]. Block copolymers were prepared via a nitroxide mediated living radical polymerization.

Block copolymers assemble into microor nano-phase separated structures with various domain shapes such as lamella, cylinder, or sphere. Exploiting nanostructures of block copolymers with appropriate designs can improve performance of applications due to allocation of functionality to each domain [6]. The other groups also reported several block copolymers for PLED applications [7,8].

More recently Tan et al. synthesized the different type of block copolymers for EL applications via the Suzuki coupling polymerization followed by the C-N coupling polymerization [9,10], which consisted of light emitting and electron transporting polyfluorene (PF) unit and hole transporting polytriarylamine (PTAA) unit. It was revealed that the introduction of PTAA increased emission efficiencies compared with PF homopolymer. This is due to the facile hole injection from the anode and/or the efficient electron block by PTAA moieties, which are located in the vicinity of the PEDOT/PSS coated on the anode through the hydrogen bonding of trioxyethylene group with PSS [10].

Here we proposed the novel methodology for the increase the efficiency in PLED. That is the novel molecular design of full functional polymers, which are block copolymers consisting of hole transporting unit and electron transporting unit with emitting moiety at the junction point. In the previous devices utilizing block copolymers as host polymers, emitting materials are low molecular weight phosphorescent dyes dispersed in host materials. In this case, emitting parts are randomly distributed in the active layer. Emission process is resulted from the recombination of holes and electrons, followed by the energy transfer from recombination centers to emitting moieties and/or the charge trap predominately occurred at the emitting sites. Therefore if the emitting parts are located in the vicinity of the interface between hole and electron transporting domains, more efficient energy transfer and carrier trap are anticipated. In order to attain the situation, a new molecular design is proposed. Target polymers are block copolymers consisting of hole transporting unit and electron transporting unit with emitting moiety at the junction point. If an ideal phase separation occurs, the emitting moiety exists at the interface between both domains.

In this study, the synthetic strategy we established for PF-b-PTAA (Suzuki coupling followed by C-N coupling polymerization) is modified to prepare a block copolymer consisting of PF and PTAA functionalized with green emitting phenoxazine moiety at the junction point of two blocks. Phenoxazine derivatives are known as a emitting dye [11,12], and have been utilized as the component of EL copolymers [13,14]. EL characteristics were preliminarily investigated for the comparison with random copolymer.

2. EXPERIMENTAL

2.1. Materials

Figure 1 illustrates the synthetic route of targeted polymers. n-Butylphenylphenoxazine (1) [14], 2,7-dibromo- 9-(4-methylphenyl)-9-(4-octylphenyl)-fluorene (4) [9], 2,7-bis(4,4,5,5-tetramethyl-1,3,2-dioxaborolan-2-yl)-9- (4-methylphenyl)-9-(4-octylphenyl)-fluorene (5) [9], 4-(4'-bromophenyl)-4"-butyl-diphenylamine (6) [15] were synthesized according to reported procedures. All reagents and solvents were used without further purification unless stated otherwise. Tetrahydrofuran (THF) was distilled over sodium and benzophenone, and stored under nitrogen atmosphere. Toluene was distilled over calcium hydrine, and stored under nitrogen atmosphere. The other regents and solvents were obtained commercially and were used as received.

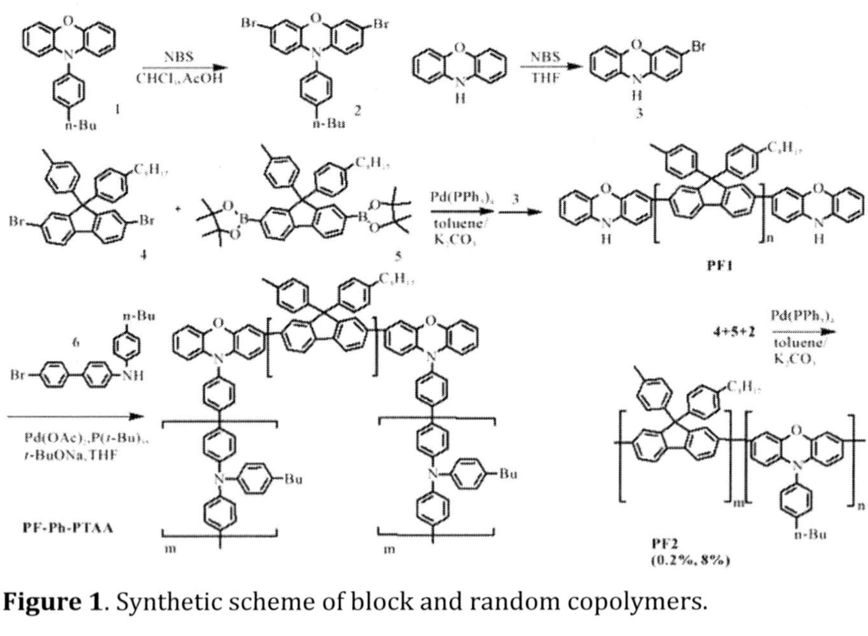

Figure 1. Synthetic scheme of block and random copolymers.

2.2. Characterization

^{1}H and ^{13}C NMR spectra were obtained on a JEOL ALPHA300 instrument at 25°C at 300 and 75 MHz, respectively. Deuterated chloroform was used as a solvent with tetramethylsilane as an internal standard. Numberand weight-average molecular weights (M_n and M_w) and polydispersity index (PDI) were determined by gel permeation chromatography (GPC) analysis calibrated by standard polystyrene samples with a JASCO RI-2031 detector eluted with chloroform at a flow rate of 0.5 ml/min at room temperature. Differential scanning calorimetry (DSC) analysis was performed on a Rigaku DSC-8230 under nitrogen atmosphere at heating and cooling rates of 10 °C/min. UV-vis absorption spectra were obtained on a JASCO V-570 spectrophotometer and fluorescence spectra were obtained with JASCO FP-6500 spectrophotometer with an excitation at 380 nm. Cyclic voltammetry (CV) was conducted on a Niko Keisoku Model NPGFZ- 2501-A potentiogalvanostat. All measurements were carried out at room temperature in a typical three-electrode cell with a working electrode (glassy carbon electrode), a reference electrode (Ag/AgCl), and a counter electrode (Pt wire) at a scanning rate of 0.1 V/s. In all measurements, acetonitrile and tetrabutylammonium tetrafluorocarbonate (Bu_4NBF_4) (0.1 M) were used as a solvent and a supporting electrolyte, respectively.

2.3. Synthesis of 3,7-Dibromo-N-butylphenylphenoxazine (2)

To a 100 mL flask equipped with a stopcock and a condenser were placed N-butylphenylphenoxazine (1) (0.44 g, 1.39 mmol), chloroform (20 mL), acetic acid (20 mL), N-bromosuccinimide (NBS) (0.52 g, 2.91 mmol) in 0°C, and the mixture was stirred for 10 h. After the reaction the solution extracted with chloroform, and organic layer washed with water and dried with $MgSO_4$. After chloroform was removed by a rotary evaporator, the crude product was purified by recrystallization with hexane. Light yellow solid was obtained. The yield was 0.49 g (75%). ^{1}H NMR (300 MHz, $CDCl_3$) δ (ppm): 0.95 (t, 3H), 1.4 (m, 2H), 1.70 (m, 2H), 2.75 (t, 2H), 5.68 (d, 2H), 5.75 (d, 2H), 6.79 (s, 2H), 7.16 (d, 2H), 7.38 (d, 2H). ^{13}C NMR (75 MHz, $CDCl_3$) δ (ppm): 14.0, 22.4, 33.5, 35.4, 112.6, 114.4, 118.5, 126.2, 129.9, 131.2, 133.4, 135.4, 143.9, 144.1.

2.4. Synthesis of 3-Bromophenoxazine (3)

To a 50 mL flask equipped with a stopcock were placed phenoxazine (1.00 g, 5.50 mmol) and distilled THF (20 mL). THF solution (15 mL) of NBS (0.97 g, 5.5 mmol) was added dropwise to the mixture and stirred for 1 h. After the filtration and the evaporation of THF, the product was resolved in ethyl acetate and washed with water 3 times. After dried with $MgSO_4$, and solvent was removed with a rotary-evaporator. The crude product was purified by column chromatography on silica gel with hexane:ethyl acetate = 1:1 as an eluent. The yield was 0.62 g (42%). ^1H NMR (300 MHz, $CDCl_3$) δ (ppm): 6.66 - 6.81 (m, 7H)

2.5. Synthesis of PF-Ran-PPh (PF2, 0.2%)

To a 10 mL flask equipped with a stopcock and a condenser were placed 4 (0.499 g, 0.143 mmol), 5 (0.500 g, 0.143 mmol), 2 (0.001 g, 0.00028 mmol), $Pd(PPh_3)_4$ (1 mol%), K_2CO_3 (2 M, 2 mL) and toluene (3 mL) under nitrogen atmosphere. The mixture was heated up to 90°C for 24 h. After the reaction the product was extracted by chloroform and washed with water and the organic layer was dried with $MgSO_4$. Solvent was removed with a rotary-evaporator. The crude product was purified by reprecipitation into methanol and Soxhlet extraction with acetone. The yield was 0.126 g (97%). PF2 with 8 mol% of 2 was synthesized with a similar manner (98%).

2.6. Synthesis of PF with Phenoxazine Moieties at Chain End (PF1)

To a 20 mL three-necked round-bottom flask equipped with a stopcock and a condenser were added 4 (0.166 g, 0.2757 mmol), 5 (0.226 g, 0.323 mmol), $Pd(PPh_3)_4$ (3.46 mg, 0.003 mmol, 0.5 mol%), K_2CO_3 (2 M, 2 mL), and toluene (4 mL) under nitrogen atmosphere, and the mixture was refluxed for 48 h. Then, 3 (0.078 g, 0.3 mmol) was added and the reflux was continued for 12 h. After the reaction the product was extracted with chloroform, and washed with salt NaCl and water. Then the organic layer was dried with $MgSO_4$. After the removal of the solvents, the product was obtained by precipitation into methanol. The yield was 0.20 g (79%). ^1H NMR (300 MHz, $CDCl_3$) δ (ppm): 0.83 - 0.88 (m, 3H), 1.24 - 1.28 (m, 10 H), 1.50 - 1.61 (m, 2H), 2.19 - 2.34 (m, 3H), 2.51 - 2.56 (m, 2H), 6.63 - 6.67 (m, 14H), 7.02 - 7.18 (m, 8H), 7.41 - 7.65 (m, 4H), 7.67 - 7.81 (m, 2H).

2.7. Synthesis of Block Copolymer with Phenoxazine Moiety at the Junction (PF-Ph-PTAA)

To a 10 mL two-necked round-bottom flask equipped with a stopcock and a condenser were added PF1 (0.131 g, 0.01 mmol of the terminal), 8 (0.114 g, 0.3 mmol), $P(t\text{-}Bu)_3$ (7.5 μL, 0.01 mmol), $Pd(OAc)_2$ (0.0014 g, 0.006 mmol), t-BuONa (0.032 g, 0.33 mmol), and THF (3 mL) under nitrogen atmosphere and the mixture was stirred under reflux for 24 h. After removing the solvent by a rotary evaporator, the product was obtained by precipitation into methanol and acetone. The yield was 0.191 g (60%).

2.8. Device Fabrication

Perior to preparation of device, a glass slide with indium tin oxide (ITO) patterns was washed by an alkaline cleaner under sonication and rinsed with deionized water. The substrate was subsequently washed by 2-propanol under sonication, and rinsed with clean 2-propanol, and dried with nitrogen. PEDOT:PSS with 30 nm of thickness was spin-coated on the substrate at 2500 rpm for 30 s from the dispersion in water filtered by 0.2 μm of membrane filter followed by annealing at 200°C for 1 h. Polymer layer was laminated on PEDOT:PSS by spincoating at 500 rpm for 30 s from chlorobenzene solution (12 mg/mL) filtered by 0.45 μm of membrane filter, and annealed at 120°C for 1 h under nitrogen atmosphere. On the polymer, 2,9-dimethyl-4,7-diphenyl-1,10-phenanthroline (BCP) with 50 nm thickness was deposited by thermally evaporation in carbon pot at 240°C under vacuum with a rate of 1.5 Å/s. As a cathode, lithium fluoride with 0.5 nm of thickness followed by a Al with 100 nm of thickness was deposited on the organic layer with a rate of 0.1 Å/s and a rate of 4.5 Å/s using tantalum and tungsten boats, respectively. Finally the polymer device was set into a metal can with barium oxide as a drying agent under a nitrogen atmosphere, and passivated with an epoxy resin (XNR 5570-B1, Nagase Chemtex Corporation) irradiated by UV light.

3. RESULTS AND DISCUSSION

3.1. Synthesis

In order to prepare the PF homopolymer with phenoxadine moieties (PF1 in **Figure 1**), Suzuki coupling polymerization of 4 and 5 in non-stoichiometric conditions (4/5 = 0.85/1.0) was carried out resulting in an end-functional polymer with boronic ester moieties, followed by the reaction with monobrominated phenoxazine. The successful introduction of phenoxazine moieties was confirmed by ^{1}H NMR spectrum. As shown in **Figure 2**(a), signals for phenoxazine protons appeared around 6.6 ppm. In order to obtain a target block copolymer (PF-Ph-PTAA) consisting of PF and PTAA functionalized with green emitting phenoxazine moiety at the junction point of two blocks, C-N coupling polymerization of 6 was conducted in the presence of PF1 using palladium acetate and tri(tert-butyl) phosphine ligand, and tert-butoxide as a base in THF for 24 h in the similar manner we reported [9,15]. As shown in **Figure 2**(b), two types of methyl signals around 0.85 ppm, and N-H signal from polymer ends (5.7 ppm) were observed indicating the introduction of TAA unit as expected.

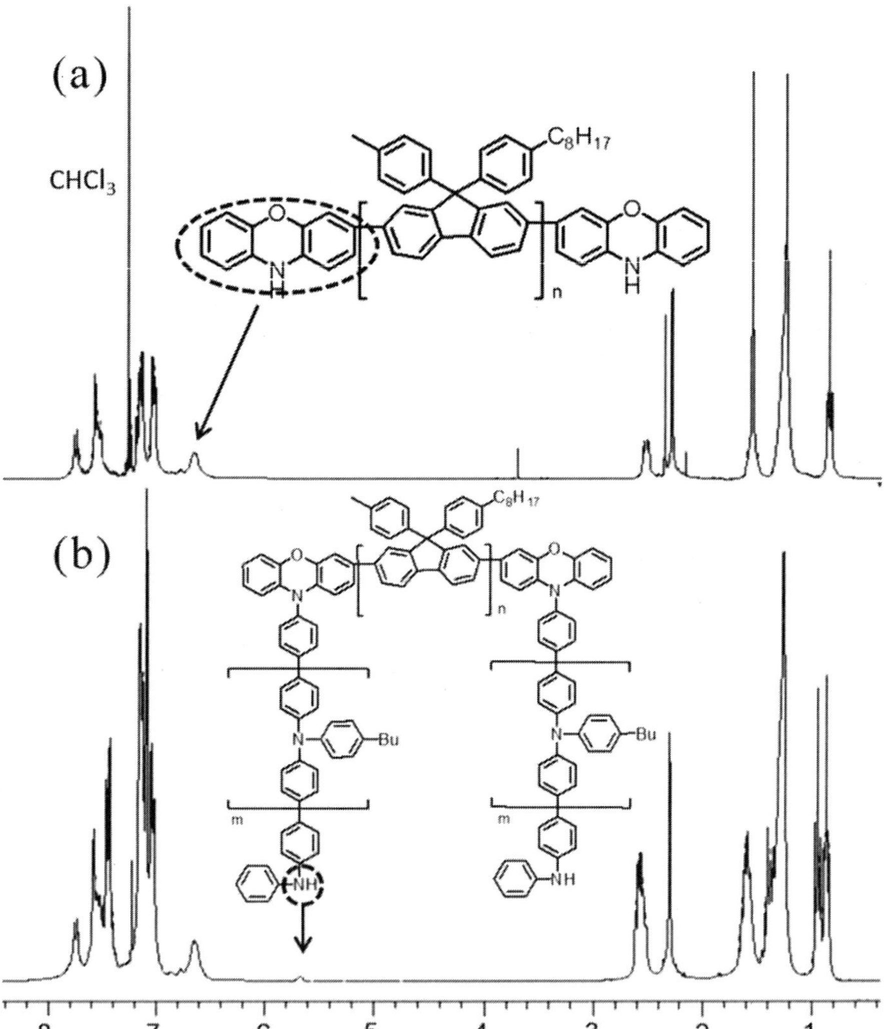

Figure 2. ^{1}H-NMR of synthesized polymers measured in $CDCl_3$ at 500 MHz and 25°C.

Figure 3 shows GPC profiles for PF1 and PF-PhPTAA. GPC trace of PF-Ph-PTAA was monomodal and shifted toward the higher molecular weight direction compared with PF1 homopolymer. Both NMR and GPC results suggest the successful formation of the designed block copolymer. Random copolymers, PF2s, were also synthesized by Suzuki coupling polymeriztion of 2 and 4 with 5. **Table 1** lists the characteristics of prepared polymers together with the data for PF-bPTAA, which is a previously prepared block copolymer without phenoxazine units [9,10].

Glass transition temperature of PF-Ph-PTAA was determined to be 148°C by DSC, and this value is almost the same as that of PF-b-PTAA. All the polymers were soluble in common organic solvents and readily formed thin films by a solution processing.

3.2. Optical and Electrochemical Properties

The optical properties of prepared polymers were investigated with UV-vis absorption and photoluminescent (PL) spectroscopies. In chloroform solutions, all polymers we examined possess similar optical properties regardless of the existence of phenoxazine moiety as shown in **Table 2**. In a solution state, all copolymers showed $\lambda_{abs, max}$ of 382 - 385 nm, and exhibited similar PL spectra in solution state, having distinct vibronic bands at about 418 and 445 nm and a shoulder around 470 nm upon excitation at 380 nm. As we previously reported [9,10], PTAA segments show negligible contribution to PL spectrum. It is also found that phenoxazine units existing at the junction point, or in the fluorene sequence of random copolymers also have almost no effects on the UV and PL spectra in the solution state.

Figure 4 shows UV-vis (a) and PL (b) spectra in a film state. Only PF2 (8%) showed different absorption spectrum as shown in **Figure 4**(a). The observation of a tail over 420 nm is probably due to the interchain interaction between phenoxazine rich sequences. The existence of phenoxazine significantly influenced PL spectra in the film state (**Figure 4**(b)). PF-b-PTAA shows similar PL spectrum to that of solution indicating that interchain interaction is negligible because of the bulky substituent. However, only 0.2% of phenoxazine moiety drastically changed the PL spectrum, i.e., the spectrum showed emission peaks at 426, and 450 nm with a tail. In the case of PF2 (8%), further red shift was observed

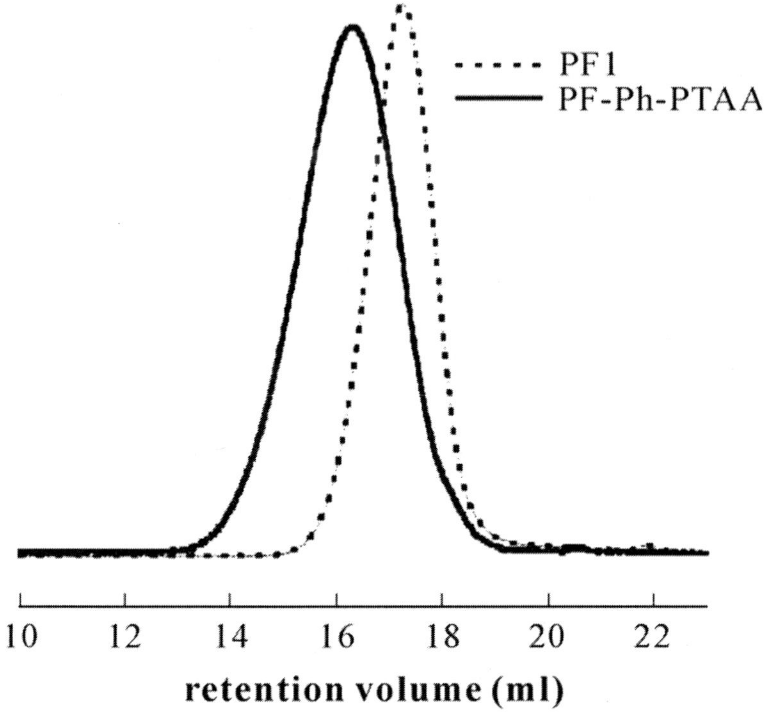

Figure 3. GPC profiles of PF1 and PF-Ph-PTAA. Effluent was monitored with UV detector (254 nm).

Table 1. Characteristics of synthesized polymers.

Polymer	Mn/10[3a]	PDI[a]	FL content[b] (mol%)	Phenoxazine content (mol%)
1PF	4.1	4.0	78	22
PF-Ph-PTAA	11	3.4	44	13
PF2 (0.2%)	8.0	2.4	99	0.2
PF2 (8%)	25	3.2	92	8.0
PF-b-PTAA[c]	19	2.4	57	0

[a] Determined by GPC; [b] determined by ^{1}H-NMR; [c] Ref. [10].

Table 2. Optical properties of synthesized polymers

Polymer	λ_{max}(UV)/nm		λ_{max}(PL)[a]/nm	
	Solution[b]	Film	Solution[c]	Film
PF-Ph-PTAA	382	380	419	426, 453
PF2 (0.2%)	385	380	419	426, 453
PF2 (8%)	385	380	419	469
PF-b-PTAA	385	385	419	419

[a] Exited at 380 nm; [b] $CHCl_3$ solution, 10^{-2} mg/mL; [c] 10^{-4} mg/mL.

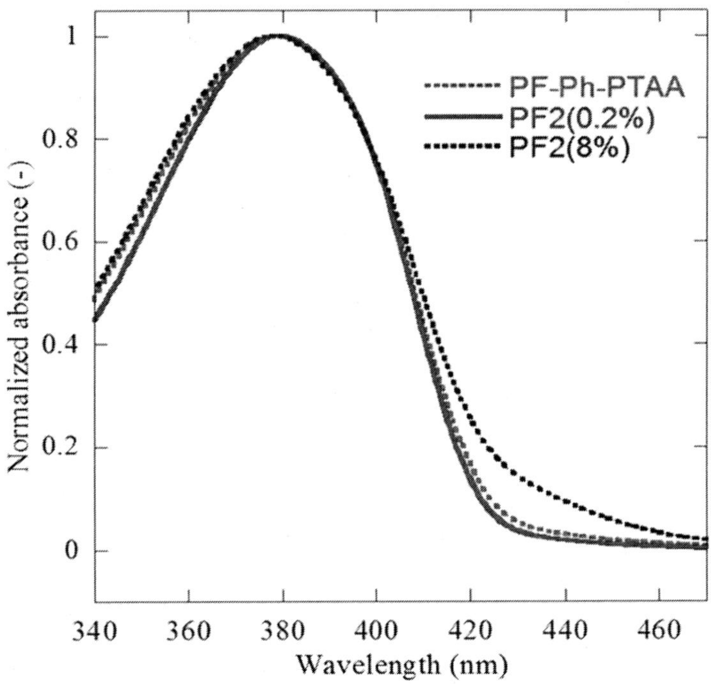

(a)

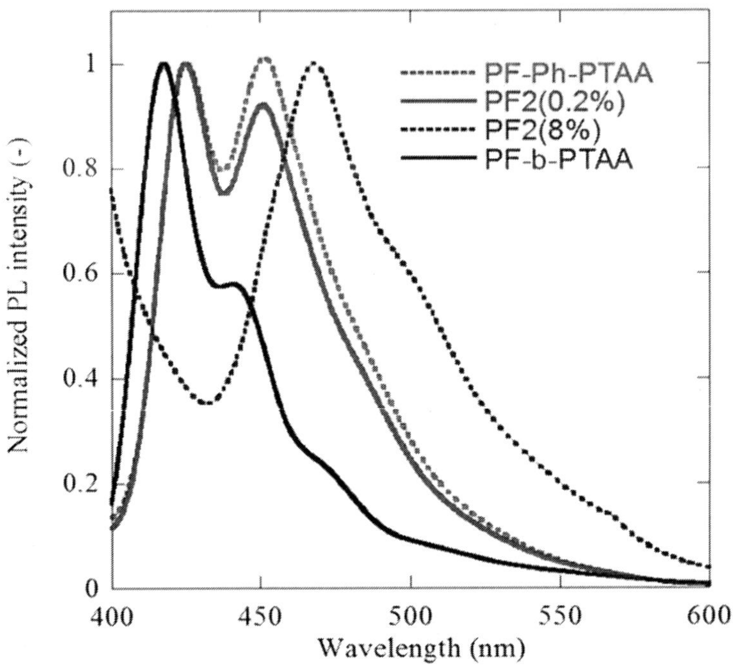

(b)

Figure 4. UV-vis (a) and PL (b) spectra of synthesized polymers in a film state. Excitation wavelength: 380 nm.

showing emission maximum only at 463 nm, indicating the effective energy transfer. Although the content of phenoxazine unit in PF-Ph-PTAA was relatively high (17%) as a result of the low molecular weight nature, it showed similar PL spectrum to that of PF2 (0.2%). This is because phenoxazine unit is isolated in single polymer chain nevertheless the high content, and/or microphase separated structure is established in PF-Ph-PTAA film.

The electrochemical properties of the copolymers were investigated by cyclic voltammetry (CV) with a reference electrode of Ag/AgCl. The oxidation process of PF-PhPTAA started around 0.84 V, and the reversible redox peak in the reduction process appeared at around −0.71 V. The energy levels of the highest occupied molecular orbital (HOMO) of the PF-Ph-PTPA was estimated from its oxidation potential according to an empirical formula, $E_{HOMO} = -e(E_{ox} + 4.4)$ V. HOMO level of the block copolymer was in -5.24 eV. The levels of the lowest unoccupied molecular orbital (LUMO) for the polymer were calculated with the reduction potential Ered from the empirical formula $E_{LUMO} = -e(E_{red} + 4.4)$ V. LUMO level of the block copolymer was in -3.69 eV. This result was almost consistent with that of PF-b-PTAA [10].

3.3. EL Properties

The performance of the EL devices based on PF-PhPTAA with the device structure of ITO/PEDOT:PSS/ polymer/LiF/Al was evaluated. The device with PF-PhPTAA blended with 20 wt% of PF-b-PTAA was also evaluated in order to elucidate the chemical environment where phenoxazine moiety locates. **Figure 5** shows luminance vs. voltage (a), and current efficiency vs. current density (b) characteristics, respectively. Although the maximum luminance for the device based on PF-PhPTPA with 20 wt% of PF-b-PTPA (1710 cd/cm^2) is slightly higher than neat PF-Ph-PTPA, both devices afforded the similar current efficiency profiles. As discussed in PL section, phenoxazine unit is isolated in single polymer chain nevertheless the high content. Therefore, the environment of emitting centers in both devices is similar each other. Compared with the device based on PF-b-PTAA [10], EL performance is slightly worse. It is speculated that this is due to the low molecular weight nature of PF-Ph-PTAA, which results in the unclear microphase separation.

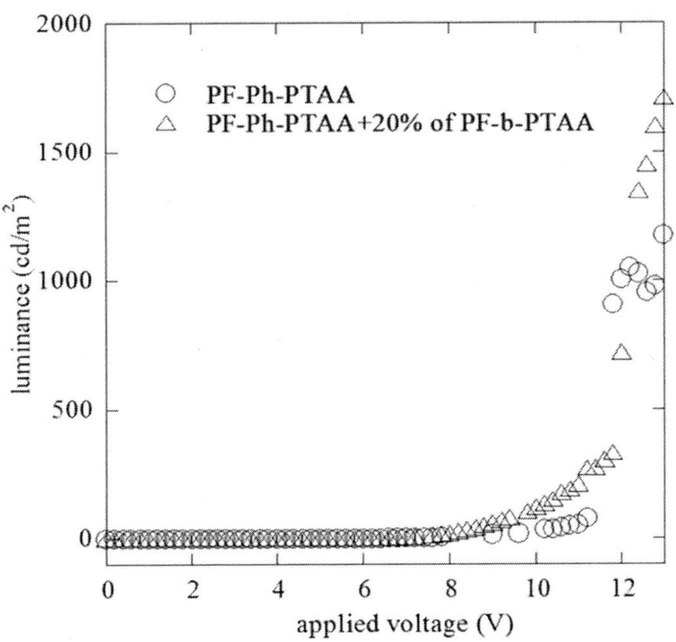

(a)

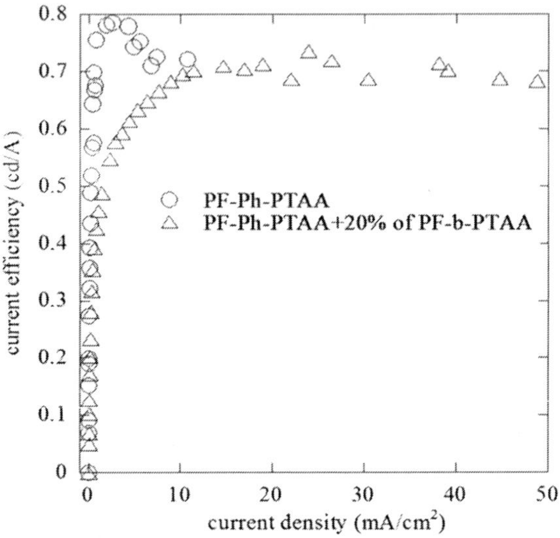

(b)

Figure 5. EL characteristics of device based on PF-PhPTAA (circle) and PF-Ph-PTAA + 20 wt% of PF-b-PTAA (triangle). (a) Luminance vs. voltage; (b) Current efficiency vs. current density.

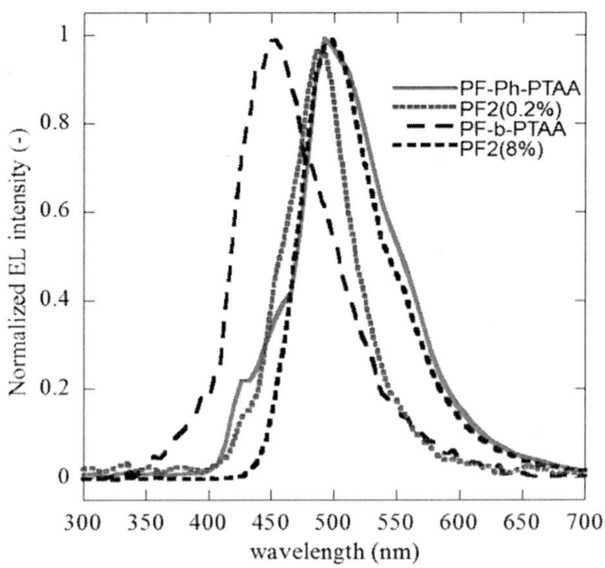

Figure 6. Electroluminescent spectra for devices based on synthesized polymers obtained at 10 V.

Figure 6 shows EL spectra for fabricated devices. Compared with the device based on PF-b-PTAA, all the devices based on phenoxazine containing polymers show ca. 50 nm red-shift. It is indicated that emission sites predominantly located in the vicinity of phenoxazine moiety. Because of shallow HOMO level of phenoxazine unit [14], it works as a hole trap resulting in the efficient recombination of hole and electron. Contrary to PL spectra, PF-Ph-PTAA showed similar EL spectrum to PF2 (8%), where a tail was observed in the long wavelength region. It is possible that the interaction of excitons exists because of the local high concentration of phenoxazine moiety.

4. CONCLUSION

A block copolymer consisting of electron transporting PF and hole transporting PTAA functionalized with green emitting phenoxazine moiety at the junction point of two blocks was successfully prepared via the combination of Suzuki, and C-N coupling polymerization based on the novel molecular design of full functional polymers for the increase the efficiency in PLED. Prepared polymers exhibited similar optical properties in chloroform solutions. On the other hand, in a film state, the existence of phenoxazine unit drastically changed PL spectra. Although the content of phenoxazine unit in PF-Ph-PTAA was relatively high (13%), it showed similar PL spectrum to that of PF2 (0.2%) indicating that phenoxazine unit is isolated in single polymer chain nevertheless the high content. EL device based on PF-Ph-PTAA showed green-emission, suggesting that emission sites predominantly located in the vicinity of phenoxazine moiety because of its shallow HOMO level. Our molecular design presented here makes it possible to control the emission site, and to afford more efficient materials for EL applications.

REFERENCES

1. J. H. Burroughes, D. D. C. Bradley, A. R. Brown, R. N. Marks, K. Mackay, R. H. Friend, P. L. Burns and A. B. Holmes, "Light-Emitting Diodes Based on Conjugated Polymers," Nature, Vol. 347, No. 6293, 1990, pp. 539- 541.

2. L. Akcelrud, "Electroluminescent Polymers," Progress in Polymer Science, Vol. 18, No. 6, 2003, pp. 875-962.

3. K. Tsuchiya, K. Sakaguchi, H. Kasuga, A. Kawakami, H. Taka, H. Kita and K. Ogino, "Synthesis of Charge Transporting Block Copolymers

Containing 2,7-Dimethoxycarbazole Units for Light Emitting Device," Polymer, Vol. 50, No. 3, 2010, pp. 616-622.

4. K. Tsuchiya, H. Kasuga, A. Kawakami, H. Taka, H. Kita and K. Ogino, "Synthesis of Bipolar Charge Transporting Block Copolymers and Characterization for Organic Light-Emitting Diode," Journal of Polymer Science Part A: Polymer Chemistry, Vol. 48, No. 7, 2010, pp. 1461-1468.

5. K. Tsuchiya, K. Sakaguchi, A. Kawakami, H. Taka, H. Kita, T. Shimomura and K. Ogino, "Charge Transporting Block Copolymer for Morphological Control in Light Emitting Device Based on Polymer Blends," Synthetic Metals, Vol. 160, No. 15-16, 2010, pp. 1679-1682.

6. J. K. Kim, S. Y. Yang, Y. Lee and Y. Kim, "Functional Nanomaterials Based on Block Copolymer Self-Assembly," Progress in Polymer Science, Vol. 35, No. 11, 2010, pp. 1325-1349.

7. L. Deng, P. T. Furuta, S. Garon, J. Li, D. Kavulak, M. E. Thompson and J. M. J. Fréchet, "Living Radical Polymerization of Bipolar Transport Materials for Highly Efficient Light Emitting Diodes," Chemistry of Materials, Vol. 18, No. 2, 2006, pp. 386-395.

8. B. Ma, B. J. Kim, L. Deng, D. A. Poulsen, M. E. Thompson and J. M. J. Fréchet, "Bipolar Copolymers as Host for Electroluminescent Devices: Effects of Molecular Structure on Film Morphology and Device Performance," Macromolecules, Vol. 40, No. 23, 2007, pp. 8156-8161.

9. Y. Tan, Z. Gu, K. Tsuchiya and K. Ogino, "Synthesis and Luminescent Properties of Block copolymers Based on Polyfluorene and Polytriphenylamine," Polymer, Vol. 53, No. 7, 2012, pp. 1444-1452.

10. Y. Tan, K. Tsuchiya and K. Ogino, "Synthesis of Polyfluorene Block Copolymers and Effect of Side Chain Group on Electroluminescent Device Performance," Chemistry Letters, Vol. 41, No. 3, 2012, pp. 257-259.

11. A. Nowakowska-Oleksy, J. Sołoducho and J. Cabaj, "Phenoxazine Based Units-Synthesis, Photophysics and Electrochemistry," Journal of Fluorescence, Vol. 21, No. 1, 2011, pp. 169-178.

12. H. Tanaka, K. Shizu, H. Miyazaki and C. Adachi, "Efficient Green Thermally Activated Delayed Fluorescence (TADF) from a Phenoxazine-Triphenyltriazine (PXZTRZ) Derivative," Chemical Communications, Vol. 48, No. 93, 2012, pp. 11392-11394.

13. J.-H Park, N. S. Cho, Y. K. Jung, H.-J. Cho, H.-K. Shim, H. Kim and Y. S. Lee, "Polymeric Light Emitting Properties and Structural Relationships of Fluorene-Based Conjugated Copolymers Containing Various Hole Transporting Derivatives," Organic Electronics, Vol. 8, No. 2-3, 2007, pp. 272-285.

14. Y. Pei, M. Otake, M. Vacha and H. Sato, "Synthesis and Characterization of a Novel Electroluminescent Polymer Based on Phenoxazine and Fluorene Derivatives," React. Funct. Polym., Vol. 67, No. 11, 2007, pp. 1211-1217.

15. K. Tsuchiya, T. Shimomura and K. Ogino, "Preparation of Diblock Copolymer Based on Poly(4-n-butyltriphenyl amine) via Palladium Coupling Polymerization," Polymer, Vol. 50, No. 1, 2009, pp. 95-101.

11

CHAPTER

Phase-Segregated Dendrigraft Copolymer Architectures

Lorena-Eugenia Sanchez Cadena [1] and Mario Gauthier [2,*]

[1]*DICIVA, Department of Environmental Sciences, University of Guanajuato/Ex-Hda. El Copal, Km. 9, Carretera Irapuato-Silao, Mexico*

[2]*Department of Chemistry, University of Waterloo, 200 University Ave. West, Waterloo, Ontario N2L 3G1, Canada*

ABSTRACT

Dendrigraft polymers have a multi-level branched architecture resulting from the covalent assembly of macromolecular building blocks. Most of these materials are obtained in divergent (core-first) synthetic procedures whereby the molecule grows outwards in successive grafting reactions or generations. Two main types of dendrigraft polymers can be identified depending on the distribution of reactive sites over the grafting substrate: Arborescent polymers have a large and variable number of more or less uniformly distributed sites, while dendrimer-like star polymers have a lower but well-defined number of grafting sites strictly located at the ends of the substrate chains. An overview of the synthesis and the characterization of dendrigraft copolymers with phase-segregated morphologies is provided in this review for both dendrigraft polymer families. The tethering of side-chains with a different composition onto branched substrates confers unusual physical properties to these copolymers, which are highlighted through selected examples.

Keywords: graft copolymers; dendritic; dendrigraft; arborescent; dendrimer-like; core-shell; core-shell-corona

1. INTRODUCTION

Different types of macromolecules with well-defined branched architectures have been synthesized over the past 30–40 years including star-branched polymers and polymacromonomers [1,2,3], dendrimers and hyperbranched polymers [4,5], and dendrigraft polymers [4,6,7]. Among these the dendritic architecture, represented by the latter three categories, has had a significant impact on polymer science because it encompasses highly branched macromolecules with a controllable molecular weight, number of branches, and terminal groups. The most distinctive characteristic of these polymers is a multi-level branched architecture, resulting from coupling reactions of either small molecule monomers (dendrimers and hyperbranched polymers), or macromolecular building blocks (dendrigraft polymers, referring collectively to arborescent and dendrimer-like star polymer systems).

The dendritic branching concept was first applied to graft polymers synthesized from polystyrene [8] and polyethylenimine [9] segments in 1991. The synthesis of these materials typically involves cycles of substrate functionalization and grafting reactions in a generation-based divergent (core-first) scheme as represented in **Scheme 1(a)**, whereby the molecule grows outwards from the substrate. Grafting linear chains onto a linear polymer substrate suitably functionalized with coupling sites yields a comb-branched polymer structure, also called a generation zero (G0) arborescent polymer. Provided that the grafted side-chains can be further functionalized with coupling sites, the process can be repeated to obtain successive generations of arborescent polymers. Thus the second functionalization and grafting cycle yields the first generation of arborescent polymer (G1, corresponding to the first generation with a dendritic or multi-level branched architecture), and so on. Geometric increases in molecular weight and branching functionality are typically observed over successive reaction cycles (generations) of arborescent polymers, while a narrow molecular weight distribution (polydispersity index $M_w/M_n < 1.1$) is maintained in most cases.

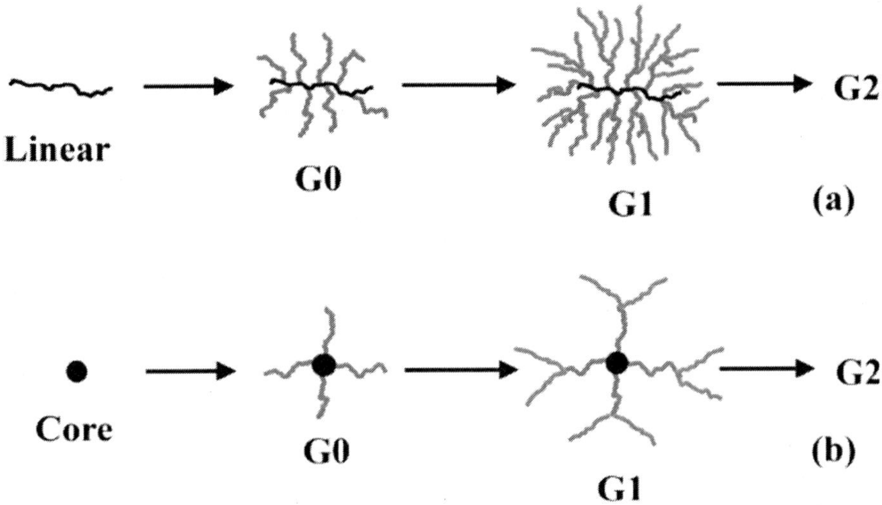

Scheme 1. Graphical representation of the synthesis of **(a)**arborescent and **(b)** dendrimer-like star polymers by successive grafting reactions of polymer segments.

Dendrimer-like star polymers [**Scheme 1(b)**], on the other hand, have a smaller but well-defined number of coupling sites strictly located at the chain ends of the previous generation, and are thus the dendrigraft polymers closest to dendrimers from a structural viewpoint [10]. Since the branching multiplicity of the chain ends (number of branching points) is relatively low (typically 2–4) in dendrimer-like polymers, the rate of molecular weight increase is also lower for these systems than for arborescent macromolecules. It should be noted that, to ensure consistency of the structure nomenclature used herein, dendrimer-like architectures with a single branching level (star polymers) are identified as G0 structures, while polymers with a second level of branching (branches upon branches) are G1 molecules, and so on.

Dendrigraft copolymers are unimolecular, covalently bonded species with well-defined structures (as opposed to multimolecular or randomly cross-linked assemblies), and branching functionalities ranging from a few to thousands of side-chains, whose basic characteristics (side-chain molecular weight and composition, branching functionality/density, and uniformity) can be accurately controlled in their synthesis. These features distinguish them from other heterogeneous macromolecular assemblies such as star-block copolymers with relatively low branching functionalities [11], metastable micelles obtained by the self-assembly of block copolymers in selective solvents [12,13] or stabilized through random cross-linking as in knedel-like structures [14], and core-shell polymer particles prepared by microemulsion polymerization [15,16]. This paper reviews the different methods developed for the synthesis of dendrigraft polymer structures, with emphasis on copolymer systems and the unusual properties arising from these heterogeneous morphologies.

2. ARBORESCENT POLYMERS WITH RANDOMLY DISTRIBUTED BRANCHING POINTS

2.1. Arborescent Polystyrene

Arborescent polystyrene is of particular importance, because it served as grafting substrate in the synthesis of many phase-separated copolymers. Consequently the synthesis of the homopolymers will be discussed first, but it will be shown that even these macromolecules display characteristics typical of heterogeneous copolymer systems due to differences in segmental density within the core and the outer shell (corona) of the molecules. The synthesis of arborescent polystyrene starts with the nucleophilic attack of "living" polystyryl anions onto a linear polystyrene substrate, randomly functionalized with suitable electrophilic substituents on a fraction (typically 25–30%) of the structural units, to generate the G0 polymer. Both chloromethyl [8] and acetyl [17] functionalities have been successfully applied as electrophilic coupling sites in the synthesis of arborescent structures (**Scheme 2**). In each case it was important to match the reactivity of the macroanions and the electrophile, in order to minimize side reactions and maximize the yield of the grafting reaction. It was thus necessary to "cap" the polystyryl anions with a single 1,1-diphenylethylene unit to suppress metal-halogen exchange reactions competing with coupling for the chloromethyl sites. The grafting yield (defined as the fraction of side-chains generated in the polymerization reaction becoming attached to the substrate) was maximized when grafting was carried out in tetrahydrofuran (THF) at −30 °C [8]. Acetyl groups, on the other hand, are susceptible to proton abstraction by the macroanion acting as a strong base. In this case the yield was maximized in THF in the 0–25 °C range, after capping the chains with a few units of relatively unhindered or more reactive monomers such as isoprene or 2-vinylpyridine, and the addition of LiCl to the reaction to suppress chain end ionization [17]. Irrespective of the grafting path selected, the functionalization and grafting reaction cycle can be repeated to obtain arborescent polystyrene structures of generations G1 and above.

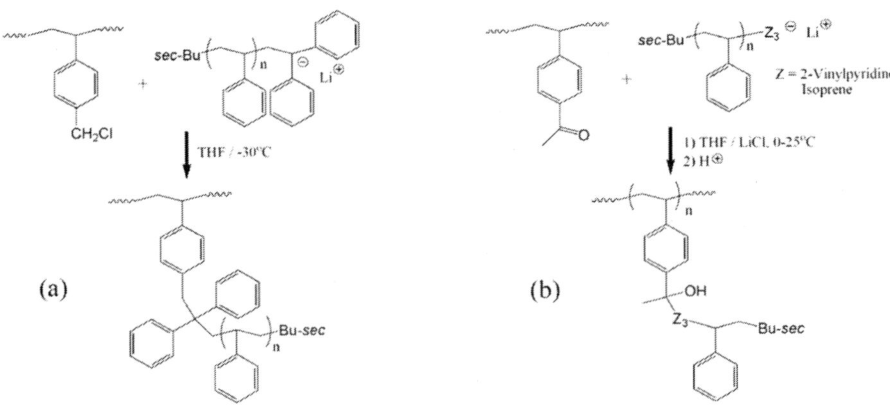

Scheme 2. Synthesis of arborescent polystyrene by grafting polystyryl anions onto **(a)** chloromethylated and **(b)** acetylated polystyrene substrates.

An important characteristic of arborescent polymers is that their apparent molecular weight, determined by size exclusion chromatography (SEC) analysis using a linear polystyrene standards calibration curve, is strongly underestimated (**Table 1**) [17]. Comparison of these values with the absolute weight-average molecular weights (M_w) determined from static light scattering measurements reveals discrepancies between both sets of results, the magnitude of the discrepancies increasing for the higher generations. This is a direct consequence of the very compact structure of arborescent polymers, which display an increase in average segmental density over successive generations, in contrast to randomly coiled linear polymers for which the hydrodynamic density decreases exponentially as the molecular weight increases. The absolute molecular weight and branching functionality of the molecules (determined from light scattering measurements) otherwise follow the expected trend, with roughly geometric increases over successive generations, while a low polydispersity index is maintained.

Table 1. Characterization data for a series of arborescent styrene homopolymers synthesized from acetylated polystyrene substrates.[a]

Sample	Generation	M_w[b]	M_w/M_n[b]	f_w[c]	M_w[d]
PS-PS5	0	5.3×10^4	1.08	11	3.3×10^4
G0PS-PS5	1	4.3×10^5	1.08	84	1.3×10^5
G1PS-PS5	2	3.9×10^6	1.09	690	4.5×10^5
G2PS-PS5	3	2.5×10^7		3800	

[a] Adapted from Reference 17 by permission from the American Chemical Society. All polymers synthesized from $M_w \approx 5{,}000$ side-chains with an acetylation level of *ca.* 25 mole% and 5 equiv of LiCl per "living" end. [b] Absolute values from SEC analysis with a light scattering detector or from batchwise light scattering measurements. [c] Branching functionality (number of side-chains added in the last grafting reaction). [d] Apparent M_w from SEC analysis using a linear polystyrene standards calibration curve.

Arborescent macromolecules, including even the homopolymers, may be regarded as having a dual phase morphology. This is because of the "diffuse layer" growth mechanism of the molecules over successive generations, represented in a cartoon fashion in **Scheme 1(a)**: A high segmental density (low chain mobility) region is expected in the core portion of the molecules, due to the coupling reactions leading to the formation of branching points, while the chains added in the last grafting cycle are only tethered to the substrate at one end. These remain relatively flexible and form a mobile shell (corona) of linear chains on the outside of the molecule. The corona should be rather diffuse not only because of the higher mobility of the chains, but also due to fluctuations in the position of the coupling sites along the side-chains of the substrate as shown in **Scheme 1(a)**. To a first approximation, the growth of arborescent molecules over successive generations may thus be represented by the addition of successive layers of chains onto a rigid core substrate (**Figure 1**).

The heterogeneous character of arborescent polystyrene molecules was first quantified by labeling a series of polymers with short branches ($M_w \approx 5{,}000$) with pyrene and observing their fluorescence quenching behavior when exposed to small molecule (nitrobenzene) and macromolecular (nitrated polystyrene) quenchers in solution [18]. The diffusion coefficients measured for the small molecule quencher within the arborescent polystyrene molecules decreased as a function of increasing generation number. Quenching experiments with linear nitrated polystyrene led to downward curvature in Stern-Volmer plots characteristic for protective quenching, *i.e.*, reduced accessibility of a portion of the chromophores to the quencher groups. The application of a fractional quenching model allowed the determination of the quenching rate constant and the fraction of accessible chromophores for each sample. These parameters decreased for the upper generation polymers, in agreement with the diffuse layer growth model of **Figure 1**: When adding successive layers of constant thickness (corresponding to the chains added in the last grafting reaction) on the surface of a sphere, the volume fraction of the core (fraction of inaccessible material, f_c) should increase over successive generations.

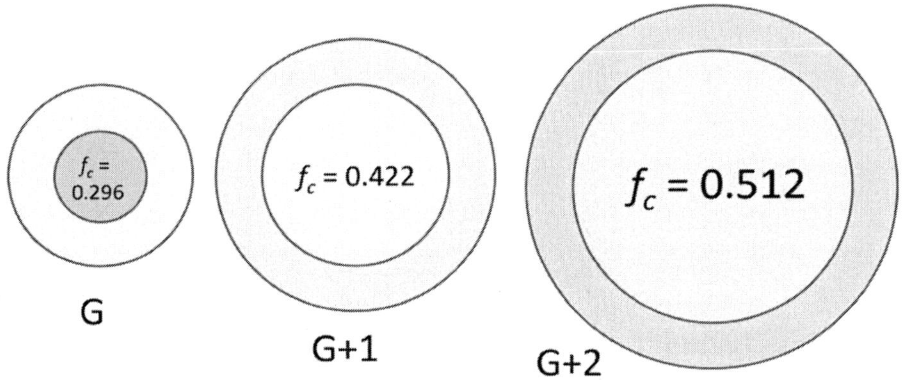

Figure 1. Schematic representation of the diffuse layer growth model for arborescent polymers. The volume fraction of inaccessible (core) material in the molecules, for a core of initial radius R = 2 onto which layers of thickness d = 1 are added, is represented by f_c.

The morphology of arborescent polystyrene molecules was also the topic of a detailed investigation using small-angle neutron scattering (SANS) experiments with contrast matching for polymers derived from arborescent polystyrene substrates grafted with deuterated polystyrene chains forming the shell [19]. Most interestingly, radial SANS contrast density profiles (proportional to the local concentration of scattering segments) could be generated for the core and the shell portions of the molecule under matched contrast conditions, and for polymers of two different generations (G3 *vs.* G4). It is clear from the profiles obtained (**Figure 2**) that the core and the shell components are segregated, even for systems expected to display minimal phase separation such as copolymers with a hydrio-polystyrene core and a deutero-polystyrene shell. The core-shell interface is nonetheless diffuse due to significant mixing between the core and the shell components, in agreement with the diffuse layer growth model proposed above.

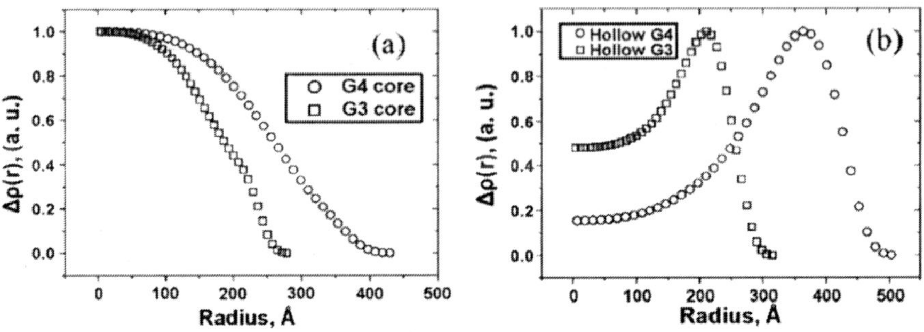

Figure 2. Radial SANS contrast profiles generated from measurements on arborescent polystyrenes with a deuterated polystyrene shell under **(a)** shell and **(b)** core contrast matching conditions. Adapted from Reference 19 by permission from the American Chemical Society.

2.2. Poly(ethylene oxide) Copolymers

The first example of an arborescent copolymer structure with a clearly heterogeneous morphology, published in 1996, was for polymers incorporating an arborescent polystyrene core grafted with poly(ethylene oxide) (PEO) segments at the chain termini [20]. The synthesis of micelle-like core-shell macromolecules with different core sizes (generations) and varying PEO contents or shell thicknesses was demonstrated.

The preparation of the arborescent polystyrene core relied on the grafting technique with chloromethyl coupling sites [8], but the addition of a PEO shell required a modification of the procedure in the last step. The synthesis of a G1 copolymer is illustrated in **Scheme 3** as an example. A comb-branched (G0) polystyrene substrate was prepared in the usual fashion, by initiating the polymerization of styrene with *sec*-butyllithium and capping with 1,1-diphenylethylene, followed by titration of the "living" anions with a solution of chloromethylated linear polystyrene. A twice-grafted (G1) substrate with protected hydroxyl end-groups was obtained by using (6-lithioexyl)acetaldehyde acetal (LHAA in**Scheme 3**) to initiate the polymerization of styrene, followed by capping and coupling with the chloromethylated G0 polymer. The acetal functionalities were then hydrolyzed and the hydroxyl end-groups were deprotonated with potassium naphthalide before adding ethylene oxide to the substrate. To maintain a narrow molecular weight distribution in the reaction, it was necessary to eliminate residual chloromethyl sites on the substrate via metal-halogen exchange prior to shell growth. The core-

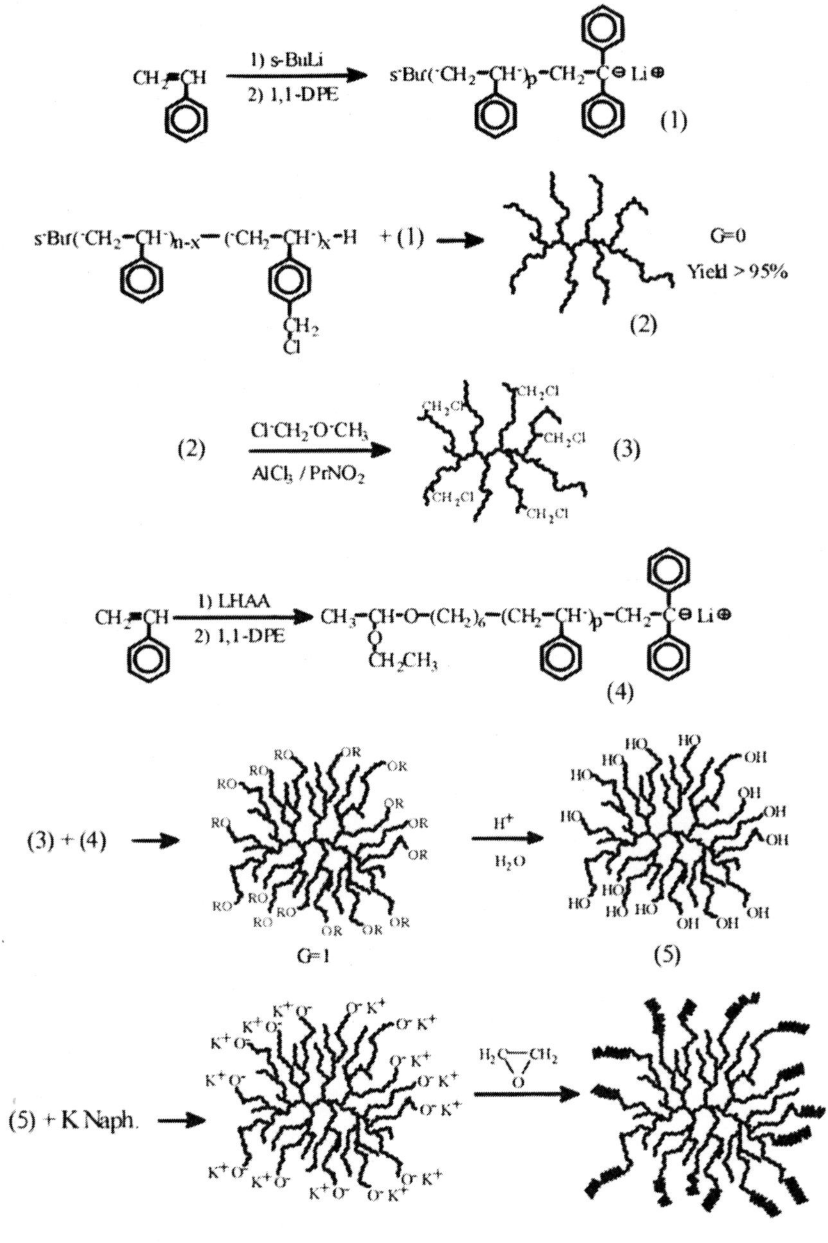

Scheme 3. Synthesis of arborescent polystyrene-*graft*-poly(ethylene oxide) amphiphilic copolymers. Reproduced from Reference 20 by permission from the American Chemical Society.

shell copolymers derived from a G1 core with $M_w = 7 \times 10^5$ had PEO contents of 19% and 66% by weight, and apparent polydispersities $M_w/M_n = 1.07–1.21$. Another sample containing 36% PEO by weight was obtained from a G4 core having $M_w \sim 10^8$.

Comparison of the hydrodynamic radii determined from dynamic light scattering measurements for the core homopolymers and the core-shell copolymers demonstrated that the hydrophilic poly(ethylene oxide) chains on the surface of the molecules essentially adopted a randomly coiled conformation, with minimal stretching of the chains. The solubility characteristics of the macromolecules were consistent with a core-shell morphology: The copolymers were freely soluble in methanol, in contrast to the bare polystyrene cores.

The self-assembly of the amphiphilic arborescent polystyrene-*graft*-poly(ethylene oxide) copolymers at the air-water interface was investigated as a function of the composition and the structure of the molecules [21,22]. The types of superstructures formed by the amphiphiles strongly depended on characteristics including their composition, branching functionality, and structural rigidity. Thus copolymers with low PEO contents simply dewetted from the water surface and formed large island-like clusters, while molecules with a high PEO content had little tendency to self-assemble in the absence of compression [**Figures 3(a,c)**] [21]. However copolymers of intermediate compositions (*ca.* 20–35% PEO content by weight) self-assembled into ribbon-like superstructures [**Figure 3(b)**] mainly via end-to-end aggregation, the extent of aggregation (ribbon length) being determined not only by the composition but also by the flexibility of the molecules (branching functionality and molecular weight of the polystyrene side-chains within the core). This end-to-end aggregation phenomenon was rationalized through sideways rearrangement of the linear PEO chains in the shell leading to an increase in the magnitude of van der Waals interactions between the hydrophobic cores along the long axis of the ribbons, while thickening of the stabilizing PEO layer on the sides of the ribbons hindered further aggregation of the molecules [**Figure 3(d)**]. It was also shown that the length of the ribbon-like superstructures increased under the influence of compression in Langmuir force balance experiments, or through temperature variations influencing the degree of hydration of the stabilizing PEO chains [22].

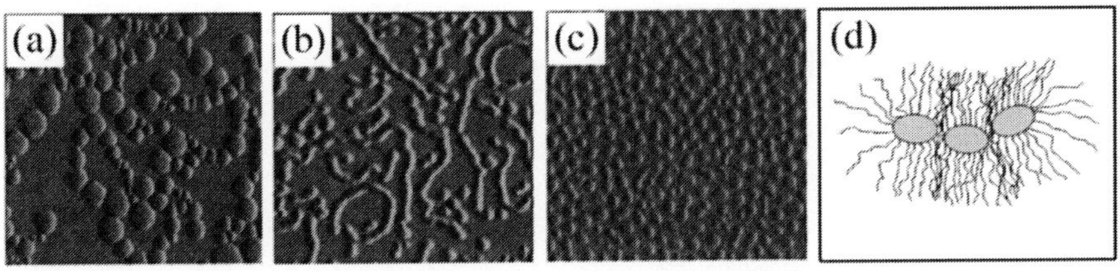

Figure 3. Self-assembly of arborescent polystyrene-*graft*-poly(ethylene oxide) copolymer molecules at the air-water interface without compression: **(a)** low branching functionality, low PEO content G1-30PS-LB-15, **(b)** low branching functionality, intermediate PEO content G1-30PS-LB-31, and **(c)** high branching functionality, high PEO content G1-30PS-HB-43. Each picture has a width of 1.5 μm. **(d)** End-to-end aggregation mechanism proposed to explain the formation of ribbons in (b). Adapted from Reference 21 by permission from Wiley-VCH.

2.3. Poly(2-vinylpyridine) Copolymers

The synthesis of arborescent polystyrene-*graft*-poly(2-vinylpyridine) copolymers was initially achieved by coupling poly(2-vinylpyridinyl)lithium with chloromethylated polystyrene substrates of generations up to G2 in tetrahydrofuran with N,N,N′,N′-tetramethylethylenediamine (TMEDA) [23]. This reaction is analogous to **Scheme 2(a)**, where the 1,1-diphenylethylene-capped polystyryllithium species are replaced with the "living" poly(2-vinylpyridinyl)lithium chains without capping. Copolymers with branching functionalities ranging from 14–3,880 and molecular weights of 8.2×10^4 to 6.7×10^7 were thus obtained while maintaining low apparent polydispersity indices ($M_w/M_n \approx 1.06$–1.15). The polystyrene content of these copolymers was rather low and varied from *ca.* 3 to 19% by weight, depending on the arborescent polystyrene substrate generation and the molecular weight of the side-chains used in the reaction. Comparable results were subsequently obtained when coupling poly(2-vinylpyridinyl)lithium with acetylated polystyrene substrates in the presence of LiCl, similarly to**Scheme 2(b)** (in this case the side-chains correspond to a long segment of "Z" units without styrene) [24].

Dynamic light scattering measurements demonstrated that arborescent poly(2-vinylpyridine) (P2VP) copolymers, when protonated with excess HCl (at high ionic strengths), expand much more in solution than their linear homologues (**Figure 4**). This effect was attributed to the higher charge density (charge/volume) attained in these branched polyelectrolytes [25], which can be viewed as P2VP homopolymers to a first approximation due to their low polystyrene contents. Nevertheless the expansion ratio of the copolymers, expressed as the ratio of hydrodynamic volumes in the ionized and the neutral states, decreased as the generation number (structural rigidity) of the copolymer increased. It was also larger for copolymers containing longer, more flexible side-chains.

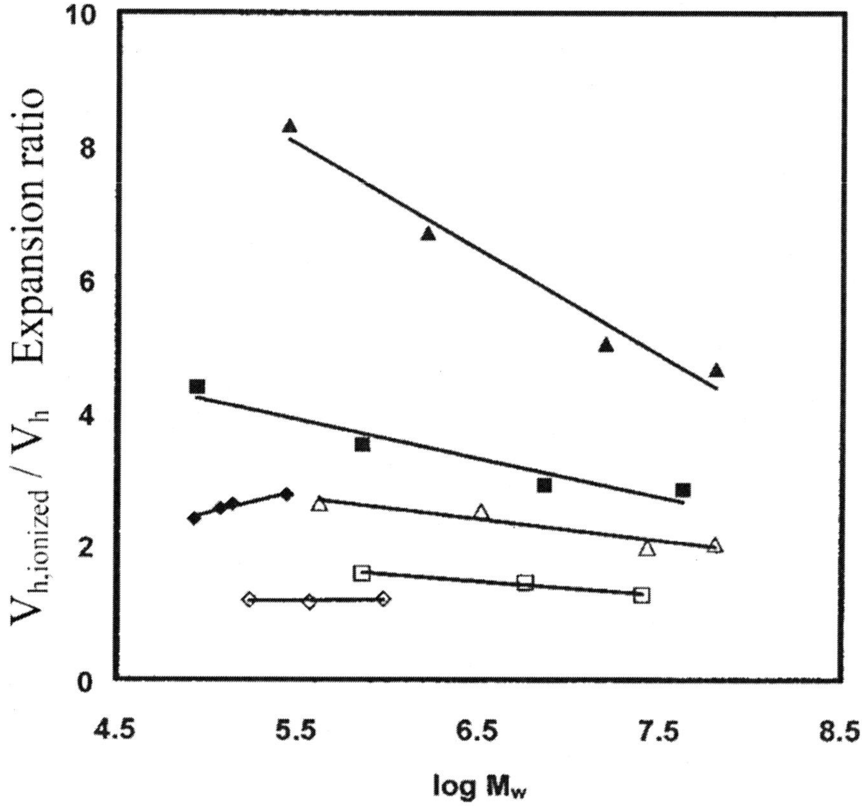

Figure 4. Hydrodynamic volume expansion for arborescent polystyrene-*graft*-poly(2-vinylpyridine), polystyrene-*graft*-poly(methacrylic acid), and their linear homologues upon ionization. Linear P2VP (◊), arborescent P2VP copolymers with short ($M_w \sim$ 5,000; □) and long ($M_w \sim$ 30,000; Δ) side-chains dissolved in MeOH/H_2O 95/5 with 0.10 N HCl. Linear PMAA (♦), arborescent PMAA copolymers with short ($M_w \sim$ 5,000; ■) and long ($M_w \sim$ 30,000; ▲) side-chains dissolved in MeOH/H_2O 95/5 with 0.05 N NaCl and neutralized with NaOH. Adapted from Reference 25 by permission from Wiley.

An investigation on the dilute-solution structure of aggregated, partially ionized arborescent P2VP copolymer molecules at low ionic strengths was completed by Yun *et al.* using SANS measurements [26]. Solutions of charged G1 arborescent polystyrene-*graft*-poly(2-vinylpyridine) copolymers in methanol-d4 and in D_2O were characterized in the dilute concentration regime (mass fraction f = 0.005–0.05). Upon addition of less than one equivalent of HCl per 2-vinylpyridine unit (degree of ionization $0 < \alpha < 0.4$), a peak appeared in plots of the scattering intensity $I(q)$ *vs.* q (where $q = \sin(\theta)4\pi/\lambda$ is the scattering vector, θ is the observation angle, and λ is the wavelength) due to aggregation of the molecules. The intermolecular distance within the aggregates, calculated from the peak position, corresponded to the expected value for the formation of superstructures having a uniform liquid-like distribution of molecules. The lower dielectric constant of methanol-d4 resulted in long-range Coulombic interactions persisting to lower polymer concentrations than in D_2O. Dynamic light scattering measurements displayed two diffusive relaxation processes under these conditions, the slow diffusion mode reflecting aggregate formation within the solutions. The SANS scattering peak and the slow diffusion mode both disappeared upon addition of NaCl or excess HCl to the solutions, due to screening of the electrostatic interactions promoting aggregate dissociation. Furthermore, the G1 copolymer grafted with long P2VP side-chains (M_w = 30,000) formed a gel upon addition of HCl for volume fractions $\phi > 0.01$. This result is consistent with enhanced molecular expansion promoting gelation for these branched polyelectrolytes, the aggregated network breaking up upon addition of NaCl or excess HCl due to screening of the electrostatic interactions [26].

The ability to control the solution properties of arborescent polyelectrolytes by varying parameters such as the length and the number of P2VP segments in the molecules is interesting for the design of pH-sensitive reversible gels with tunable properties such as the sol-gel transition point and the gel modulus [23].

The arborescent polystyrene-*graft*-poly(2-vinylpyridine) copolymers were also investigated for their potential application as unimolecular micelles, to take advantage of their heterogeneous morphology combining a hydrophobic core and a corona of polyelectrolyte chains. The copolymers derived from a linear polystyrene substrate aggregated when dissolved in aqueous HCl solutions, but those incorporating G0–G2 substrates yielded unimolecular (non-aggregated) micellar solutions with interesting solubilization properties for various polycyclic aromatic hydrophobes [27]. It was shown that the solubilization kinetics and capacity of the micelles varied with their composition (polystyrene content), but also with their structure (generation number, molecular weight of the P2VP side-chains). The kinetics for pyrene solubilization were found to obey a first order kinetic model where the time-dependent concentration of pyrene within the micelles, $[PyM]_t$, could be expressed in terms of the equilibrium concentration $[PyM]_{eq}$ as $[PyM]_t = [PyM]_{eq}(1 - e^{-kt})$ (**Figure 5**), where the solubilization rate constant k varied *ca.* 3-fold among the different samples investigated. While the copolymers used in the investigation were not specifically designed to serve as drug delivery vehicles (being non-biocompatible and having low polystyrene weight fractions), the study still demonstrates that the solubilization or release characteristics of arborescent unimolecular micelles can be tailored through variations in structure and composition. The solubilization capacity of the micelles also varied with the type of hydrophobic probe solubilized, significantly higher capacities being observed for probes with non-specific solubilization within the polystyrene core and the P2VP shell of the molecules rather than exclusively within the core [27]. Similar observations were reported in a release study using lidocaine and indomethacin as model drugs, a higher loading capacity being observed for indomethacin due to strong interactions between its carboxylic acid group and the P2VP component of the copolymers in the side-chains [28].

2.4. Poly(methacrylic acid) Copolymers

The grafting of poly(*tert*-butyl methacrylate) chain segments onto arborescent polystyrene was achieved by a procedure analogous to **Scheme 2(a)**, albeit it was necessary to replace the chloromethylated polystyrene substrates with their bromomethylated analogues to achieve reasonable grafting yields [25]. The resulting copolymer molecules had poly(*tert*-butyl methacrylate) contents ranging from 81 to 98% by weight. The *tert*-butyl ester group was easily hydrolyzed by treatments with trimethylsilyl iodide and HCl, and neutralization of the methacrylic acid units with a base generated the corresponding polyelectrolytes. These molecules displayed enhanced molecular expansion in solution upon neutralization, the effect being even more important than for arborescent polystyrene-*graft*-poly(2-vinylpyridine) copolymers (**Figure 4**).

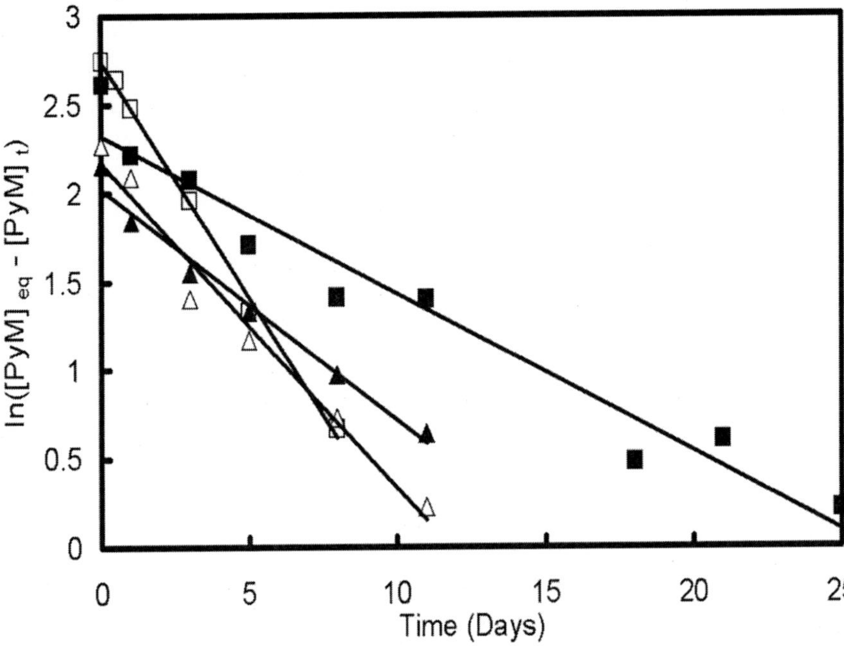

Figure 5. Analysis of the micellar solubilization kinetics of pyrene by arborescent polystyrene-*graft*-poly(2-vinylpyridine) copolymers. The following rate constants were obtained: Δ G0PS-P2VP5-12, k = (0.18 ± 0.01) d^{-1}; ▲ G1PS-P2VP5-13, k = (0.13 ± 0.01) d^{-1}; □ G1PS-P2VP5-20, k = (0.266 ± 0.006) d^{-1}; ■ G2PS-P2VP5-18, k = (0.089 ± 0.007) d^{-1}. Reprinted from Reference 27 by permission from Elsevier.

2.5. Polyisoprene Copolymers

Arborescent copolymers containing polyisoprene (PIP) segments were obtained by coupling polyisoprenyllithium "living" chains with either chloromethylated [29] or acetylated [30] polystyrene substrates according to the procedures outlined in **Scheme 2**. The synthesis from chloromethylated substrates proceeded most efficiently after capping of the chains with 1,1-diphenylethylene and coupling at −30 °C. For grafting on acetylated polystyrene substrates the yield was maximized at 25 °C, in the presence of 5 equiv of LiCl per "living" end to attenuate the reactivity of polyisoprenyllithium. Isoprene was polymerized with *sec*-butyllithium either in cyclohexane or in tetrahydrofuran, to yield predominantly *cis*-1,4 or mixed microstructures, respectively. However when the polymerization was carried out in cyclohexane, the addition of tetrahydrofuran before the coupling step was still preferable to increase the grafting yield. Arborescent copolymers were synthesized by grafting PIP side-chains having a M_w of either 5,000 (PIP5) or 30,000 (PIP30) onto linear, G0, G1, and G2 polystyrene. The copolymers with short (PIP5) side-chains had a polyisoprene content of 80–90% by weight, increasing to 94–98% for long (PIP30) side-chains. The expected geometric increases in branching functionality and molecular weight were observed, while a narrow molecular weight distribution was maintained over successive generations.

Film formation by the isoprene copolymers on mica surfaces was investigated using tapping mode atomic force microscopy (AFM) after spin-casting from different solvents [30]. Measurements in the phase contrast mode are ideal to probe phase separation in these systems, because the glassy polystyrene core and the rubbery polyisoprene shell should lead to significant phase changes in the signal. A heterogeneous morphology was indeed detected for these copolymer systems (**Figure 6**) and, surprisingly, even for the copolymers with very low polystyrene contents (down to *ca.* 3% by weight). It was interesting that when heptane, a solvent selective for the polyisoprene segments, was used to prepare the films, phase separation between the polystyrene core and the polyisoprene shell was significantly enhanced in phase contrast imaging, particularly for copolymers with longer PIP side-chains (low polystyrene contents). Conversely, in nonselective solvents (toluene and chloroform) the phase contrast was reduced, presumably due to enhanced mixing of the polystyrene and polyisoprene components. This clearly confirms the occurrence of phase separation between the polystyrene core and the polyisoprene shell of the molecules on the nanometric scale, but also that the extent of phase separation achieved strongly depends on the selectivity of the solvent used in film preparation [30].

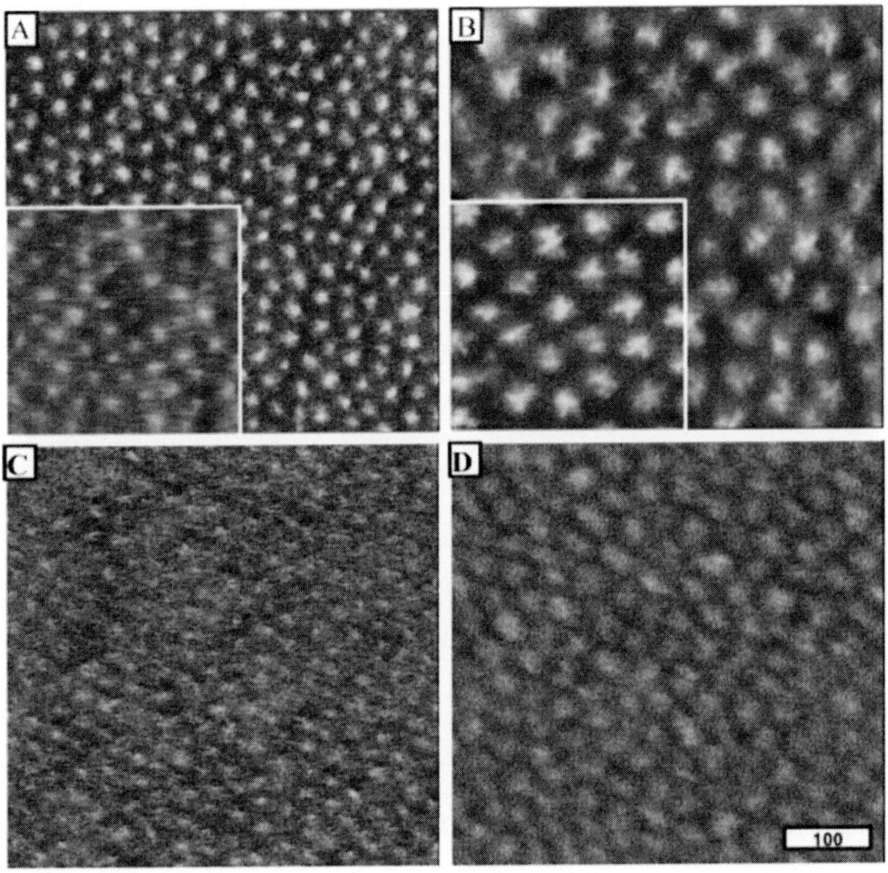

Figure 6. AFM phase contrast images for films of arborescent polystyrene-*graft*-polyisoprene obtained by spin-casting (concentration ~1 mg/mL): G0PS-PIP30 in heptane (**A**) and chloroform (**C**); G1PS-PIP30 in heptane (**B**) and chloroform (**D**). The insets of (A) and (B) are the height images, showing the topology of the monolayers, and the 100-nm scale bar shown in (D) is the same for all the images. Reprinted from Reference 30 by permission from the American Chemical Society.

While AFM imaging provides direct evidence for phase segregation within the individual styrene-isoprene graft copolymer molecules, characterization results obtained by other methods were also consistent with a heterogeneous morphology. Thus breakdown of the time-temperature superposition principle was observed in an investigation of the rheological properties of arborescent polystyrene-*graft*-polyisoprene copolymers, when constructing modulus-frequency master curves for copolymers with sufficiently large (G1 and G2) polystyrene cores [31]. This type of behavior is typical for heterogeneous polymer systems.

The arborescent polystyrene-*graft*-polyisoprene copolymers were very recently investigated for their potential use as polymer processing additives in linear low density polyethylene (LLDPE) resins [32]. The copolymers, after modification with fluorinated substituents on a fraction of the isoprene units, were shown to suppress sharkskin formation and delay the onset of cyclic melt fracture when used at a concentration of 0.5% by weight in the extrusion of LLDPE monofilaments. Pressure reductions reaching up to 29% as compared with the unmodified LLDPE resin were also observed in the extrusion experiments.

2.6. Layered Copolymer Architectures

All the examples discussed so far are for copolymers with core-shell morphologies, derived from arborescent polystyrene substrates grafted with side-chains having a different chemical composition and forming the shell. Multilayered core-shell-corona systems have also been obtained recently by a similar approach. Thus the coupling reaction of "living" poly(2-vinylpyridine)-*block*-polystyrene carrying the macroanion at the P2VP chain end yielded a layered morphology with a polystyrene core, an inner shell of P2VP, and a corona of polystyrene chains on the outside (**Figure 7**) [33]. The usefulness of these complex macromolecular architectures lays in their ability to complex various transition metal salts within the P2VP shell, while the polystyrene corona ensures good solubility of the metal-loaded micelles with minimal aggregation in non-polar solvents such as toluene. These copolymers may thus serve as templates for the formation of metallic nanoparticles, in analogy to block copolymer micelles [34], but are potentially much more powerful: Spherical micelles are only formed by block copolymers under specific conditions (solvent selective for the shell component, appropriate volume fraction of each block), but these restrictions are relaxed for arborescent core-shell-corona copolymers. The covalently bonded architecture of these molecules also precludes extensive rearrangements of the side-chains, including their dynamic exchange (dissociation) from the micellar structure. Consequently, the micellar morphology of the arborescent copolymer molecules is preserved even under non-selective solvency conditions (e.g., tetrahydrofuran, chloroform) that hinder the self-assembly of poly(2-vinylpyridine)-*block*-polystyrene copolymers into micelles. A preliminary investigation on the core-shell-corona copolymer systems demonstrated that intricate metallic nanoparticle morphologies could be obtained upon loading metallic compounds such as tetrachloroauric acid ($HAuCl_4$) in these template molecules (**Figure 7**).

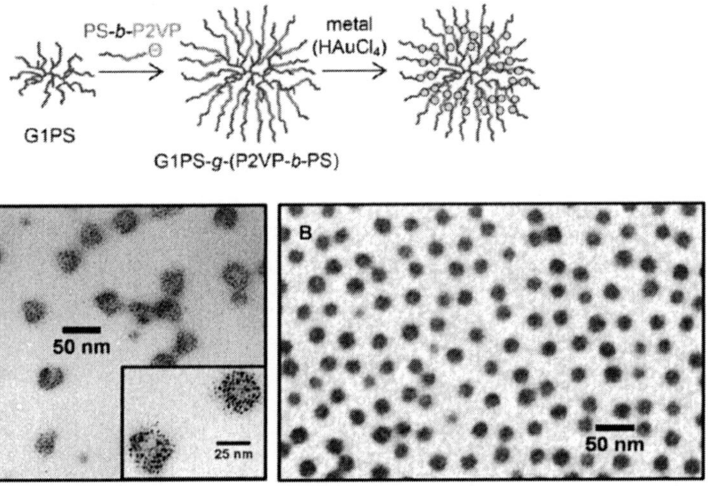

Figure 7. Top: Schematic representation of the synthesis of a core-shell-corona arborescent copolymer structure by grafting "living" block copolymer chains onto an arborescent polystyrene substrate. Bottom: Examples of TEM micrographs obtained for G2 (**A**) and G0 (**B**) templates loaded with $HAuCl_4$ at 0.25 and 0.50 equiv relative to 2VP residues, respectively. Adapted from Reference 33 by permission from the American Chemical Society.

2.7. One-pot Synthetic Procedures

The stepwise or generation-based synthesis is clearly advantageous in terms of the extent of control achieved over the reaction and the characterization of the products obtained: The molecular weight of the side-chains and the substitution level of the coupling substrates can be easily controlled. A sample of side-chains can be removed from the reactor prior to the grafting reaction and analyzed by SEC to determine their exact molecular weight and size uniformity (polydispersity index). Finally, since the absolute molecular weight of the substrate and the graft polymer can be measured, the exact number of side-chains grafted onto the substrate can be determined for each generation. On the down side, product isolation after functionalization of the substrate, and the generation of "living" chains for each grafting cycle are time-consuming and relatively expensive procedures. To circumvent these problems, a one-pot process has been developed by Yuan and Gauthier for the synthesis of arborescent polymers which does not require the isolation and purification of reaction intermediates [35,36]. The procedure is applicable to the preparation of G0 and G1 homo- and copolymer architectures from successive functionalization and grafting reactions carried out sequentially in the same reactor, and yields products with fairly low polydispersity indices ($M_w/M_n \approx 1.2$–1.3 typically).

The one-pot method starts with the anionic copolymerization of styrene and *m*-diisopropenylbenzene as shown in **Scheme 4**. To compensate for the unfavorable copolymerization behavior of the two monomers, a semi-batch monomer addition protocol was developed to minimize segregation of the *m*-diisopropenylbenzene units along the chains. After the preparation of the linear copolymer, the "living" chains are terminated by titration of the solution with methanol. The pendent isopropenyl moieties of the copolymer are then activated with *sec*-butyllithium to generate initiating sites along the linear substrate, and a G0 styrene-*m*-diisopropenylbenzene copolymer is obtained by successive additions of a styrene-*m*-diisopropenylbenzene monomer mixture. Upon termination of the "living" G0 copolymer, the pendent isopropenyl groups on the side-chains are again activated and used to initiate the polymerization of styrene to obtain a G1 styrene homopolymer, or other monomers such as 2-vinylpyridine or *tert*-butyl methacrylate (as shown in **Scheme 4**) to give the corresponding copolymer. This approach, while not yielding products as well defined as the stepwise method, is much more convenient when large samples are required for less demanding applications. The formation of linear homopolymer in the procedure outlined in **Scheme 4** (due to incomplete reaction of the *sec*-butyllithium used to activate the initiating sites) varied from 6–34% of the total sample weight, depending on the monomer and the substrate used in the reaction. Unfortunately, the exact characterization of the branched polymer structure obtained (branching functionality and molecular weight) is difficult in this case, as for all anionic *grafting from* procedures in general. It was indeed pointed out that in these anionic *grafting from* procedures, the branching functionality is usually approximated from the molecular weight of the graft polymer and the linear contaminant formed. Unfortunately this calculation method can lead to an underestimated branching functionality, because chain growth is faster for the linear polymer contaminant than for the graft polymer [35,36].

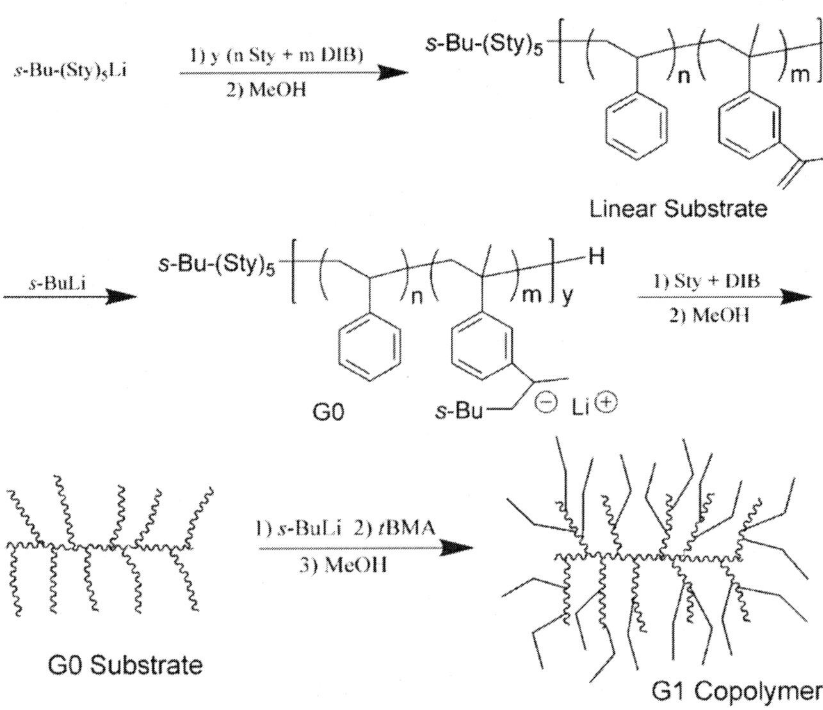

Scheme 4. Synthesis of arborescent polystyrene-*graft*-poly(*tert*-butyl methacrylate) copolymers by a one-pot process. Reprinted from Reference 36 by permission from Wiley-VCH.

The one-pot synthesis of arborescent polyisobutylene structures by cationic polymerization in the presence of inimers (monomers containing a latent initiating site within their structure) has received much attention by the group of Puskas [37]. This technique, initially developed for the preparation of arborescent isobutylene homopolymers, was subsequently extended to the preparation of copolymers with polystyrene or poly(*p*-methylstyrene) segments as shown in **Scheme 5**. For example, the cationic copolymerization of isobutylene with 4-(2-methoxyisopropyl)styrene serving as inimer produces "living" macromonomers which can self-condense *in situ and* copolymerize with isobutylene in a*grafting through* scheme to give dendritic branched structures carrying multiple cationic active sites at the chain ends. The addition of styrene or *p*-methylstyrene to the "living" branched substrate leads to the growth of glassy styrenic segments (*grafting from* process). The main interest in these copolymers lays in their elastomeric properties, as the glassy segments aggregate into phase-separated domains, giving rise to thermoplastic elastomers (TPE) with excellent tensile properties [38,39,40]. The biomedical application of these systems as coatings for the sustained release of drugs in stents has also received attention [41], considering that the analogous linear block copolymer TPE are already FDA-approved.

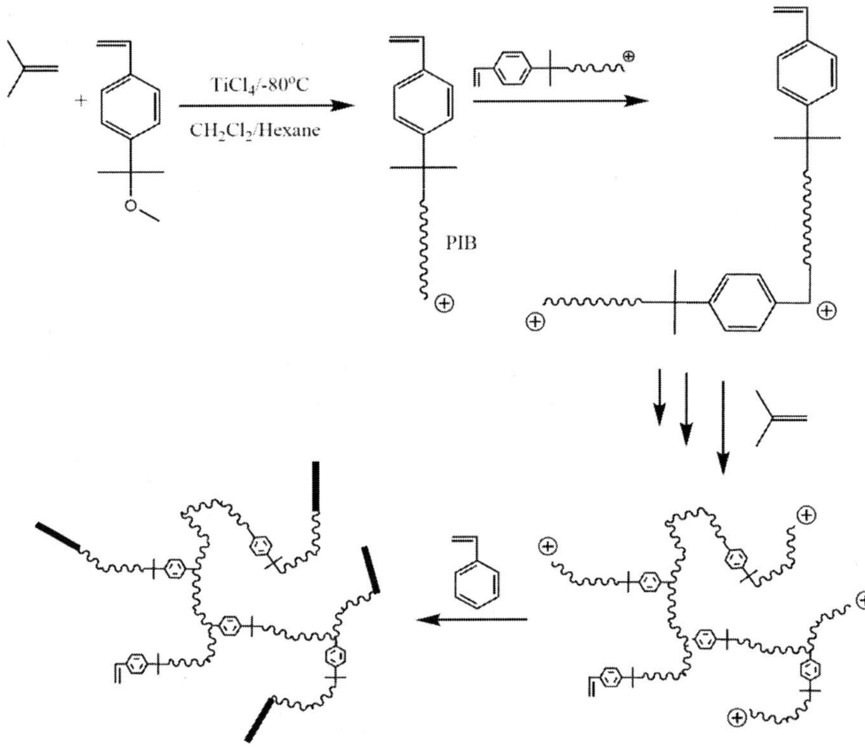

Scheme 5. One-pot synthesis of arborescent polyisobutylene with polystyrene end-blocks by the cationic inimer method according to Puskas.

3. ARBORESCENT POLYMERS WITH SEGREGATED BRANCHING POINTS

The arborescent polymer syntheses discussed in the previous section depend (with the exception of the inimer technique) on the random introduction of coupling sites on the substrate in a post-polymerization modification reaction for each grafting cycle. An alternate approach avoiding the functionalization step was suggested by Deffieux *et al.*, by growing short polymer segments providing a variable but controllable number of coupling sites grouped at the end of the substrate chains. The synthesis of an amphiphilic copolymer structure incorporating an arborescent polystyrene substrate and a corona of poly(methyl vinyl ether) chains is illustrated in **Scheme 6** as an example [42]. A linear poly(chloroethyl vinyl ether) segment was first synthesized by cationic polymerization and side-chains, generated by the anionic polymerization of styrene initiated with 3-lithiopropionaldehyde diethyl acetal, were *grafted onto* the substrate to obtain a G0 polymer. After conversion of the acetal moieties to α-iodo end-groups with trimethylsilyl iodide (TMSI), the substrate was activated for the cationic polymerization of a new portion of chloroethyl vinyl ether to extend (*graft from*) the side-chains of the G0 polymer. An additional anionic grafting cycle with acetal-capped polystyryllithium provided a G1 substrate, that was again activated with TMSI and $ZnCl_2$ before adding methyl vinyl ether to grow a shell of water-soluble poly(methyl vinyl ether) segments from the end of the polystyrene core chains. In a closely related technique, amphiphilic structures were obtained by successive additions of 2-((*tert*-butyldimethylsilyl)-oxy)ethyl vinyl ether (SiVE) and 1,2:3,4-di-O-isopropylidene-6-O-(2-vinyloxy ethyl)-D-galactopyranose monomers, followed by deprotection of the silyl ether units to generate a hydrophilic block copolymer corona of poly(2-hydroxyethyl vinyl ether) and saccharide-functionalized poly(2-hydroxyethyl vinyl ether) segments (**Figure 8**) [43,44]. These amphiphilic copolymers were shown to display micellar properties, enabling the solubilization of hydrophobic compounds such as pyrene and manganese tetraphenyl porphyrin in water [44]. Another strategy investigated for the generation of amphiphilic macromolecules from the arborescent polystyrene substrates involved a transacetalization reaction of the acetal-capped polystyrene substrate (G1 structure in **Scheme 6**) with α-bis(hydroxymethyl)-functionalized poly(ethylene oxide) chains [45].

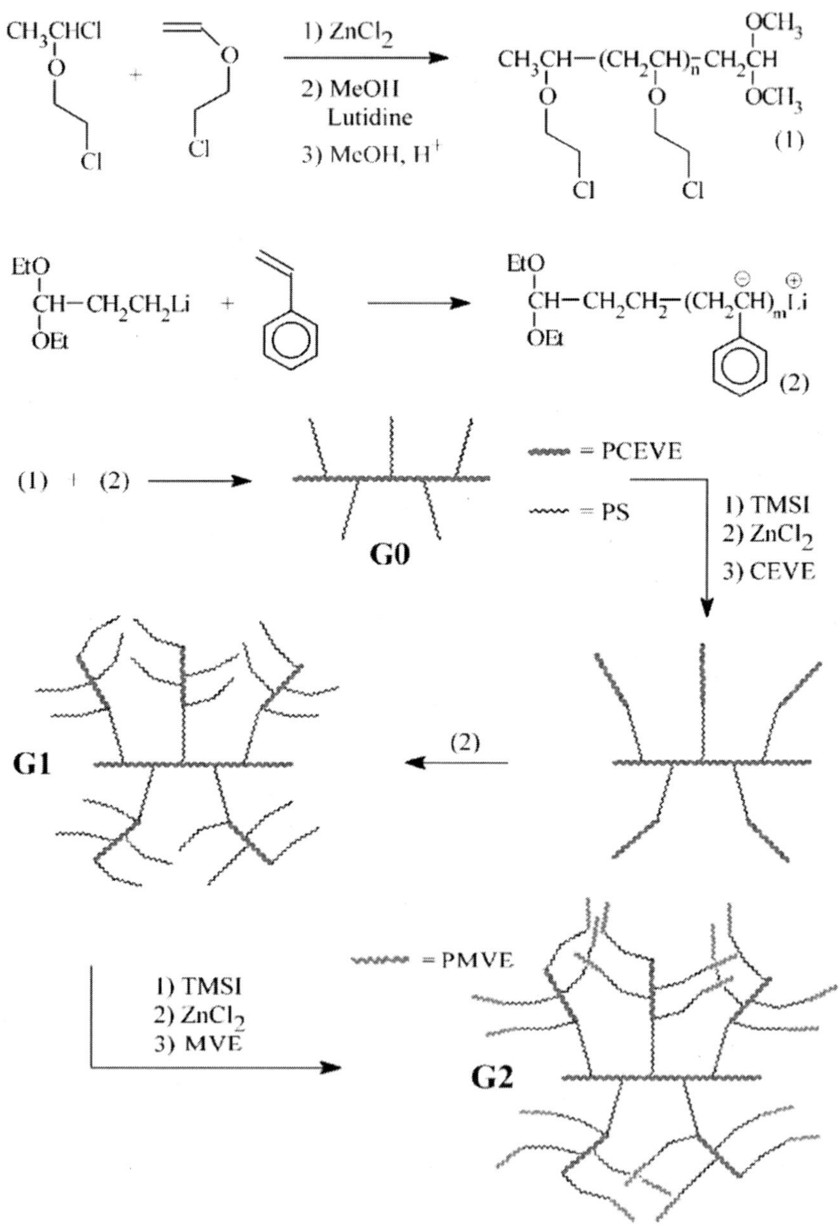

Scheme 6. Arborescent polystyrene-*graft*-poly(methyl vinyl ether) copolymer synthesis according to Deffieux *et al.*

A similar method was developed by Dworak and Wałach for the synthesis of double amphiphilic arborescent copolymer structures with poly(ethylene oxide) and polyglycidol segments [46]. In this case an anionic "living" 4-arm poly(ethylene oxide) star-branched substrate was used to initiate the polymerization of a protected glycidol monomer (1-ethoxyethyl glycidyl ether) and extend the arms with a short poly(glycidol acetal) end-block. After hydrolysis of the acetal protecting groups, the free glycidol moieties of the G0 polymer were activated with a base and used for a second cycle of sequential polymerizations of ethylene oxide and glycidol acetal monomers. The G1 polymer did not display a phase-segregated morphology in differential scanning calorimetry (DSC) measurements due to the short (number-average degree of polymerization X_n = 4–6) polyglycidol segments employed, but crystallization of the poly(ethylene oxide) phase was observed for copolymers with longer (X_n = 30–50) side-chains. It was shown that the copolymers could serve as transfer (solubilizing) agents for calmagite (a polar azo dye) in nonpolar solvents such as CH_2Cl_2, the solubilization capacity reaching up to 4% dye by weight of dendritic copolymer. A variation of this procedure starting from a triblock copolymer of poly(ethylene oxide) with X_n = 90 flanked with two short (X_n = 5) polyglycidol segments led to a pom-pom dendritic architecture [47].

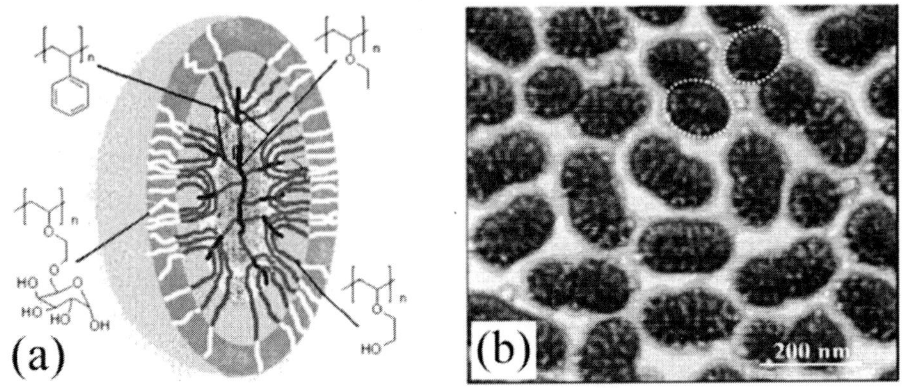

Figure 8. (**a**) Example of an amphiphilic core-shell-corona architecture with segregated branching points, and (**b**) AFM image in the phase contrast tapping mode showing a nodular core morphology and confirming the presence of a hydrophilic shell in the core-shell-corona molecules. Adapted from Reference 43 by permission from the American Chemical Society.

4. DENDRIMER-LIKE STAR POLYMERS

The synthesis of dendritic polymer architectures from polymeric building blocks through a strictly controlled terminal grafting process, to yield dendrigraft polymer molecules that are the closest analogues of dendrimers, was introduced in 1995 by Six and Gnanou [48]. The scope of the terminal grafting methodology was initially limited to the synthesis of homopolymers of ethylene oxide with low branching functionalities (the synthesis of a G1, twice-grafted structure with 6 side-chains was reported in the original paper), but it has meanwhile been expanded considerably by Gnanou and others to include much higher branching functionalities and the synthesis of copolymers. Distinct strategies have been developed for different monomers, and consequently they will be considered separately. The discussion on dendrimer-like star polymer structures is organized according to the type of grafting methodology used for their synthesis, namely *grafting from* (with a substrate acting as a polyfunctional initiator), *grafting onto* (by coupling a substrate with side-chains), and *grafting through* (by the incorporation of macromonomers into growing chains).

4.1. Grafting from Methodologies

The chain-end derivatization method developed by Gnanou has been applied mainly to the synthesis of copolymers incorporating a branched poly(ethylene oxide) core and a corona of chains with a different composition. The preparation of a dendrimer-like star polymer incorporating three poly(ethylene oxide) and six polystyrene side-chains is shown in **Scheme 7** to illustrate this strategy [49]. The synthesis starts from 2-ethyl-2-hydroxymethyl-1,3-propanediol serving as a trifunctional anionic initiator for the polymerization of ethylene oxide in dimethylsulfoxide (DMSO) after partial (20%) deprotonation with diphenylmethylpotassium (DPMK). The use of DMSO represents a key step in this procedure as it allows fast proton hopping between the hydroxyl and the alcoholate chain ends, which ensures the uniform growth of all the side-chains, while partial deprotonation minimizes alcoholate moiety aggregation leading to poor solubility at high ion concentrations. The hydroxyl functionalities of the star polymer are then capped with 2,2-bis(2-(bromomethyl)propionato)propionyl chloride to introduce two secondary bromide functionalities at each chain end. These sites, when activated with CuBr and 2,2'-bipyridine in the presence of styrene, produce two polystyrene segments from each chain end of the triarm star by atom transfer radical polymerization (ATRP).

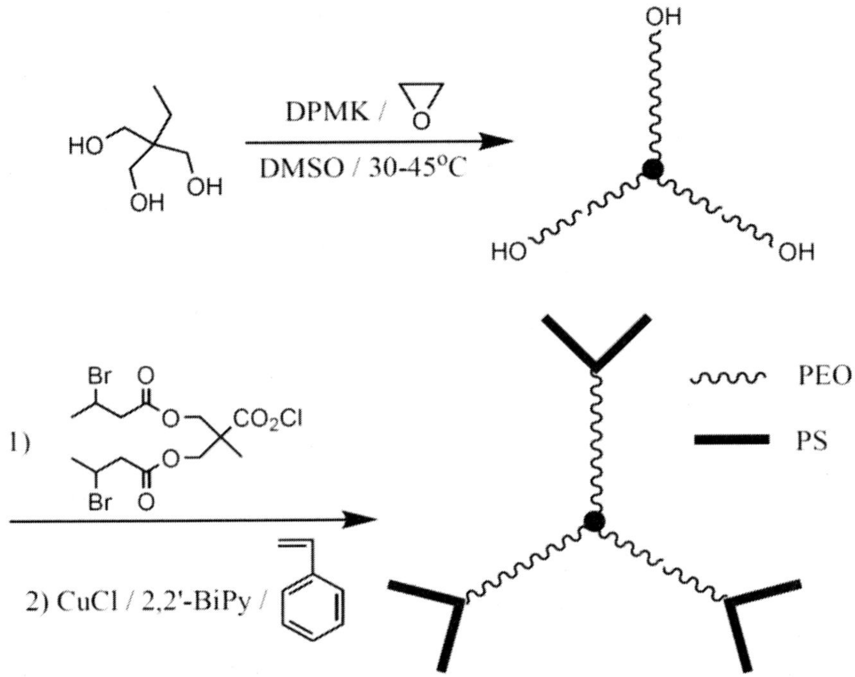

Scheme 7. Synthesis of a dendrimer-like star copolymer with a poly(ethylene oxide) core and polystyrene side-chains according to Gnanou *et al.*

Variations in the basic method outlined in **Scheme 7** have enabled the synthesis of various types of copolymers through analogous *grafting from* schemes using different branching agents and combinations of anionic, cationic, and ATRP polymerization techniques. This includes copolymers with polystyrene cores and a corona of poly(ethylene oxide) [50] or poly(*tert*-butyl acrylate)/poly(acrylic acid) [51,52] chains, copolymers derived from poly(ethylene oxide) cores with poly(*tert*-butyl acrylate)/poly(acrylic acid) segments in their corona [53], as well as more complex structures such as dendrimer-like poly(ethylene oxide) containing up to 48 poly(ethylene oxide) segments on the outside and 21 side-chains of poly(*tert*-butyl acrylate)/poly(acrylic acid) covalently bonded to the branch junctions of the preceding generations [54]. The self-assembly of some of the *tert*-butyl acrylate copolymers was investigated by the Langmuir force balance technique [51,52], while dynamic light scattering measurements in aqueous solutions showed that the copolymers with poly(acrylic acid) segments grafted at the branch junctions displayed pH-dependent size variations depending on the degree of ionization of the polyelectrolyte segments [54].

Hedrick *et al.* also achieved the synthesis of different dendrimer-like star polymers through a terminal*grafting from* methodology, but starting from ε-caprolactone and polyfunctional hydroxylated substrates activated with stannous 2-ethylhexanoate ($Sn(Oct)_2$) [55]. The basic technique, initially developed for the synthesis of dendrimer-like ε-caprolactone homopolymers, was subsequently extended to the synthesis of different copolymers incorporating substituted polylactone, polylactide, poly(2-hydroxyethyl methacrylate), and poly(methyl methacrylate) segments [56,57]. The synthesis of a G1 dendrimer-like copolymer with 6 poly(ε-caprolactone) and 24 poly(methyl methacrylate) chain segments is provided in **Scheme 8** as an example [57]. A core molecule containing 6 hydroxyl groups is first activated with $Sn(Oct)_2$ in the presence of ε-caprolactone to produce a six-arm star polymer with hydroxyl chain ends. Branching points are introduced at the ends of the star-branched structure by coupling with a dendritic fragment (**1**) containing four protected hydroxyl groups in the presence of diisopropyl azodicarboxylate (DIAD) and triphenylphosphine (TPP). These are subsequently deprotected and converted to tertiary bromide moieties serving in the ATRP growth of poly(methyl methacrylate) chain segments.

The copolymers were verified to possess phase-segregated morphologies on the basis of dynamic mechanical analysis and differential scanning calorimetry measurements [56,57]. The morphology of dendrimer-like molecules was investigated in SANS contrast matching experiments for a series of isomeric copolymers incorporating isomeric branched poly(ε-caprolactone) cores varying from star-branched to G2 dendrimer-like structures, with deuterated poly(methyl methacrylate) segments end-grafted on the last generation of poly(ε-caprolactone) segments [58]. The main finding was that cores with more highly branched structures (of increasing generation number) had a more uniform distribution of chain segments, presenting similarities to uniformly branched polymer networks.

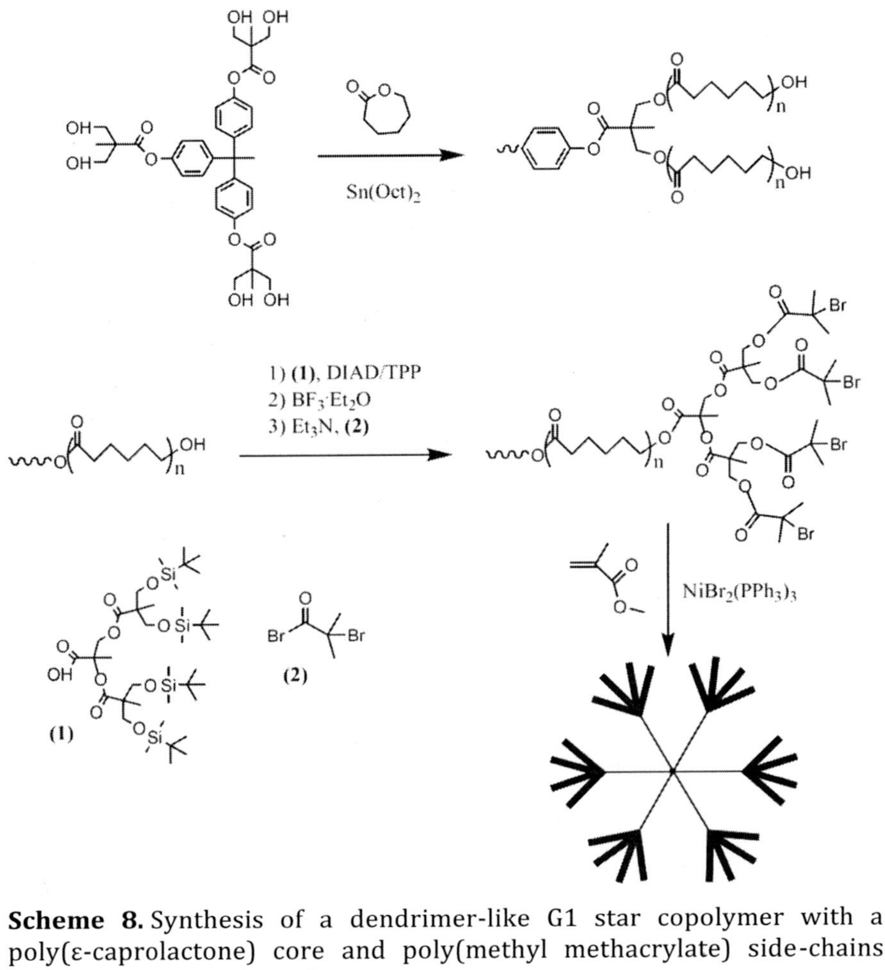

Scheme 8. Synthesis of a dendrimer-like G1 star copolymer with a poly(ε-caprolactone) core and poly(methyl methacrylate) side-chains according to Hedrick *et al.*

4.2. Grafting onto Methodologies

Different strategies have been developed by the group of Hirao for the synthesis of dendrimer-like star polymers incorporating segments of poly(methyl methacrylate) and polystyrene building blocks, as well as other polymers as corona materials. These techniques are based on the introduction of latent or protected coupling sites as chain ends in the synthesis of the side-chains, using suitable protected organolithium initiators, and coupling of the "living" polymers with a substrate. The method presented in 2004 for the synthesis of dendrimer-like star homopolymers of poly(methyl methacrylate) (PMMA) is outlined in **Scheme 9** [59]. The synthesis starts with the addition of *sec*-butyllithium to 1,1-bis(3-*tert*-butyldimethylsilyloxymethylphenyl)ethylene, a 1,1-diphenylethylene derivative containing two benzylic groups protected as silyl ethers. The resulting organolithium compound, when used to initiate the polymerization of methyl methacrylate, introduces latent coupling sites at the end of the PMMA chain opposite to the propagating centre. The "living" chains are coupled with α,α'-dibromoxylene and treated with trimethylsilyl chloride/LiBr to convert the four silyl ether moieties to benzylic bromide coupling sites. Further cycles of grafting with "living" PMMA side-chains containing silyl ether-protected coupling sites and conversion to benzylic bromide sites yields the subsequent generations of dendritic PMMA-based homopolymers. Variations of this technique have been applied to the synthesis of dendrimer-like PMMA copolymers with tetrafunctional branching points and a corona of poly(methacrylic acid), poly((2,3-dihydroxy)propyl methacrylate), or poly(2-vinylpyridine) chain segments [60]. Copolymers having layered morphologies, with alternating polystyrene and PMMA segments, have also been reported [61].

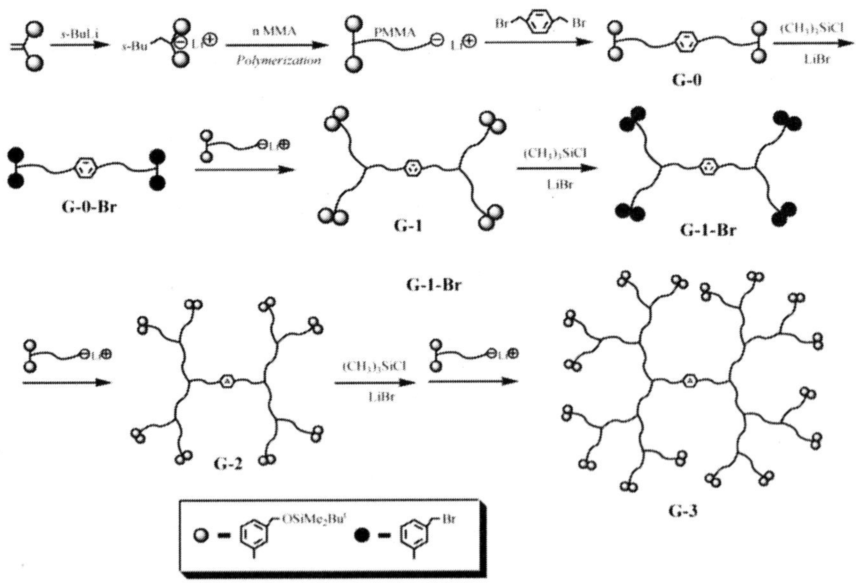

Scheme 9. Synthesis of dendrimer-like star poly(methyl methacrylate) according to Hirao *et al.* Reprinted from Reference 59 by permission from the American Chemical Society.

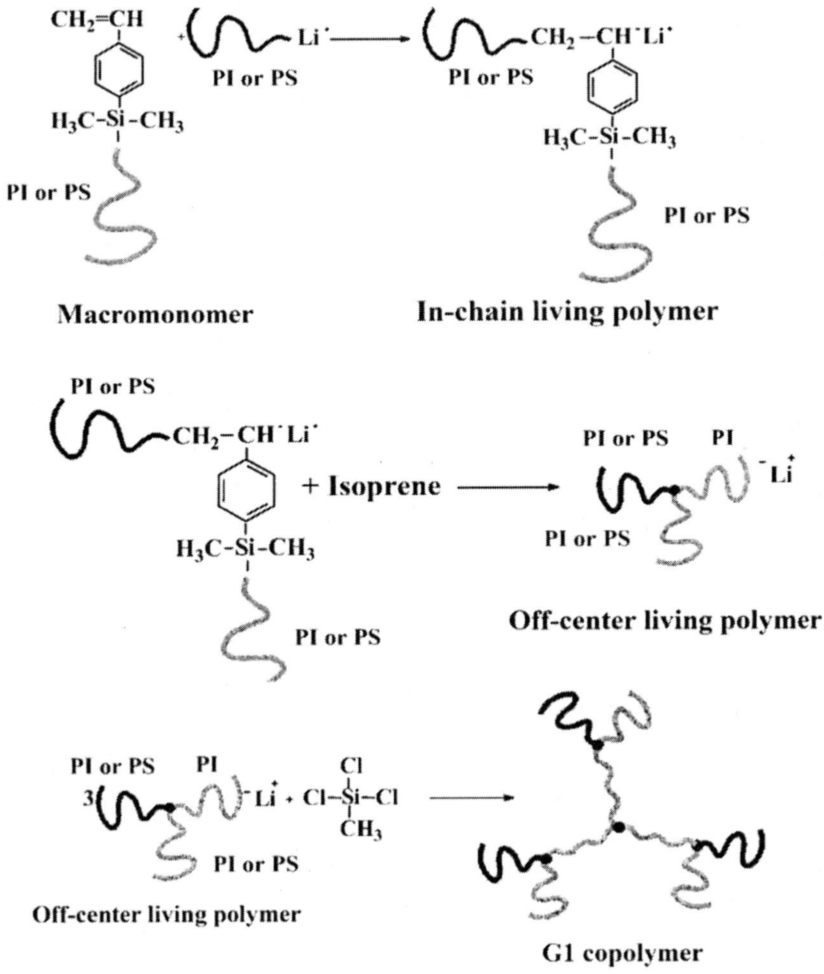

Scheme 10. Stepwise synthesis of dendrimer-like star copolymers with polystyrene and polyisoprene segments according to Hadjichristidis *et al.* Adapted from Reference 62 by permission from Wiley.

4.3. Grafting through Methodology

The synthesis of dendrimer-like star copolymers of low branching functionalities (G1 structures) has been achieved by Hadjichristidis *et al.* by a stepwise *grafting through* technique summarized in **Scheme 10** [62]. Macromonomers were generated by coupling polyisoprenyllithium or polystyryllithium with one equivalent of 4-

(chlorodimethylsilyl)styrene and reacted with a "living" polymer (polyisoprenyllithium or polystyryllithium) to generate a polymer with a propagating center at the junction point (in-chain "living" polymer in **Scheme 10**). This macroanion was used to initiate the polymerization of isoprene and give an off-center "living" polymer, which was ultimately coupled with trichloromethylsilane to yield the G1 dendrimer-like copolymer structure. The formation of multimolecular micelles by self-assembly of these copolymers in selective solvents has been the topic of an investigation [63].

5. CONCLUSIONS

The papers reviewed herein show that a wide range of heterogeneous dendrigraft copolymer architectures with core-shell and core-shell-corona morphologies can be synthesized, and at a much lower cost than for typical dendrimer syntheses. The very high molecular weight and larger size of dendrigraft polymers as compared to dendrimers is another important feature distinguishing these two families of dendritic macromolecules. Furthermore, the uniform size/molecular weight distribution of dendrigraft polymers in comparison to hyperbranched systems can be advantageous for certain applications. The methods available for the synthesis of dendrigraft polymers include stepwise cycles of substrate functionalization and grafting, and one-pot methods not requiring the isolation and purification of reaction intermediates. The second approach is particularly attractive, because it enables the synthesis of dendrigraft polymers on a large scale and opens up the path to commercial applications for these materials. A few of these potential applications have already been demonstrated, but it is clear that much work remains to be done to realize the full potential of dendrigraft polymers.

ACKNOWLEDGEMENTS

The financial support of the University of Guanajuato is gratefully acknowledged by L.-E. S. C.

REFERENCES AND NOTES

1. Tezuka, Y.; Oike, H. Topological polymer chemistry. *Prog. Polym. Sci.* **2002**, *27*, 1069–1122.
2. Hadjichristidis, N.; Pitsikalis, M.; Iatrou, H.; Pispas, S. The strength of the macromonomer strategy for complex macromolecular architecture: Molecular characterization, properties and applications of polymacromonomers. *Macromol. Rapid Commun.* **2003**, *24*, 979–1013.
3. Taton, D. From stars to microgels. In *Macromolecular Engineering*; Matyjaszewski, K., Gnanou, Y., Leibler, L., Eds.; Wiley-VCH: Weinheim, Germany, 2007; Volume 2, pp. 1007–1056.
4. Fréchet, J.M.J.; Tomalia, D.A. *Dendrimers and Dendritic Polymers*; Wiley: West Sussex, UK, 2001.
5. Vögtle, F.; Richardt, G.; Werner, N. *Dendrimer Chemistry: Concepts, Syntheses, Properties, Applications*; Wiley-VCH: Weinheim, Germany, 2009.
6. Teertstra, S.J.; Gauthier, M. Dendrigraft polymers: Macromolecular engineering on a mesoscopic scale. *Prog. Polym. Sci.* **2004**, *29*, 277–327.
7. Gauthier, M. Arborescent polymers and other dendrigraft polymers: A journey into structural diversity. *J. Polym. Sci. A Polym. Chem.* **2007**, *45*, 3803–3810.
8. Gauthier, M.; Möller, M. Uniform highly branched polymers by anionic grafting: Arborescent graft polymers. *Macromolecules* **1991**, *24*, 4548–4553.
9. Tomalia, D.A.; Hedstrand, D.M.; Ferritto, M.S. Comb-burst dendrimer topology: New macromolecular architecture derived from dendritic grafting. *Macromolecules* **1991**, *24*, 1435–1438.
10. For a general review on the synthesis of dendrimer-like star polymers, see for example, Hirao, A.; Sugiyama, K.; Matsuo, A.; Tsunoda, Y.; Watanabe, T. Synthesis of well-defined dendritic hyperbranched polymers by iterative methodologies using living/controlled polymerizations.*Polym. Int.* **2008**, *57*, 554–570.
11. Ishizu, K.; Uchida, S. Synthesis and microphase-separated structures of star-block copolymers.*Prog. Polym. Sci.* **1999**, *24*, 1439–1480.

12. Gohy, J.-F. Block copolymer micelles. *Adv. Polym. Sci.* **2005**, *190*, 65–136.

13. Denkova, A.G.; Mendes, E.; Coppens, M.-O. Non-equilibrium dynamics of block copolymer micelles in solution: Recent insights and open questions. *Soft Matter* **2010**, *6*, 2351–2357.

14. O'Reilly, R.K.; Hawker, C.J.; Wooley, K.L. Cross-linked block copolymer micelles: Functional nanostructures of great potential and versatility. *Chem. Soc. Rev.* **2006**, *35*, 1068–1083.

15. Norakankorn, C.; Pan, Q.; Rempel, G.L.; Kiatkamjornwong, S. Factorial experimental design on synthesis of functional core/shell polymeric nanoparticles via differential microemulsion polymerization. *J. Appl. Polym. Sci.* **2010**, *116*, 1291–1298.

16. Pascual, S.; Urbani, C.N.; Monteiro, M.J. Functionalization of polymer nanoparticles formed by microemulsion RAFT-mediated polymerization. *Macromol. React. Eng.* **2010**, *4*, 257–263.

17. Li, J.; Gauthier, M. A novel synthetic path to arborescent graft polystyrenes. *Macromolecules* **2001**,*34*, 8918–8924.

18. Frank, R.S.; Merkle, G.; Gauthier, M. Characterization of pyrene-labeled arborescent polystyrenes using fluorescence quenching techniques. *Macromolecules* **1997**, *30*, 5397–5402.

19. Yun, S.I.; Lai, K.-C.; Briber, R.M.; Teertstra, S.J.; Gauthier, M.; Bauer, B.J. Conformation of arborescent polymers in solution by small-angle neutron scattering: Segment density and core-shell morphology. *Macromolecules* **2008**, *41*, 175–183.

20. Gauthier, M.; Tichagwa, L.; Downey, J.S.; Gao, S. Arborescent graft copolymers: Highly branched macromolecules with a core-shell morphology. *Macromolecules* **1996**, *29*, 519–527.

21. Njikang, G.N.; Cao, L.; Gauthier, M. Self-assembly of arborescent polystyrene-*graft*-poly(ethylene oxide) copolymers at the air/water interface. *Macromol. Chem. Phys.* **2008**, *209*, 907–918.

22. Njikang, G.N.; Cao, L.; Gauthier, M. Pressure- and temperature-induced association of arborescent polystyrene-*graft*-poly(ethylene oxide) copolymers at the air-water interface.*Langmuir* **2008**, *24*, 12919–12927.

23. Kee, R.A.; Gauthier, M. Arborescent polystyrene-*graft*-poly(2-vinylpyridine) copolymers: Synthesis and enhanced polyelectrolyte effect in solution. *Macromolecules* **2002**, *35*, 6526–6532.

24. Gauthier, M.; Li, J.; Dockendorff, J. Arborescent polystyrene-*graft*-poly(2-vinylpyridine) copolymers as unimolecular micelles. Synthesis from acetylated substrates. *Macromolecules* **2003**,*36*, 2642–2648.

25. Kee, R.A.; Gauthier, M. Arborescent polystyrene-*graft*-poly(*tert*-butyl methacrylate) copolymers: Synthesis and enhanced polyelectrolyte effect in solution. *J. Polym. Sci. A Polym. Chem.* **2008**, *46*, 2335–2346.

26. Yun, S.I.; Briber, R.M.; Kee, R.A.; Gauthier, M. Dilute-solution structure of charged arborescent graft polymer. *Polymer* **2006**, *47*, 2750–2759.

27. Njikang, G.; Gauthier, M.; Li, J. Arborescent polystyrene-*graft*-poly(2-vinylpyridine) copolymers as unimolecular micelles: Solubilization studies. *Polymer* **2008**, *49*, 1276–1284.

28. Njikang, G.N.; Gauthier, M.; Li, J. Sustained release properties of arborescent polystyrene-*graft*-poly(2-vinylpyridine) copolymers. *Polymer* **2008**, *49*, 5474–5481.

29. Kee, R.A.; Gauthier, M. Arborescent polystyrene-*graft*-polyisoprene copolymers. *Macromolecules***1999**, *32*, 6478–6484.

30. Li, J.; Gauthier, M.; Teertstra, S.J.; Xu, H.; Sheiko, S.S. Synthesis of arborescent polystyrene-*graft*-polyisoprene copolymers using acetylated substrates. *Macromolecules* **2004**, *37*, 795–802.

31. Teertstra, S.J.; Gauthier, M. Viscoelastic properties of arborescent polystyrene-*graft*-polyisoprene copolymers. *Macromolecules* **2007**, *40*, 1657–1666.

32. Gauthier, M.; Lin, W.-Y.; Teertstra, S.J.; Tzoganakis, C. Fluorine-containing arborescent polystyrene-*graft*-polyisoprene copolymers as polymer processing additives. *Polymer* **2010**, *51*, 3123–3129.

33. Dockendorff, J.; Gauthier, M.; Mourran, A.; Möller, M. Arborescent amphiphilic copolymers as templates for the preparation of gold nanoparticles. *Macromolecules* **2008**, *41*, 6621–6623.

34. Mössmer, S.; Spatz, J.P.; Möller, M.; Aberle, T.; Schmidt, J.; Burchard, W. Solution behavior of poly(styrene)-*block*-poly(2-vinylpyridine) micelles containing gold nanoparticles. *Macromolecules***2000**, *33*, 4791–4798.

35. Yuan, Z.; Gauthier, M. One-pot synthesis of arborescent polystyrenes. *Macromolecules* **2006**, *39*, 2063–2071.

36. Yuan, Z.; Gauthier, M. Synthesis of arborescent copolymers by a one-pot method. *Macromol. Chem. Phys.* **2007**, *208*, 1615–1624.

37. Puskas, J.E.; Chen, Y.; Dahman, Y.; Padavan, D. Polyisobutylene-based biomaterials. *J. Polym. Sci. A Polym. Chem.* **2004**, *42*, 3091–3109.

38. Puskas, J.E.; Kwon, Y.; Anthony, P.; Bhowmick, A.K. Synthesis and characterization of novel dendritic (arborescent, hyperbranched) polyisobutylene-polystyrene block copolymers. *J. Polym. Sci. A Polym. Chem.* **2005**, *43*, 1811–1826.

39. Puskas, J.E.; Kwon, Y. Biomacromolecular engineering: Design, synthesis and characterization. One-pot synthesis of block copolymers of arborescent polyisobutylene and polystyrene. *Polym. Adv. Technol.* **2006**, *17*, 615–620.

40. Puskas, J.E.; Dos Santos, L.M.; Kaszas, G.; Kulbaba, K. Novel thermoplastic elastomers based on arborescent (dendritic) polyisobutylene with short copolymer end sequences. *J. Polym. Sci. A: Polym. Chem.* **2009**, *47*, 1148–1158.

41. Puskas, J.E.; Hoerr, R.A. Drug release from novel rubbery coatings. *Macromol. Symp.* **2010**, *291-292*, 326–329.

42. Schappacher, M.; Putaux, J.L.; Lefebvre, C.; Deffieux, A. Molecular containers based on amphiphilic PS-b-PMVE dendrigraft copolymers: Topology, organization, and aqueous solution properties. *J. Am. Chem. Soc.* **2005**, *127*, 2990–2998.

43. Schappacher, M.; Deffieux, A.; Putaux, J.-L.; Viville, P.; Lazzaroni, R. Synthesis and characterization of water-soluble amphipatic polystyrene-based dendrigrafts. *Macromolecules***2003**, *36*, 5776–5783.

44. Schappacher, M.; Deffieux, A. Encapsulation of metallo-porphyrins into water-soluble amphipatic dendrigrafts. *Polymer* **2004**, *45*, 4633–4639.

45. Bernard, J.; Schappacher, M.; Viville, P.; Lazzaroni, R.; Deffieux, A. Synthesis and properties of PS-PEO core-shell amphiphilic dendrigrafts. *Polymer* **2005**, *46*, 6767–6776.

46. Dworak, A.; Wałach, W. Synthesis, characterization and properties of functional star and dendritic block copolymers of ethylene oxide and glycidol with oligoglycidol branching units. *Polymer***2009**, *50*, 3440–3447.

47. Walach, W.; Trzebicka, B.; Justynska, J.; Dworak, A. High molecular arborescent polyoxyethylene with hydroxyl containing shell. *Polymer* **2004**, *45*, 1755–1762.

48. Six, J.-L.; Gnanou, Y. From star-shaped to dendritic poly(ethylene oxide)s: Toward increasingly branched architectures by anionic polymerization. *Macromol. Symp.* **1995**, *95*, 137–150.

49. Angot, S.; Taton, D.; Gnanou, Y. Amphiphilic stars and dendrimer-like architectures based on poly(ethylene oxide) and polystyrene. *Macromolecules* **2000**, *33*, 5418–5426.

50. Taton, D.; Cloutet, E.; Gnanou, Y. Novel amphiphilic branched copolymers based on polystyrene and poly(ethylene oxide). *Macromol. Chem. Phys.* **1998**, *199*, 2501–2510.

51. Matmour, R.; Lepoittevin, B.; Joncheray, T.J.; El-Khouri, R.J.; Taton, D.; Duran, R.S.; Gnanou, Y. Synthesis and investigation of surface properties of dendrimer-like copolymers based on polystyrene and poly(*tert*-butylacrylate). *Macromolecules* **2005**, *38*, 5459–5467.

52. Joncheray, T.J.; Bernard, S.A.; Matmour, R.; Lepoittevin, B.; El-Khouri, R.J.; Taton, D.; Gnanou, Y.; Duran, R.S. Polystyrene-b-poly(*tert*-butyl acrylate) and polystyrene-*b*-poly(acrylic acid) dendrimer-like copolymers: Two-dimensional self-assembly at the air-water interface. *Langmuir* **2007**, *23*, 2531–2538.

53. Hou, S.; Chaikof, E.L.; Taton, D.; Gnanou, Y. Synthesis of water-soluble star-block and dendrimer-like copolymers based on poly(ethylene oxide) and poly(acrylic acid). *Macromolecules* **2003**, *36*, 3874–3881.

54. Feng, X.; Taton, D.; Borsali, R.; Chaikof, E.L.; Gnanou, Y. pH Responsiveness of dendrimer-like poly(ethylene oxide)s. *J. Am. Chem. Soc.* **2006**, *128*, 11551–11562.

55. Trollsas, M.; Hedrick, J.L. Dendrimer-like star polymers. *J. Am. Chem. Soc.* **1998**, *120*, 4644–4651.

56. Trollsas, M.; Kelly, M.A.; Claesson, H.; Siemens, R.; Hedrick, J.L. Highly branched block copolymers: Design, synthesis, and morphology. *Macromolecules* **1999**, *32*, 4917–4924.

57. Hedrick, J.L.; Trollsas, M.; Hawker, C.J.; Atthoff, N.; Claesson, H.; Heise, A.; Miller, R.D.; Mecerreyes, D.; Jérôme, R.; Dubois, P. Dendrimer-like star block and amphiphilic copolymers by combination of ring opening and atom transfer radical polymerization. *Macromolecules* **1998**, *31*, 8691–8705.

58. Stancik, C.M.; Pople, J.A.; Trollsas, M.; Lindner, P.; Hedrick, J.L.; Gast, A.P. Impact of core architecture on solution properties of dendrimer-like star copolymers. *Macromolecules* **2003**, *36*, 5765–5775.

59. Matsuo, A.; Watanabe, T.; Hirao, A. Synthesis of well-defined dendrimer-like branched polymers and block copolymer by the iterative approach involving coupling reaction of living anionic polymer and functionalization. *Macromolecules* **2004**, *37*, 6283–6290.

60. Hirao, A.; Watanabe, T.; Ishizu, K.; Ree, M.; Jin, S.; Jin, K.S.; Deffieux, A.; Schappacher, M.; Carlotti, S. Precise synthesis and characterization of fourth-generation dendrimer-like star-branched poly(methyl methacrylate)s and block copolymers by iterative methodology based on living anionic polymerization. *Macromolecules* **2009**, *42*, 682–693.

61. Yoo, H.-S.; Watanabe, T.; Hirao, A. Precise synthesis of dendrimer-like star-branched polystyrenes and block copolymers composed of polystyrene and poly(methyl methacrylate) segments by an iterative methodology using living anionic polymerization. *Macromolecules* **2009**, *42*, 4558–4570.

62. Chalari, I.; Hadjichristidis, N. Synthesis of well-defined second-generation dendritic polymers of isoprene (I) and styrene (S): $(S_2I)_3$, $(SI'I)_3$, $(I''I'I)_3$, and $(I'_2I)_4$. *J. Polym. Sci. A Polym. Chem.* **2002**, *40*, 1519–1526.

63. Fernyhough, C.M.; Chalari, I.; Pispas, S.; Hadjichristidis, N. Micellar behavior of a well-defined dendritic polymer $(PS_2PI)_3$: The effects of architecture and solvent selectivity. *Eur. Polym. J.* **2004**,*40*, 73–79.

12

CHAPTER

Control of Microdomain Orientation in Block Copolymer Thin Films by Electric Field for Proton Exchange Membrane

Joonwon Bae

Department of Applied Chemistry, Dongduk Women's University, Seoul, South Korea

ABSTRACT

Owing to the recent push toward efficient energy storage/conversion devices, fuel cells have become a strong candidate for energy conversion equipments. On the other hand, block copolymer polyelectrolytes are interesting materials for proton exchange membranes in fuel cells. Thus a considerable attention has been paid to the development of block copolymer polyelectrolyte membranes. In this study, the microdomains in block copolymer polyelectrolytes were controlled by external electric fields to develop high performance membranes with improved proton conductivity. The microdomain alignments in sulfonated polystyrene-b-hydrogenated poly butadiene-b-polystyrene block copolymer electrolyte were monitored by cross-sectional transmission electron microscopy analysis. The proton conductivities of the block copolymer electrolyte membranes were measured before and after exposure to electric field. In addition, the morphological features of the block copolymer electrolyte were observed with small angle x-ray scattering and atomic force microscopy.

Keywords: Proton Conducting Membrane, Block Copolymer Polyelectrolyte, Microdomain Alignment, Electric Field, Fuel Cell, Proton Conductivity

1. INTRODUCTION

Recently, energy storage/conversion devices are attracting a considerable amount of interest from all fields of science and technology. These equipments include batteries, capacitors, fuel cells, and solar cells. Among these novel devices, fuel cells are the most strong candidate for the energy conversion systems, which can convert chemical energy directly into electric energy [1] .

In order to improve the overall performances and efficiencies of fuel cells, it is critical to incorporate high performance proton conducting membranes (PCMs), especially in two kinds of popular fuel cells such as proton exchange membrane fuel cell (PEMFC) and direct methanol fuel cell (DMFC) [2] [3] . To date, the PCM based on sulfonated fluorocarbon polymer, for example, Nafion® is the most widely used one [4] . Even if Nafion has high proton conductivity and good thermal stability, its methanol permeability is too high to use in practical applications [4] . Therefore, several attempts have been made in order to develop Nafion® membranes exhibiting reduced methanol permeability while maintaining proton conductivity [4] -[8] .

On the other side, numerous studies have been conducted on developing new PCMs using block copolymers (BCPs) over the past decades [9] -[19] . Attention to the block copolymer polyelectrolytes (BCPEs) as a candidate for PCMs has been paid, because they exhibit unique morphologies such as sphere, cylinder, lamellar, and gyroid due to the self assembled-nanostructures. It has been found that the microdomain structure in BCPEs is strongly dependent on the solvent [20] , annealing condition [21] , shear force [22] , and film thickness [23] . Also, it is very important to understand transport properties such as proton conductivity and methanol permeability through BCPE membranes when the size, shape, and ordering directions of microdomains in the membranes are different [19] [24] .

Among various BCPEs, the most extensively studied material for PCMs is BCPEs based on the thermoplastic elastomers such as styrene-b-ethylene/butylene-b-styrene (SEBS), which is derived from the hydrogenation reaction of triblock copolymer, styrene-b-butadiene-b-styrene [25] -[39] . The synthesis [25] , properties [27] [28] [31] [32] , and morphologies [37] -[39] of the SEBS BCPEsare well documented. Kim's group has reported the transport properties and morphological characteristics of sulfonated SEBS extensively [10] [38] [39] .

Even if the chemical and physical properties of sulfonated SEBS BCPEs have been extensively investigated, however, the effect of microdomain alignments in SEBS BCPEs on the proton conductivity of PCMs has been scarcely investigated yet [40] . In this study, the microdomain orientations in sulfonated SEBS BCPEs were controlled by external electric fields. The dependence of proton conductivity on microdomain alignments was examined by transmission electron microscopy (TEM) and conductivity measurements. It is certain that this work provides fundamental information for understanding the structure-property relation in BCPE PCMs.

2. EXPERIMENTAL

2.1. Materials

A block copolymer, styrene-b-ethylene/butylene-b-styrene (SEBS, 80 Kg/mol, 28 wt.% polystyrene) was supplied by Kuraray (SEPTON). The block copolymer was sulfonated to prepare sulfonated SEBS (sSEBS) according to the well-established procedure in the previous literature [25] . The sulfonating reagent was acetyl sulfate. The degree of sulfonation was found 22 mol%/hard segment by elemental analysis. Figure 1 exhibits the chemical structure of sSEBS BCPE. A solvent, tetrahydrofuran (THF) was dried by a common method.

2.2. Preparation of Thin sSEBS Films

A solution of sSEBS in THF was prepared and filtered. Concentration of the solution was 3 wt.%. A sSEBS film was spin-coated onto silicon wafer at 2000 rpm for 60 s from the solution. Then the films were dried under ambient conditions until complete evaporation of solvent.

2.3. Preparation of Thick sSEBS Films for SAXS

A 10 wt.% solution of sSEBS in THF was prepared and poured into a glass dish. Then the film was placed at room temperature until complete evaporation of solvent. The membrane was recovered as a flat-sheet type and the thickness was measured at dry state using a digital micrometer.

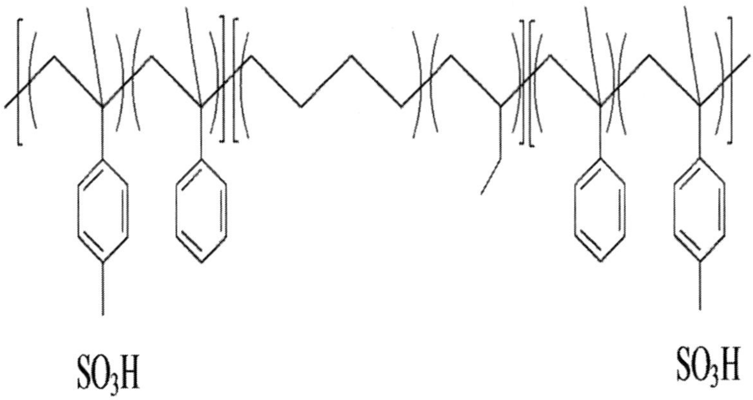

Figure 1. Chemical structure of the sulfonated SEBS electrolyte.

2.4. Application of Electric Field

Figure 2 shows the diagram of experimental apparatus for application of electric field to the prepared sSEBS film. An aluminized Kapton film served as the top electrode and asilicon wafer covered with Al foil as the other. A thin layer (20 - 25 mm) of cross-linked PDMS (Sylgard) was used as a buffer layer between the Kapton electrode and the copolymer thin film. The PDMS layer conforms to the electrode surface, eliminates air gaps between the top electrode and the copolymer film, and maintains asmooth surface of the copolymer film. The copolymer films were heated to 130°C under N_2 with an applied electric fieldof 20 V/mm for 6 ~ 12 h and then quenched to room temperature before removing the electric field.

2.5. Proton Conductivity and Water Uptake Measurement

The membranes were immersed in deionized water at least 6 h before measurements.Proton conductivity was measured using a complex impedance analyzer (ZAH-NER IM-6). AsSEBS membrane was sandwiched between two stainless electrodes and an ac perturbation of 1 V was applied to the cell.

The membranes were soaked for at least 3 days in deionized water and weighed to determine the uptake contents. Weight of a dried membrane was measured after drying the sample at 60°C under vacuum overnight. The uptake content was calculated by

$$\text{Uptake content}(\%) = \frac{w_{wet} - w_{dry}}{w_{dry}} \times 100 \tag{1}$$

where w_{dry} and w_{wet} are the mass of dried and wet samples, respectively.

2.6. Instrumentations

Small angle x-ray scattering (SAXS) patterns were obtained with a Bruker AXS Nanostar small angle X-ray scattering spectrometer with a generator voltage of 40 KV and a current of 35 mA. Scanning force microscopy (SFM) images were obtained in both the height and phase-contrast mode using a Digital Instruments Dimension 3100 scanning force microscope in a tapping mode. Transmission electron microscopy (TEM) experiments were performed with a JEOL 200CX TEM operated at an accelerating voltage of 200 kV. To prepare cross-sectional TEM specimens, a thin layer of carbon was evaporated onto the film surface before embedding in epoxy resin to prevent the diffusion of the epoxy resin into hybrid thin film. The thin film was then embedded in an epoxy resin and cured at 60°C for 12 h. The films were removed from the substrate by dipping into liquid N_2. Ultrathin sections (60 nm) were collected at room temperature using a Leica Ultracut Microtome, equipped with a diamond knife. The thin sections were exposed to RuO_4 for 10 min to enhance the contrast.

3. RESULTS & DISCUSSION

In determining sulfonation degree from the elemental analysis result, it must be assumed that the sulfonation takes place exclusively on the polystyrene end blocks. It should be mentioned that the sulfonation degree actually obtained is significantly lower than the expected value [25] . The efficiency of sulfonation employed in this study is approximately 22%.

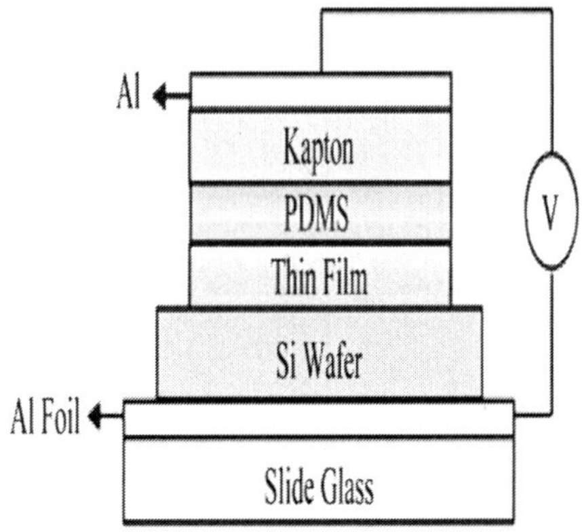

Figure 2. Schematic diagram of the experimental apparatus for application of electric field.

In addition, it is well known that proton conducts through the ionic channels or the water clusters where water or methanol molecules are likely to swell a polymer matrix. The membranes prepared in this study have hydrophilic sulfonic acid groups enough to show the high proton conductivity because of their morphologies with percolated ionic channels [10] . In general, the proton conductivity is primarily dependent on the concentration of sulfonic acid groups. More specifically, the proton conductivity is closely related to the average number of water molecules per fixed sulfonic acid group [39] . Therefore, the water uptake content was measured for sSEBS membranes using Equation (1). The obtained water uptake for pristine sSEBS film was approximately 20 wt.% regardless of film thickness. It was also revealed that the exposure to electric field did not significantly change the water uptake amount.

Figure 3 illustrates SAXS profiles of sSEBS film cast from THF before (solid line) and after (dotted line) thermal annealing at 130°C under N_2 for 48 h. THF was selected as solvent in this experiment because it is a good solvent for both constituent blocks. In addition, the molecular architecture of sSEBS in THF was clearly examined [28] . The thermal annealing has been performed well above the glass transition temperature (T_g) of both blocks, 100 and −40°C for PS and PEB, respectively. It is shown that the sSEBS membrane have lamellar microdomain morphology as indicated by the presence of maxima at positions with ratios of 1:1.95. The peak positions up to the second correlation peak are well resolved for the curve. It is known that the microscopically arranged lamellae of block copolymers exhibit the position of maximum intensity at q values o 1, 2, 3, 4, etc. It was also found that the apparent spacing (D) by using Bragg's equation, $D = 2p/q_{max}$, was estimated to be 35.1 nm from the position of first-order peak. It is accepted that the d-spacing is remarkably influenced by the solvent, water uptake, and sulfonation degree [38] [39] .

Figure 4 displays the tapping mode phase image of the sSEBS sample. The unique morphology consisting of bright and dark microdomain is observed remarkably. For sSEBS (14/72/14), an intermediate between the cylinder and lamellar structure is formed. It has been reported that a further slight increase in PS weight ratio changed the microstructure to a well-ordered lamellar [37] . The obtained morphology appears worm-like with lamellar regions dispersed throughout, and might be termed "frustrated" due to the relative disorder [41] .

Figure 5 exhibits the cross-sectional TEM images of sSEBS membrane obtained before and after exposure to electric field (20 V/mm). Figure 5(a) shows the equilibrium microstructure of sSEBS after thermal annealing at 130°C under N_2 for 48 h. The microdomain morphology is apparently lamellar throughout the entire film, where the bright and dark regions correspond to elastomer and polystyrene, respectively. This phenomenon can be explained by the following reasoning.

As for block copolymer thin film, the mean field theory (including Flory-Huggins) offers a fundamental concept of phase separation. In particular, the lamellar phases of triblock copolymer are well-described by strong segregation theory [42] . It has been shown that the domain spacing (D) for the lamellar structures of block copolymer is dependent on three physical parameters such as volume fraction of chain (f), molecular weight (N), and interaction parameters (c) between blocks when the morphology shows lamellar phases. However, it is often difficult to evaluate the exact values of interaction parameters, because the three parameters are strong dependent on temperature and compositions. Instead of interaction parameters, the solubility parameters are of use to understand the interaction between polymers. In particular, the solubility parameter is a useful measure of interaction between blocks, when polymers are partially functionalized with a group as seen in the BCPE employed in this study. Alternatively, the interaction parameter is given by

$$\chi_{sPS-EB} = \left(\delta_{sPS} - \delta_{EB}\right)^2 / \rho_0 k_b T$$

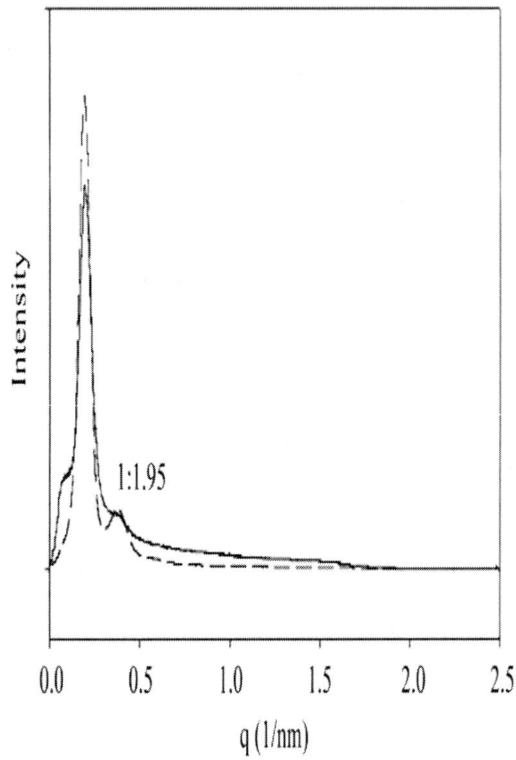

Figure 3. Small angle x-ray scattering patterns of the sulfonated SEBS membrane having an equilibrium microstructure obtained before (solid) and after (dotted) thermal annealing at 130°C under N_2 for 48 h.

where d_i is the solubility parameter of block i, r_0 the number density of the mono-unit, and k_b the Boltzmann constant [43] . The reference values for d_{PS}, d_{sPS}, and d_{EB} are 9.1, 16.6, and 7.8 $cal^{1/2}/cm^{2/3}$, respectively [44] . It was calculated that the value of cN for the sSEBS sample is approximately 200 at room temperature. According to this value, the morphology of sSEBS is likely to form a lamellar structure, as confirmed by Figure 5(a). It is commonly accepted that the cN values were influenced by the intermolecular ionic aggregation within the partially sulfonated polystyrene domains. Furthermore, strong ionic interactions like hydrogen bondings play role of physical crosslinking, which might induce an effect of increasing molecular weight.

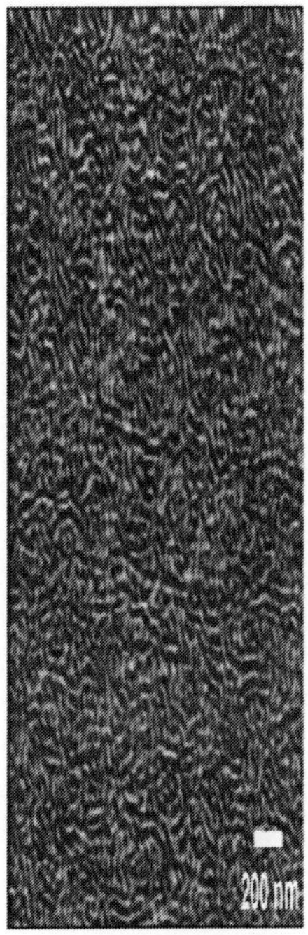

Figure 4. Scanning force microscopy phase image of the as-spun sulfonated SEBS membrane obtained at tapping mode.

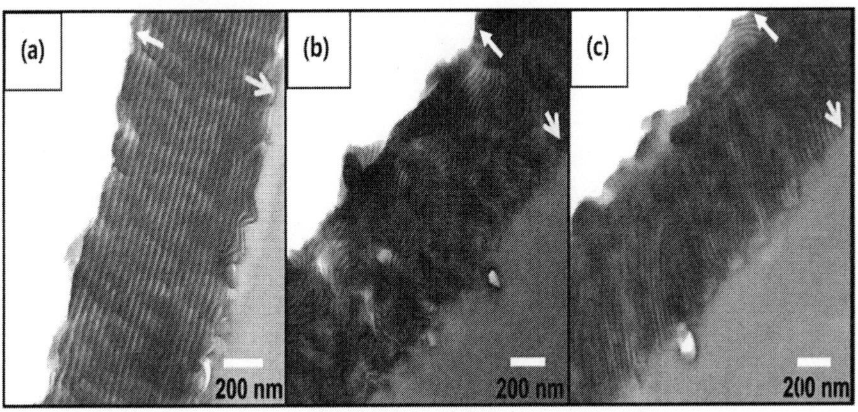

Figure 5. Cross-sectional TEM images of the sulfonated SEBS membranes (a) thermally annealed at 130°C under N_2 for 48 h, exposed to electric field of 20 V/mm for (b) 6 h and (c) 12 h at 130°C under N_2. The white and yellow arrows indicate the film/substrate and the film/air interfaces, respectively.

Figure 5(b) and Figure 5(c) present the cross-sectional TEM images of sSEBS films exposed to an electric field of 20 V/mm as a function of exposure time. It is conspicous that the microdomains oriented normal to the surface, adapted to the direction of applied electric field (yellow arrow). When an applied electric field exceeds a critical value to overcome interfacial interaction, the surface/interface becomes unstable. The electric field aligns interfaces separating two dielectric blocks parallel to the electric field vector. This leads to the formation of arrays of microdomains normal to the surface to minimize the moment of electrostatic forces exerted onto the interface between two blocks, driven by differences in the dielectric constants of the two bodies at the interface [45] .

It is also discernible that the alignments started locally after 6 h exposure and then extended to the overall range in the middle of BCP thin film under an electric field. It is notable that the vertical alignment extends completely to the film/substrate interface (white arrow). This is a strikingly different phenomenon from that seen in most block copolymer thin films. This is because the presence of ionic functional groups in polystyrene block can promote the response of PS microdomain to the direction of electric field even at the vicinity of film/substrate interface. In this case, the electrostatic force on sulfonated polystyrene domain exerted by the electric field might easily surpass the interfacial energy [46] .

Table 1. Summary of the proton conductivities of the membranes.

Sample	Proton conductivity (S/cm)
As spun (<1 μm)	0.018
Thin film (<1 μm) exposed to e-field	0.021
Thick film (30 μm) exposed to e-field	0.029

The proton conductivities of sSEBS samples are summarized in Table 1. The proton conductivity is intimately dependent on the various parameters such as sulfonation degree, water uptake, and methanol permeability. In this experiment, the effect of those parameters is excluded, because the other experimental parameters are fixed. For thin sSEBS membranes (<1 mm), the proton conductivity slightly increased after application of electric field. On the contrary, the conductivity increased significantly by exposure to electric field for thick membrane. This is attributed to the facilitated formation of percolated conducting channels inside membrane through microdomain alignments. Meanwhile, the increased water uptake amount in thicker membrane might be of help to elevate the proton conductivity.

However, the mere application of electric field did not elevate the proton conductivity significantly. This is because conductivity requires connectivity of microdomainsacross macroscopic length scales and obtaining this in the aligned orientation appears difficult, especially for thin films.

4. CONCLUSION

The dependence of proton conductivities on microdomain alignments in sulfonated SEBS block copolymer electrolytes was revealed. It could be concluded that the proton conductivities of sulfonated SEBS membranes were elevated by the application of electric field to promote microdomain alignments. In addition, a schematic pathway to microdomain orientation in sulfonated SEBS under electric field was presented. A more detailed study regarding this topic is underway. This article can provide an interesting outcome for understanding the structure-property relation in proton exchange membranes.

ACKNOWLEDGEMENTS

This study was supported by Dongduk Women's University Grant [2012-02204].

REFERENCES

1. Baldauf, M. and Preidel, W. (1999) Status of the Development of Direct Methanol Fuel Cell. Journal of Power Sources, 84, 161-166.
2. Rikukawa, M. and Sanui, K. (2000) Proton-Conduction Polymer Electrolyte Membranes Based on Hydrocarbon Polymers. Progress in Polymer Science, 25, 1463-1502.
3. Elabd, Y.A. and Hickner, M.A. (2011) Block Copolymers for Fuel Cells. Macromolecules, 44, 1-11.
4. Heitner-Wirguin, C. (1996) Recent advances in Perfluorinated Ionomer Membranes: Structure, Properties and Applications. Journal of Membrane Science, 120, 1-33.
5. Costamagna, P. and Srinivasan, S. (2001) Quantum Jumps in the PEMFC Science and Technology from the 1960s to the Year 2000;

Part I. Fundamental Scientific Aspects. Journal of Membrane Science, 102, 242-252.

6. Vielstich, W. and Ives, D.J.G. (1970)Fuel cells; Modern processes for the Electrochemical Production of Energy. Wiley, London.

7. Tricoli, V. (1998) Proton and Methanol Transport in Poly (Perfluorosulfonate) Membranes Containing Cs^+ and H^+ Cations. Journal of the Electrochemical Society, 145, 3798-3801.

8. Tricoli, V., Carretta, N. and Bartolozzi, M. (2000) A Comparative Investigation of Proton and Methanol Transport in Fluorinated Ionomeric Membranes. Journal of the Electrochemical Society, 147, 1286-1290.

9. Hickner, M.A., Ghassemi, H., Kim, Y.S., Einsla, B.R. and McGrath, J.E. (2004) Alternative Polymer Systems for Proton Exchange Membranes (PEMs). Chemical Reviews, 104, 4587-4612.

10. Kim, J., Kim, B. and Jung, B. (2002) Proton Conductivities and Methanol Permeabilities of Membranes Made from Partially Sulfonated Polystyrene-Block-Poly (Ethylene-Ran-Butylene)-Block-Polystyrene Copolymers. Journal of Membrane Science, 207, 129-137.

11. Elabd, Y.A., Napadensky, E., Sloan, J.M., Crawford, D.M. and Walker, C.W. (2003) Triblock Copolymer Ionomer Membranes: Part I. Methanol and Proton Transport. Journal of Membrane Science, 217, 227-242.

12. Shi, Z.Q. and Holdcroft, S. (2005) Synthesis and Proton Conductivity of Partially Sulfonated Poly ([Vinylidene Difluoride-co-Hexafluoropropylene]-b-Styrene) Block Copolymers. Macromolecules, 38, 4193-4201.

13. Park, M.J., Downing, K.H., Jackson, A., Gomez, E.D., Minor, A.M., Cookson, D., Weber, A.Z. and Balsara, N.P. (2007) Increased Water Retention in Polymer Electrolyte Membranes at Elevated Temperatures Assisted by Capillary Condensation. Nano Letters, 7, 3547-3552.

14. Xu, K., Li, K., Khanchaitit, P. and Wang, Q. (2007) Synthesis and Characterization of Self-Assembled SulfonatedPoly (Styrene-b-Vinylidene Fluoride-b-Styrene) Triblock Copolymers for Proton Conductive Membranes. Chemistry of Materials, 19, 5937-5945.

15. Saito, T., Moore, H.D. and Hickner, M.A. (2010) Synthesis of Midblock-Sulfonated Triblock Copolymers. Macromolecules, 43, 599-601.

16. Chen, L., Hallinan, Jr., D.T., Elabd, Y.A. and Hillmyer, M.A. (2009) Highly Selective Polymer Electrolyte Membranes from Reactive Block Polymers. Macromolecules, 42, 6075-6085.

17. Lee, H.-S., Roy, A., Lane, O., Dunn, S. and McGrath, J.E. (2008) Hydrophilic-Hydrophobic Multiblock Copolymers Based on Poly (Arylene Ether Sulfone) via Low-Temperature Coupling Reactions for Proton Exchange Membrane Fuel Cells. Polymer, 49, 715-723.

18. Bae, B., Miyatake, K. and Watanabe, M. (2010) Sulfonated Poly (Arylene Ether Sulfone Ketone) Multiblock Copolymers with Highly Sulfonated Block. Synthesis and Properties. Macromolecules, 43, 2684-2691.

19. Csernica, J., Baddour, R.F. and Cohen, R.E. (1987) Gas Permeability of a Polystyrene-Polybutadiene Block Copolymer with Oriented Lamellar Domains. Macromolecules, 20, 2468-2471.

20. Brinkmann, S., Stadler, R. and Thomas, E.L. (1998) New Structural Motif Inhexagonally Ordered Cylindrical Ternary (ABC) Block Copolymer Microdomains. Macromolecules, 31, 6566-6574.

21. Hajduk, D.A., Gruner, S.M., Rangarajan, P., Register, R.A., Fetters, L.J., Honeker, C., Albalak, R.J. and Thomas, E.L. (1994) Observation of a Reversiblethermotropic Order-Order Transition in a Diblock Copolymer. Macromolecules, 27, 490-501.

22. Morrison, F.A. and Winter, H.H. (1989) The Effect of Unidirectionalshear on the Structure of Triblock Copolymers. I. Polystyrene-Polybutadiene-Polystyrene. Macromolecules, 22, 3533-3540.

23. Heck, B., Arends, P., Ganter, M., Kressler, J. and Stühn, B. (1997) SAXS and TEM Studies on Poly(strene)-blockpoly(ethene-co-but-1-ene)-block-poly(styrene) in Bulk and at Various Interfaces. Macromolecules, 30, 4559-4566.

24. Csernica, J., Baddour, R.F. and Cohen, R.E. (1989) Gas Permeability of Apolystyrene/Polybutadiene Block Copolymer Possessing a Misoriented Lamellar Morphology. Macromolecules, 22, 1493-1496.

25. Weiss, R.A., Sen, A., Willis, C.L. and Pottick, A. (1991) Block Copolymer Ionomers: 1. Synthesis and Physical Properties of Sulfonated Poly(styrene-ethylene/butylene-styrene). Polymer, 32, 1867-1874. [Paper reference:4]

26. Jacobs, P.M. and Jones, F.R. (1989) Diffusion of Moisture into Two-Phase Polymers. Journal of Materials Science, 24, 2331-2336.

27. Weiss, R.A., Sen, A., Pottick, A. and Willis, C.L. (1991) Block Copolymer Ionomers: 2.Viscoelastic and Mechanical Properties of Sulfonated Poly(styrene-ethylene/butylene-styrene). Polymer, 32, 2785-2792.

28. Wu, C., Woo, K. and Jiang, M. (1996) Light-Scattering Studies of Styrene-(ethylene-co-butylene)-Styrene Triblock Copolymer and Its Sulfonated Ionomers in Tetrahydrofuran. Macromolecules, 29, 5361-5367.

29. Adams, J.L., Quiram, D.J., Graessley, W.W., Register, R.A. and Marchand, G.R. (1998) Interaction Strengths in Styrene-Diene Block Copolymers and Their Hydrogenated Derivatives. Macromolecules, 31, 201-204.

30. Ovejero, G., Perez, P., Romero, M.D., Guzman, I. and Diez, E. (2007) Solubility and Flory Huggins Parameters of SBES, Poly(Styrene-b-butene/ethylene-b-styrene) Triblock Copolymer, Determined by Intrinsic Viscosity. European Polymer Journal, 43, 1444-1449.

31. Blackwell, R.I. and Mauritz, K.A. (2004) Dynamic Mechanical Properties of Annealed Sulfonated Poly(styrene-b-[ethylene/butylene]-b-styrene) Block Copolymers. Polymer, 45, 3457-3463.

32. Chen, H., Hassan, M.K., Peddini, S.K. and Mauritz, K.A. (2011) Macromolecular Dynamics of Sulfonated Poly(Styrene-b-ethylene-ran-butylene-b-styrene) Block Copolymers by Broadband Dielectric Spectroscopy. European Polymer Journal, 47, 1936-1948.

33. Han, C.D., Chun, S.B., Hahn, S.F., Harper, S.Q., Savickas, P.J., Meunier, D.M., Li, L. and Yalcin, T. (1998) Phase Behavior of Polystyrene/Polybutadiene and Polystyrene/Hydrogenated Polybutadiene Mixtures: Effect of the Microstructure of Polybutadiene. Macromolecules, 31, 394-402.

34. Wang, Y., Shen, J.S. and Long, C.F. (2001) The Effect of Casting Temperature on Morphology of Poly(Styrene-ethylene/butylene-styrene) Triblock Copolymer. Polymer, 42, 8443-8446.

35. Wang, L., Hong, S., Hu, H., Zhao, J. and Han, C.C. (2007) The Surface Morphology Evolution of an Ultrathin Film of Poly[styrene-b-(ethylene-co-butylene)-b-styrene] during Its Dewetting Process. Langmuir, 23, 2304-2307.

36. Wang, L., Zhao, J. and Han, C.C. (2008) Phase Separation of Polystyrene-b-(Ethylene-co-butylene)-b-styrene (SEBS) Deposited on Polystyrene Thin Films. Polymer, 49, 2153-2159.

37. Wang, D., Fujinami, S., Liu, H., Nakajima, K. and Nishi, T. (2010) Investigation of True Surface Morphology and Nanomechanical Properties of Poly(styrene-b-ethylene-co-butylene-b-styrene) Using Nanomechanical Mapping: Effects of Composition. Macromolecules, 43, 9049-9055.

38. Kim, J., Kim, B., Jung, B., Kang, Y.S., Ha, H.Y., Oh, I.H. and Ihn, K.J. (2002) Effect of Casting Solvent on Morphology and PhysicalProperties of Partially Sulfonated Polystyrene-block-poly(ethylene-ran-butylene)-block-polystyrene Copolymers. Macromolecular Rapid Communications, 23, 753-756.

39. Kim, B., Kim, J. and Jung, B. (2005) Morphology and Transport Properties of Protons and Methanol through Partially Sulfonated Block Copolymers. Journal of Membrane Science, 250, 175-182.

40. Park, M.J. and Balsara, N.P. (2010) Anisotropic Proton Conduction in Aligned Block Copolymer Electrolyte Membranes at Equilibrium with Humid Air. Macromolecules, 43, 292-298.

41. Mauritz, K.A., Blackwell, R.I. and Beyer, F.L. (2004) Viscoelastic Properties and Morphology of Sulfonated Poly(styrene-b-ethylene/butylene-b-styrene) Block Copolymers (sBCP), and sBCP/[silicate] Nanostructured Materials. Polymer, 45, 3001-3016.

42. Matsen, M.W. (2000) Equilibrium Behavior of Asymmetric ABA Triblock Copolymer Melts, Journal of Chemical Physics, 113, 5539.

43. Shibayama, M., Hashimoto, T. and Kawai, H. (1983) Ordered Structure Inblock Polymer Solutions. 5. Equilibrium and Nonequilibrium Aspects of Microdomain Formation. Macromolecules, 16, 1434-1443.

44. Krevelen, V. (1990) Properties of Polymers. 3rd Editon, Elsevier, New York, 189.

45. Xu, T., Hawker, C.J. and Russell, T.P. (2003) Interfacial Energy Effects on the Electric Field Alignment of Symmetric Diblock Copolymers. Macromolecules, 36, 6178-6182.

46. Wang, J.Y., Chen, W., Roy, C., Sievert, J.D. and Russell, T.P. (2008) Influence of Ionic Complexes on Phase Behavior of Polystyrene-b-poly(methyl methacrylate) Copolymers. Macromolecules, 41, 963-969.

13
CHAPTER

Nanoparticle Network Formation in Nanostructured and Disordered Block Copolymer Matrices

Michelle K Gaines, Steven D Smith, Jon Samseth, Saad A Khan and Richard J Spontak

Department of Materials Science & Engineering, North Carolina State University

Department of Chemical & Biomolecular Engineering, North Carolina State University

ABSTRACT

Incorporation of nanoparticles composed of surface-functionalized fumed silica (FS) or native colloidal silica (CS) into a nanostructured block copolymer yields hybrid nanocomposites whose mechanical properties can be tuned by nanoparticle concentration and surface chemistry. In this work, dynamic rheology is used to probe the frequency and thermal responses of nanocomposites composed of a symmetric poly(styrene-*b*-methyl methacrylate) (SM) diblock copolymer and varying in nanoparticle concentration and surface functionality. At sufficiently high loading levels, FS nanoparticle aggregates establish a load-bearing colloidal network within the copolymer matrix. Transmission electron microscopy images reveal the morphological characteristics of the nanocomposites under these conditions.

Keywords Block copolymer Colloidal network Nanostructured polymer Nanocomposite Silica Nanoparticles Fumed silica Colloidal silica
Presented by Michelle K Gaines

INTRODUCTION

Block copolymers remain one of the most extensively studied classes of polymers due to their innate ability to infuse the dissimilar properties of homopolymers into a single material by spontaneously self-organizing at the molecular level. Molecular self-assembly is a direct consequence of thermodynamic incompatibility between the contiguous sequences comprising a block copolymer and results in the formation of (a)periodic nanoscale morphologies that can be tailored for diverse (nano)technologies [1, 2]. Ordered morphologies observed in simple AB diblock copolymers include A(B) spheres on a body- or face-centered cubic lattice in a B(A) matrix, A(B) cylinders on a hexagonal lattice in a B(A) matrix, triply periodic bicontinuous channels or alternating lamellar sheets. Targeted addition of a selective low-molar-mass solvent or relatively low-molecular-weight homopolymer to an ordered block copolymer can be used to preferentially swell one of the domains comprising the copolymer nanostructure and ultimately yield tunable transitions to morphologies with specific mechanical or spatial properties [3, 4, 5, 6, 7, 8]. Recent studies have extended this general design paradigm by modifying ordered block copolymers with surface-functionalized inorganic nanoparticles to achieve hybrid nanocomposites. Unlike conventional nanocomposites prepared from homopolymers, block copolymer nanocomposites rely on the existing copolymer nanostructure to template—that is, spatially modulate—the nanoparticles.

Previous experimental studies [8, 9, 10, 11, 12] of block copolymer nanocomposites have focused on the precise positioning of nanoparticles within the copolymer nanostructure for use in optics, such as waveguides. Such efforts have demonstrated that, if sufficiently small with respect to the host copolymer molecules (i.e., the characteristic size of the copolymer nanostructure), nonselective nanoparticles tend to distribute uniformly throughout the copolymer matrix in much the same fashion as nonselective solvent molecules. Selective nanoparticles, on the other hand, tend to locate along the interface separating adjacent domains within the copolymer nanostructure due to interfacial energy considerations, which result in fewer contacts between A and B repeat units. Larger selective nanoparticles are enthalpically driven to the core of the compatible domains to minimize repulsive contacts with incompatible blocks. Surface-functionalized nanoparticles have likewise been incorporated into ordered block copolymers to promote changes in morphology [13, 14, 15], as well as changes in phase behavior [16, 17, 18] discerned from the order–disorder transition (ODT). We have recently demonstrated [18] that the ODT of a poly(styrene-*b*-methyl methacrylate) (SM) diblock copolymer modified with surface-functionalized fumed silica (FS) or native (hydroxyl-terminated) colloidal silica (CS) decreases monotonically with increasing nanoparticle loading. If, however, oligostyrene-functionalized CS is added to the copolymer, the ODT increases slightly before dropping, in qualitative agreement with self-consistent field predictions.

The objective of the present work is to examine how the mechanical properties of such block copolymer nanocomposites evolve as the concentration of surface-functionalized FS and native CS is systematically increased. Dynamic melt rheology is employed to investigate the mechanical properties, and transmission electron microscopy (TEM) is used to examine the morphology of one of the nanocomposites.

EXPERIMENTAL

The SM copolymer was synthesized via sequential living anionic polymerization of the S block in cyclohexane at 60°C, followed by the M block in tetrahydrofuran at −78°C, with *sec*-butyllithium as the initiator. According to proton nuclear magnetic resonance (^{1}H NMR) spectroscopy and size-exclusion chromatography (SEC), the block masses measured 13,000 each, with an overall polydispersity of 1.05. Three grades of functionalized FS were obtained in powder form from Degussa Corp. (Parsippany, NJ) and probed the effects of hydrophilicity versus hydrophobicity and block selectivity: hydroxyl-terminated (OH), methacrylate-terminated (MA) and octyl-terminated (C8). According to the manufacturer, the primary particle size in each case was ~12 nm. The CS nanoparticles with an average diameter of 10–15 nm were provided as a suspension (20% solids) in dimethylacetamide by Nissan Chemicals (Houston, TX). Specimens for dynamic melt rheology were produced by ultrasonicating the nanoparticles (at a specimen-specific concentration relative to the copolymer) for 30 min in toluene to achieve a satisfactory dispersion, followed by copolymer dissolution and further ultrasonication, and then air- and vacuum-drying, all performed at ambient temperature. No copolymer degradation due to ultrasonication was detected according to SEC analysis of the resultant films.

Dynamic rheology was performed on an ARES strain-controlled rheometer equipped with serrated 8 mm parallel plates and operated at 2% strain amplitude to ensure linear viscoelasticity. Disks measuring 8 mm in diameter and 1 mm thick were melt-pressed at 150°C and heated to 220°C under nitrogen. Frequency (ω) spectra were acquired at discrete temperatures above and below the ODT, while isochronal temperature sweeps were performed at ω = 1 rad/s and a cooling rate of 1°C/min under a nitrogen purge to avoid oxidative degradation. Specimens for TEMT were prepared by sectioning the glassy nanocomposites at ambient temperature.

Electron-transparent sections measuring *ca.* 150 nm thick were subjected to the vapor of 0.5% RuO_4(aq) to selectively stain the styrenic units. Serial TEM tilt images were collected on a Gatan UltraScan 4000 CCD camera at a resolution of 0.76 nm/pixel and tilt angles ranging from −69° to +69° at an angular interval of 1.5° on a Technai T20 microscope operated at 200 kV. While the full tilt series was aligned using a pre-calibrated geometric model based on a high-precision goniometer stage [19], only representative images acquired at 0°, 15° and 30° are reported herein.

RESULTS AND DISCUSSION

According to the discontinuous change in the dynamic storage modulus (G′) encountered during an isochronal temperature sweep (upon cooling) [20], the ODT of the neat copolymer is determined to be 186 ± 1°C (data not shown here). Addition of FS or CS nanoparticles up to 10 wt% reduces the ODT by as much as ~7°C, depending on surface functionality. This change in ODT is markedly different from that observed in block copolymer nanocomposites composed of a poly(styrene-*b*-isoprene) diblock copolymer modified with C_{60} buckyballs [21]. In that case, the ODT is found by dynamic rheology to decrease by 21°C upon incorporation of only 0.04 wt% C_{60}. At higher concentrations of C_{60} nanoparticles (up to 0.5 wt%), the ODT does not change further, but it does progressively broaden. Such sensitivity is indicative of chemical interactions between the copolymer molecule (specifically, the unsaturated isoprenic units) and the carbonaceous nanoparticles [22], thereby generating a cross-linked material. Such interactions are not expected in the present systems, although we suspect that hydroxyl-terminated silica nanoparticles may bind with the acrylic units upon thermal treatment. To discern the effect of siliceous nanoparticles on mechanical properties at higher concentrations, representative temperature sweeps are presented in Fig. 1 for SM nanocomposites containing 20 wt% of the FS-OH, FS-C8 and CS additives.

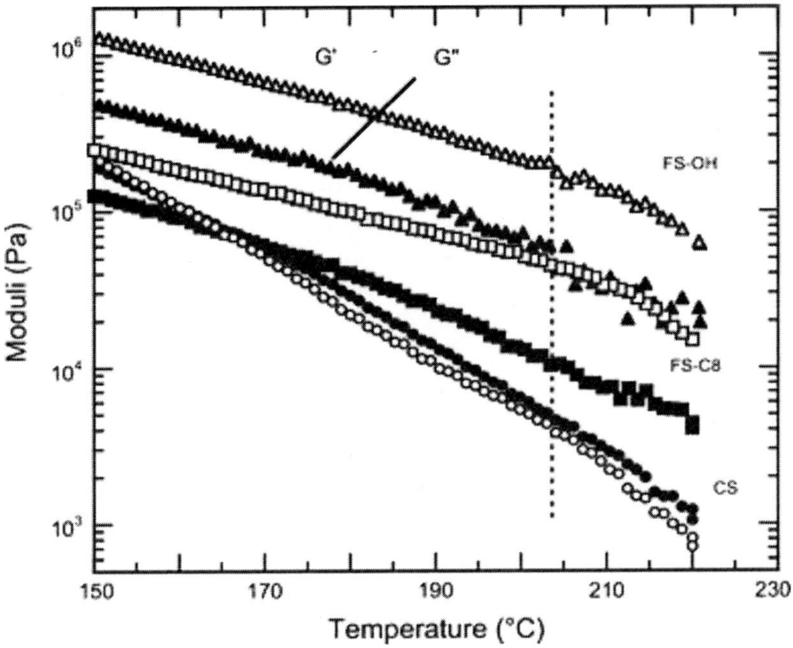

Figure 1 Temperature dependence of the dynamic shear moduli (G′, *open*; G″, *filled*) for SM nanocomposites containing 20 wt% of three different nanoscale additives: hydroxyl-terminated colloidal silica (CS, *circles*), hydroxyl-terminated fumed silica (FS-OH, *triangles*) and octyl-terminated fumed silica (FS-C8, *squares*). The *dotted vertical line* identifies an abrupt change in G′ that occurs in the case of the hydroxyl-terminated nanoparticles

These results immediately indicate that the fumed nanoparticles, which exist as branched aggregates commonly measuring on the order of hundreds of nanometers, have a more pronounced effect on the SM copolymer than do CS nanoparticles. Specifically, G′ measured for the nanocomposites with FS consistently exceeds the dynamic loss modulus (G″) over the entire temperature interval examined, even though the copolymer exists as a structureless melt. In contrast, G′ for the CS-containing nanocomposite increases beyond G″ only at low temperatures. Close examination of the data in this figure also reveals that (1) the ODT of the copolymer in the SM/CS nanocomposite persists in the vicinity of 172°C, which constitutes a 14°C reduction in the ODT of the copolymer and (2) a second, less pronounced event appears to occur in the SM/FS nanocomposites at *ca*. 204°C. This second event is absent in the neat copolymer, as well as in the system modified with CS, whereas neither

SM/FS nanocomposite exhibits a discernible ODT. Similar disappearance of the ODT is observed [21] in the case of nanocomposites modified with C_{60} buckyballs. Over the range from 160 to 200°C, the dynamic moduli measured from the SM-based nanocomposites described in Fig. 1 are well behaved and permit direct assessment of modulus enhancement as functions of nanoparticle concentration and surface chemistry below and above the copolymer ODT.

The variation of the dynamic storage modulus (G′) and dissipation factor (tanδ = G″/G′) with nanoparticle concentration is provided at a temperature below the virgin copolymer ODT (at 160°C) in Fig. 2a, 2b, respectively, and above the ODT (at 200°C) in Fig. 2c, 2d, respectively. In Fig. 2a, values of G′ measured from FS-containing nanocomposites generally increase slowly up to ~5 wt% and then increase sharply at higher FS concentrations. In contrast, G′ from the SM/CS nanocomposite remains relatively flat and does not exhibit an abrupt rise until 20 wt% CS, indicating that single CS nanoparticles are less effective at improving the rigidity of the nanocomposite at low loading levels than the aggregated FS nanoparticles. Within the family of FS nanoparticles examined here, the FS-MA variant appears to be the most effective, while the FS-C8 nanoparticles are the least effective, at improving the mechanical properties of the copolymer. This is most likely due to the presumably nonselective nature of the FS-C8 nanoparticles. The FS-MA and FS-OH nanoparticles, on the other hand, can preferentially interact with the PMMA units of the copolymer and therefore immobilize the chains. The solid and dashed lines included in the figure are meant as guides for the eye, but correspond to exponential regressions and fit the data surprisingly well. Similar behavior is observed at 200°C (cf. Fig. 2c) when the copolymer is above its ODT and disordered. It is interesting that the onset of the increase in G′ occurs at about the same concentration of both FS and CS nanoparticles.

In Fig. 2b, tanδ, a direct measure of liquid- versus solid-like behavior, is provided as a function of nanoparticle concentration and shows that, up to the concentration where G′ suddenly increases in Fig. 2a, tanδ is, for the most part, greater than unity. Since tanδ = G″/G′, this observation indicates that the nanocomposite behaves liquid-like. At higher nanoparticle concentrations, tanδ decreases below unity, and the material behaves more solid-like. While there is little systematic variation among the three surface-functionalized FS grades, the SM/CS nanocomposites exhibit the greatest liquid-like tendency, marginally behaving solid-like at 20 wt% CS. As before, similar results are seen in Fig. 2d, which displays tanδ as a function of nanoparticle concentration at 200°C, above the copolymer ODT. Solid-like behavior becomes evident at nanoparticle loading levels that correspond to the sharp rise in G′. The one series that deviates from the data previously discussed with regard to Fig. 2b consists of the SM/FS-C8 nanocomposites. According to Fig. 2d, tanδ for this series attains the highest tanδ value measured (15.5 at 0.1 wt%) in this study and thus exhibits the greatest liquid-like behavior of all the nanocomposites investigated. As the concentration of FS-C8 is increased, however, values of tanδ for this series become comparable to those measured for the other FS, as well as CS, series.

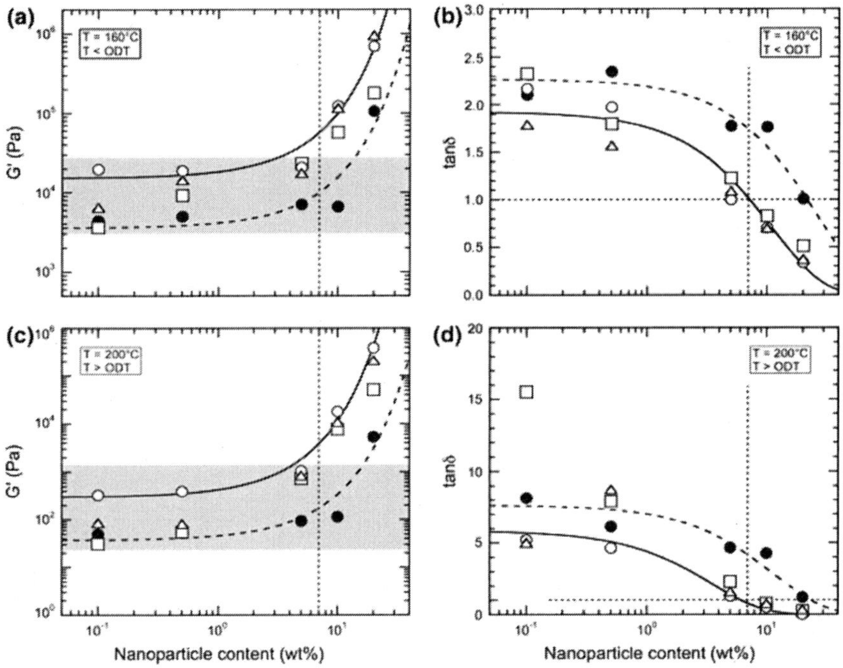

Figure 2 Dependence of G′ (**a**,**c**) and tanδ (**b**,**d**) on nanoparticle concentration for four nanoparticle species—CS (*filled circle*), FS-OH (*open triangle*), FS-C8 (*open square*) and FS-MA (*open circle*)—at two temperatures (in °C): 160 (**a**,**b**) and 200 (**c**,**d**). The *solid* and *dashed lines* serve as guides for the eye for the FS-MA and CS data, respectively; whereas the*shaded region* shows the range in G′ over which nanoparticle concentration generally has little effect on nanocomposite mechanical properties. The *vertical dotted line* identifies the concentration vicinity beyond which the nanocomposites behave solid-like, and the *horizontal dotted line* (**c**,**d** only) signifies where tanδ = 1

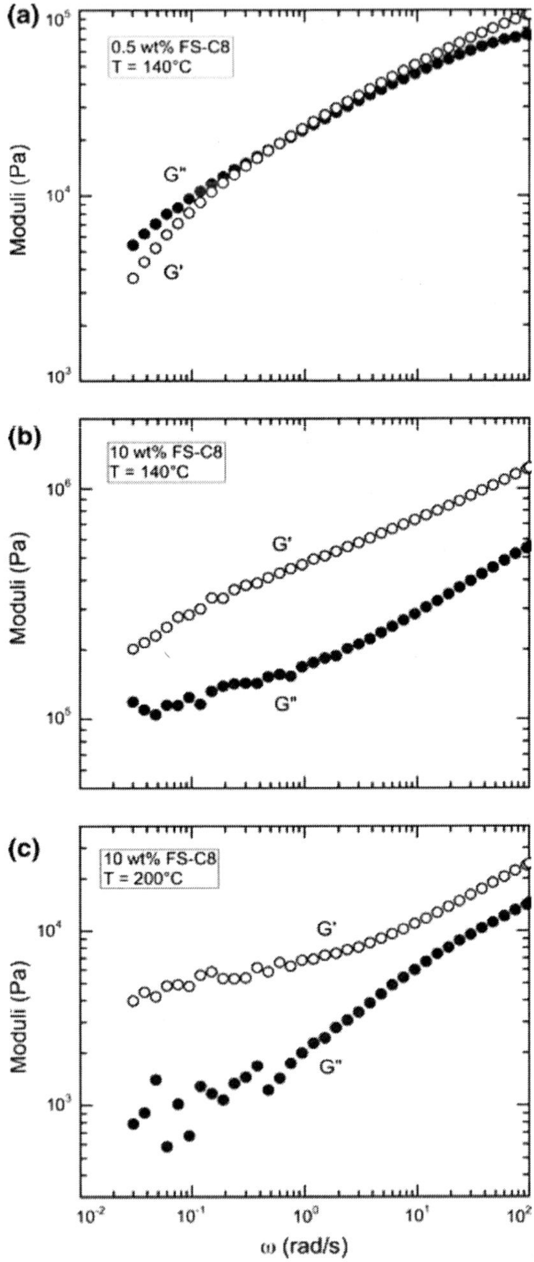

Figure 3 Isothermal frequency spectra acquired for G′ (*open circle*) and G″ (*filled circle*) at the following conditions: **a** 0.5 wt% FS-C8 at 140°C, **b** 10 wt% FS-C8 at 140°C and **c** 10 wt% FS-C8 at 200°C

Of all the additives considered here, the discrete CS nanoparticles appear to be the least effective in improving the mechanical properties of the SM copolymer (due to the formation of discrete, rather than interconnected, aggregate structures) and are not considered further. The FS-C8 nanoparticles likewise appear ineffective at low nanoparticle concentrations and temperatures above the copolymer ODT, but their efficacy progressively improves as the nanoparticle concentration increases. Since this series of nanocomposites represents the worst case of FS-based nanoparticles in terms of property-enhancing attributes, it is used to probe the ability of FS nanoparticles to form colloidal networks within the nanostructured copolymer matrix. Figure 3a shows the frequency spectra for G′ and G″ at 140°C and a nanoparticle concentration of 0.5 wt% FS-C8. Viscoelastic behavior is observed wherein G″ exceeds G′ at low ω, but G′ grows larger than G″ at high ω [23]. The crossover point at ω_c, where G′ and G″ intersect, yields a characteristic relaxation time (τ) for the material. In this case, $\tau = 1/\omega_c$ is about 1.67 s. At 200°C, similar behavior is observed (data not shown), but τ decreases by an order of magnitude to 0.13 s.

When the concentration of FS-C8 is increased to 10 wt%, the frequency spectra change dramatically from the one displayed in Fig. 3a. In Fig. 3b, G′ is greater than G″ over the entire ω range examined, although both are frequency dependent. These data are reminiscent of weak gel-like materials possessing a long relaxation time [24, 25, 26, 27]. In the disordered SM copolymer matrix at 200°C (Fig. 3c), we notice a remarkable behavior: G′ continues to exceed G″, remaining parallel at high ω (with G′ and G″ scaling as $\omega^{0.35}$) but starting to show evidence of a plateau at low ω. The presence of a low frequency plateau in G′ suggests that this material is more gel-like than any of the others [25, 27, 28, 29]. Interestingly, the modulus of this sample is lower than that of the specimen portrayed in Fig. 3b. These results taken together suggest the presence of a sample-spanning, self-supporting network within the disordered copolymer melt. In the ordered state, the nanocomposite possesses a higher modulus, but an apparently weaker FS network (due possibly to network disruption upon copolymer ordering). Since this is the least effective modifier of the FS family, it immediately follows that the other additives exhibit comparable, if not more pronounced, network behavior at nanoparticle concentrations of 10 wt% or more.

We now turn our attention to the most promising network-forming nanoparticle grade: FS-MA. Conventional TEM images acquired from relatively thick stained sections of two nanocomposites containing 5 and 20 wt% FS-MA are provided at 0°, 15° and 30° tilt in the top and bottom rows, respectively, of Fig. 4. In the case of the material with 5 wt% FS-MA, the lamellar morphology of the SM copolymer is evident and possesses a period of 20 ± 2 nm. Existence of lamellae confirms that the copolymer molecules are capable of self-organizing (albeit under frustrating conditions) and that an ODT should be observed, which it is according to dynamic rheology [18]. Discrete clusters of FS-MA nanoparticles are likewise visible and measure from ~10 to 35 nm across (which is on the same scale as the primary FS-MA nanoparticles). Recall that, of all the FS nanoparticles investigated here, the FS-MA nanoparticles are expected to be the most uniformly dispersed throughout the copolymer matrix, because they promote the greatest and most consistent property enhancement. Since the nanoparticles reside throughout the specimen and TEM images provide 2D projections of 3D objects, the precise effect of these nanoparticles/clusters on copolymer nanostructure cannot be directly assessed without the use of transmission electron microtomography [30, 31], which will be provided in a forthcoming publication.

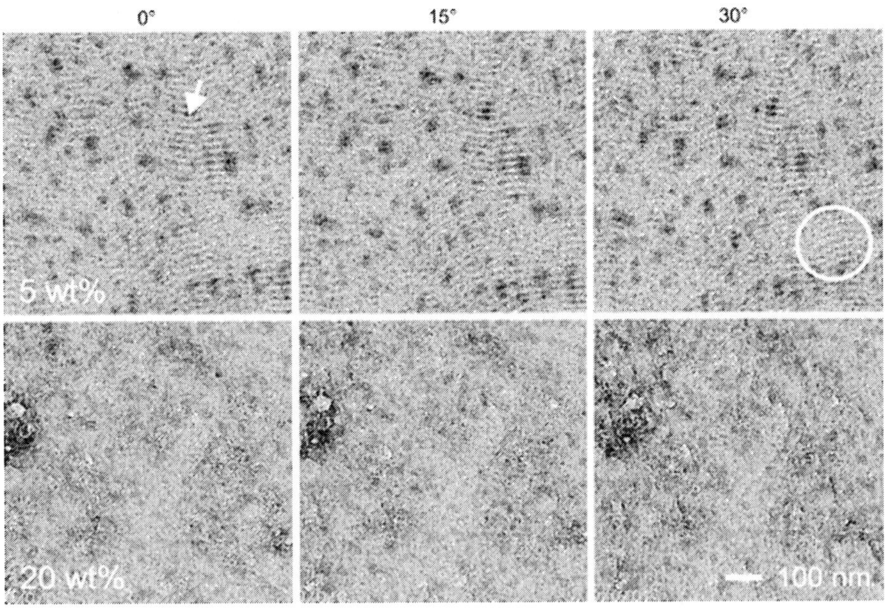

Figure 4 TEM images acquired at three different tilt angles (labeled) from SM/FS-MA nanocomposites containing 5 and 20 wt% FS-MA (*top* and *bottom* rows, respectively). In each case, the FS-MA nanoparticles appear as electron-opaque (*dark*) aggregate features, whereas the styrenic lamellae of the SM copolymer are likewise dark due to selective staining. The*arrow* shows the location of lamellae comprising the copolymer nanostructure, whereas the *circled region* highlights copolymer lamellae that appear distorted or partially discontinuous

It is apparent, however, from the tilt images corresponding to the nanocomposite with 5 wt% FS-MA that the copolymer lamellae are not highly oriented due, in large part, to the presence of the nanoparticles. Moreover, in several locations throughout the field of view, the lamellae appear distorted or even discontinuous (cf. the circled region at 30° tilt), which is consistent with our previous phase study [18] indicating that the stability of the copolymer nanostructure (discerned from the magnitude of the ODT) in this nanocomposite is lower than that of the neat copolymer. In the case of the nanocomposite containing 20 wt% FS-MA, however, the lamellar nanostructure of the copolymer is altogether eliminated, replaced by a continuous background of nearly constant optical density, whereas the FS-MA nanoparticles form a continuous network that extends throughout the material. These results agree with our findings from dynamic rheology: (1) no ODT is discernible from isochronal temperature sweeps of this nanocomposite and (2) this nanocomposite exhibits solid-like behavior both at low and high temperatures in the melt. Thus, we provide experimental evidence to demonstrate that incorporation of nanoparticles in block copolymer melts can induce sufficient molecular frustration via nanoscale confinement to completely thwart the ability of the copolymer molecules to form a periodic nanostructure.

CONCLUSIONS

Addition of native and surface-functionalized siliceous nanoparticles varying in hydrophobicity and inherent aggregation to a nanostructured block copolymer melt has little effect on the rheological properties at low nanoparticle concentrations, but promotes an abrupt increase in G' and a corresponding decrease in $\tan\delta$ (below unity) at high nanoparticle loading levels. In this latter regime, the nanocomposite melt behaves solid-like at temperatures above and below the copolymer ODT, suggesting that a colloidal network composed of nanoparticles develops. Existence of such a network is confirmed from mechanical frequency spectra acquired at different nanoparticle concentrations and temperatures. Transmission electron microscopy provides direct visual evidence of a clustered nanoparticle network [32] within the ordered copolymer nanostructure and establishes that two dissimilar nanostructures, both capable of imparting solid-like behavior to soft materials, can coexist in block copolymer nanocomposite melts [33].

NOTES

Presented by Michelle K Gaines

ACKNOWLEDGMENTS

This work was supported by the Research Council of Norway under the NANOMAT Program. M. K. G. expresses her gratitude for a GEM Fellowship and a NOBCChE Procter & Gamble Fellowship.

REFERENCES

1. Hamley IW: *The Physics of Block Copolymers.* Oxford University Press, New York; 1998.
2. Lazzari M, Liu G, Lecommandoux S (eds.) (Eds): *Block Copolymers in Nanoscience.* Wiley, Weinheim; 2006.
3. Alexandridis P, Lindman B: *Amphiphilic Block Copolymers: Self-Assembly and Applications.* Elsevier, Amsterdam; 2000.
4. Alexandridis P, Spontak RJ: *Curr. Opin. Colloid Interface Sci.* 1999, **4:** 130.
5. Lodge TP, Pudil B, Hanley KJ: *Macromolecules.* 2002, **35:** 4707.
6. Spontak RJ, Patel NP: *Developments in Block Copolymer Science and Technology.* Edited by: Hamley IW. Wiley, New York; 2004:159–212.
7. Hamley IW: *Block Copolymers in Solution: Fundamentals and Applications.* Wiley: Hoboken, NJ; 2005.
8. Spontak RJ, Shankar R, Bowman MK, Krishnan AS, Hamersky MW, Samseth J, Bockstaller MR, Rasmussen KØ: *Nano Lett.*. 2006, **6:** 2115.
9. Tsutsumi K, Funaki Y, Hirokawa Y, Hashimoto T: *Langmuir.* 1999, **15:** 5200.

10. Bockstaller MR, Lapetnikov Y, Margel S, Thomas EL: *J. Am. Chem. Soc.*. 2003, **125:** 5276.

11. Bockstaller MR, Thomas EL: *Phys. Rev. Lett.*. 2004, **93:** 166106.

12. Bockstaller MR, Mickiewicz RA, Thomas EL: *Adv. Mater.*. 2005, **17:** 1331.

13. Lo C-T, Lee B, Pol VG, Dietz Rago NL, Seifert S, Winans RE, Thiyagarajan P: *Macromolecules*. 2007, **40:** 8302.

14. Kim BJ, Chiu JJ, Yi GR, Pine DJ, Kramer EJ: *Adv. Mater.*. 2005, **17:** 2618.

15. Kim BJ, Fredrickson GH, Hawker CJ, Kramer EJ: *Langmuir*. 2007, **23:** 7804.

16. Lee KM, Han CD: *Macromolecules*. 2003, **36:** 804.

17. Jain A, Gutmann JS, Garcia CBW, Zhang Y, Tate MW, Gruner SM, Wiesner U: *Macromolecules*. 2002, **35:** 4862.

18. Gaines MK, Smith SD, Samseth J, Bockstaller MR, Thompson RB, Rasmussen KØ, Spontak RJ: *Soft Matter*. 2008, **4:**1609.

19. Zheng QXS, Braunfeld MB, Sedat JW, Agard DA: *J. Struct. Biol.*. 2004, **147:** 91.

20. Rosedale JH, Bates FS: *Macromolecules*. 1990, **23:** 2329.

21. Zhao Y, Hashimoto T, Douglas JF: *J. Chem. Phys.*. 2009, **130:** 124901.

22. Laiho A, Ras RHA, Valkama S, Ruokolainen J, Österbacka R, Ikkala O: *Macromolecules*. 2006, **39:** 7648.

23. Dealy JM, Larson RG: *Structure and Rheology of Molten Polymers: From Structure to Flow Behavior and Back Again*. Hanser, Munich; 2006.

24. English RJ, Raghavan SR, Jenkins RD, Khan SA: *J. Rheol.*. 1999, **43:** 1175.

25. Tayal A, Pai VB, Khan SA: *Macromolecules*. 1999, **32:** 5567.

26. Chiou BS, Raghavan SR, Khan SA: *Macromolecules*. 2001, **34:** 4526. C

27. Kumar R, Kalur GC, Ziserman L, Raghavan SR: *Langmuir*. 2007, **23:** 12849.

28. English RJ, Laurer JH, Spontak RJ, Khan SA: *Ind. Eng. Chem. Res.*. 2002, **41:** 6425.

29. Pai VB, Khan SA: *Carbohydr. Polym.*. 2002, **49:** 207.

30. Jinnai H, Spontak RJ, Nishi T: *Macromolecules.* 2010, **43:** 1675.

31. Abetz V, Spontak RJ, Talmon Y: *Macromolecular Engineering Precise Synthesis, Materials Properties, Applications.* Edited by: ed. by Matyjaszewski K, Gnanou Y, Leibler L. Wiley, Weinheim; (2007):1649–1685.

32. Rahedi AJ, Douglas JF, Starr FW: *J. Chem. Phys..* 2008, **128:** 024902.

33. Thompson RB, Ginzburg VV, Matsen MW, Balazs AC: *Science.* 2001, **292:** 2469.

Index